# Lecture Notes in Mathematics

Editorial Policy
for the publication of monographs

1. Lecture Notes aim to report new developments in all areas of mathematics – quickly, informally and at a high level. Monograph manuscripts should be reasonably self-contained and rounded off. Thus they may, and often will, present not only results of the author but also related work by other people. They may be based on specialized lecture courses. Furthermore, the manuscripts should provide sufficient motivation, examples and applications. This clearly distinguishes Lecture Notes from journal articles or technical reports which normally are very concise. Articles intended for a journal but too long to be accepted by most journals, usually do not have this "lecture notes" character. For similar reasons it is unusual for doctoral theses to be accepted for the Lecture Notes series.

2. Manuscripts should be submitted (preferably in duplicate) either to one of the series editors or to Springer-Verlag, Heidelberg. In general, manuscripts will be sent out to 2 external referees for evaluation. If a decision cannot yet be reached on the basis of the first 2 reports, further referees may be contacted: the author will be informed of this. A final decision to publish can be made only on the basis of the complete manuscript, however a refereeing process leading to a preliminary decision can be based on a pre-final or incomplete manuscript. For such a preliminary decision, the amount of material that will be considered should include a detailed outline describing the planned contents of each chapter, a bibliography and several sample chapters.
Authors should be aware that incomplete or insufficiently close to final manuscripts almost always result in longer refereeing times and nevertheless unclear referees' recommendations, making further refereeing of a final draft necessary.
Authors should also be aware that parallel submission of their manuscript to another publisher while under consideration for LNM will in general lead to immediate rejection.

3. Manuscripts should in general be submitted in English.
Final manuscripts should contain at least 100 pages of mathematical text and should include
- a table of contents;
- an informative introduction, with adequate motivation and perhaps some historical remarks: it should be accessible to a reader not intimately familiar with the topic treated;
- a subject index: as a rule this is genuinely helpful for the reader.

Continued on inside back-cover

# Lecture Notes in Mathematics

Edited by A. Dold, F. Takens and B. Teissier

**Editorial Policy**
for the publication of monographs

1. Lecture Notes aim to report new developments in all areas of mathematics – quickly, informally and at a high level. Monograph manuscripts should be reasonably self-contained and rounded off. Thus they may, and often will, present not only results of the author but also related work by other people. They may be based on specialized lecture courses. Furthermore, the manuscripts should provide sufficient motivation, examples and applications. This clearly distinguishes Lecture Notes from journal articles or technical reports which normally are very concise. Articles intended for a journal but too long to be accepted by most journals, usually do not have this "lecture notes" character. For similar reasons it is unusual for doctoral theses to be accepted for the Lecture Notes series.

2. Manuscripts should be submitted (preferably in duplicate) either to one of the series editors or to Springer-Verlag, Heidelberg. In general, manuscripts will be sent out to 2 external referees for evaluation. If a decision cannot yet be reached on the basis of the first 2 reports, further referees may be contacted: the author will be informed of this. A final decision to publish can be made only on the basis of the complete manuscript, however a refereeing process leading to a preliminary decision can be based on a pre-final or incomplete manuscript. The strict minimum amount of material that will be considered should include a detailed outline describing the planned contents of each chapter, a bibliography and several sample chapters.
Authors should be aware that incomplete or insufficiently close to final manuscripts almost always result in longer refereeing times and nevertheless unclear referees' recommendations, making further refereeing of a final draft necessary.
Authors should also be aware that parallel submission of their manuscript to another publisher while under consideration for LNM will in general lead to immediate rejection.

3. Manuscripts should in general be submitted in English.
Final manuscripts should contain at least 100 pages of mathematical text and should include
- a table of contents;
- an informative introduction, with adequate motivation and perhaps some historical remarks: it should be accessible to a reader not intimately familiar with the topic treated;
- a subject index: as a rule this is genuinely helpful for the reader.

Continued on back inside cover

# Lecture Notes in Mathematics 1747

**Springer**
*Berlin*
*Heidelberg*
*New York*
*Barcelona*
*Hong Kong*
*London*
*Milan*
*Paris*
*Singapore*
*Tokyo*

Lars Winther Christensen

# Gorenstein Dimensions

Springer

Author

Lars Winther Christensen
Matematisk Afdeling
Københavns Universitet
Universitetsparken 5
2100 København Ø, Danmark

E-mail: winther@math.ku.dk

Cataloging-in-Publication Data applied for

Die Deutsche Bibliothek - CIP-Einheitsaufnahme:

Winther Christensen, Lars:
Gorenstein dimensions / Lars Winther Christensen. - Berlin ; Heidelberg ; New York ; Barcelona ; Hong Kong ; London ; Milan ; Paris ; Singapore ; Tokyo : Springer, 2000
(Lecture notes in mathematics ; 1747)
ISBN 3-540-41132-1

Mathematics Subject Classification (2000): 13-02, 13C15, 13D02, 13D05, 13D07, 13D25, 13E05, 13H10, 18G25

ISSN 0075-8434
ISBN 3-540-41132-1 Springer-Verlag Berlin Heidelberg New York



Springer-Verlag Berlin Heidelberg New York
a member of BertelsmannSpringer Science+Business Media GmbH



Typesetting: Camera-ready TEX output by the author
SPIN: 10724258 41/3142-543210 - Printed on acid-free paper

# Preface

In 1995, almost five years ago, Hans–Bjørn Foxby gave me a copy of *Anneaux de Gorenstein, et torsion en algèbre commutative*, a set of notes based on lectures given by Auslander in 1966–67. I was told that the notes contained ideas about something called 'Gorenstein dimensions', a concept which had received renewed attention in the early 1990s and might prove to be an interesting topic for my Master's thesis.

I was easily convinced: Gorenstein dimensions have been part of my life ever since. I have, already, expressed my gratitude to Foxby on several occasions, however, I wish to do it again: This book is an enhanced and extended version of my Master's thesis from 1996, and I thank Hans–Bjørn Foxby for encouraging me to publish it and for his continual support during the entire process.

Among the people who have helped me complete this project, my friend and colleague Srikanth Iyengar stands out. I was about to start the project when we first met in late 1998, and Srikanth has from day one shown a genuine and lasting interest in the project: reading at least one version of every chapter and making valuable comments on my style and significant improvements to several proofs.

I also thank Anders Frankild and Mette Thrane Nielsen for reading parts of the manuscript, and Luchezar Avramov and Peter Jørgensen for their readiness to discuss specific details in some of my proofs.

Finally, I thank Line, my wife, for her endless love, support, and encouragement.

The invaluable help and support from colleagues, friends, and family notwithstanding, this book is no better than its author. I have tried to set out the text in such a way that the main features stand out clearly, and I have taken great care to supply detailed proofs; it may sometimes seem that I go to great lengths to explain the obvious, but that is how I am.

Copenhagen, June 2000

Lars Winther Christensen

# Contents

**Introduction** 1

**Synopsis** 3

**Conventions and Prerequisites** 9

Notation and Basics . . . . . 9
Standard Tools . . . . . 11
Standard Homomorphisms . . . . . 11
Homological Dimensions . . . . . 13
A Hierarchy of Rings . . . . . 14

**1 The Classical Gorenstein Dimension** 17

1.1 The G–class . . . . . 17
1.2 G–dimension of Finite Modules . . . . . 22
1.3 Standard Operating Procedures . . . . . 29
1.4 Local Rings . . . . . 32
1.5 G–dimension versus Projective Dimension . . . . . 37

**2 G–dimension and Reflexive Complexes** 41

2.1 Reflexive Complexes . . . . . 41
2.2 The Module Case . . . . . 47
2.3 G–dimension of Complexes with Finite Homology . . . . . 52
2.4 Testing G–dimension . . . . . 58

**3 Auslander Categories** 65

3.1 The Auslander Class . . . . . 65
3.2 The Bass Class . . . . . 71
3.3 Foxby Equivalence . . . . . 76
3.4 Cohen–Macaulay Rings . . . . . 83

**4 G–projectivity** 91

4.1 The G–class Revisited . . . . . 91
4.2 Gorenstein Projective Modules . . . . . 97
4.3 G–projectives over Cohen–Macaulay Rings . . . . . 99
4.4 Gorenstein Projective Dimension . . . . . 105

**5 G–flatness** **113**
5.1 Gorenstein Flat Modules . . . . . . . . . . . . . . . . . . . . 113
5.2 Gorenstein Flat Dimension . . . . . . . . . . . . . . . . . . 120
5.3 The Ultimate AB Formula . . . . . . . . . . . . . . . . . . . 127
5.4 Comparing Tor–dimensions . . . . . . . . . . . . . . . . . . 131

**6 G–injectivity** **135**
6.1 Gorenstein Injective Modules . . . . . . . . . . . . . . . . . 135
6.2 Gorenstein Injective Dimension . . . . . . . . . . . . . . . . 141
6.3 G–injective versus G–flat Dimension . . . . . . . . . . . . . 148
6.4 Exercises in Stability . . . . . . . . . . . . . . . . . . . . . 152

**A Hyperhomology** **159**
A.1 Basic Definitions and Notation . . . . . . . . . . . . . . . . 160
A.2 Standard Functors and Morphisms . . . . . . . . . . . . . . 168
A.3 Resolutions . . . . . . . . . . . . . . . . . . . . . . . . . . 171
A.4 (Almost) Derived Functors . . . . . . . . . . . . . . . . . . 175
A.5 Homological Dimensions . . . . . . . . . . . . . . . . . . . 180
A.6 Depth and Width . . . . . . . . . . . . . . . . . . . . . . . 183
A.7 Numerical and Formal Invariants . . . . . . . . . . . . . . . 185
A.8 Dualizing Complexes . . . . . . . . . . . . . . . . . . . . . 187

**Bibliography** **191**

**List of Symbols** **197**

**Index** **199**

# Introduction

Introduction

In 1967 Auslander [1] introduced a new invariant for finitely generated modules over commutative Noetherian rings: a relative homological dimension called the G–dimension. The 'G' is, no doubt, for 'Gorenstein' and chosen because the following are equivalent for a local ring $R$:

- $R$ is Gorenstein.
- The residue field $R/\mathfrak{m}$ has finite G–dimension ($\mathfrak{m}$ is the unique maximal ideal).
- All finitely generated $R$–modules have finite G–dimension.

This characterization of Gorenstein rings (rings of finite self-injective dimension) is parallel to the Auslander–Buchsbaum–Serre characterization of regular rings (rings of finite global dimension), but to make the analogy complete a fourth condition, dealing with non-finitely generated modules, is needed.

So far, the most successful approach to G–dimension for non-finitely generated modules is the one taken in the 1990s by Enochs et al. in [22–32]. At first (quoting from the abstract of [32]) "... to get good results it was necessary to take the base ring Gorenstein", but the theory of Foxby equivalence[1] has subsequently brought about good results over rings with dualizing complexes in general. In particular, Enochs' group [32] and Foxby [39] have outlined a beautiful theory for Gorenstein projective and flat dimensions (extensions of the original G–dimension) and Gorenstein injective dimension (dual to the Gorenstein projective one) over Cohen–Macaulay local rings with a dualizing module.

The purpose of this monograph is to give a detailed and up to date presentation of the theory of Gorenstein dimensions. In chapter 1 we review Auslander's G–dimension using homological algebra in the tradition of the fifties and sixties. In the second chapter we extend the G–dimension to complexes and start using hyperhomological algebra (an extension of homological algebra for modules). The Gorenstein projective, flat, and injective dimensions are treated in chapters 4, 5, and 6, and the theory of Foxby equivalence is dealt with in chapter 3. The synopsis, following immediately after this introduction, gives an overview of the principal results.

[1] Some authors call it Foxby duality.

This book is intended as a reference for Gorenstein dimensions. It is aimed at mathematicians, especially graduate students, working with homological dimensions in commutative algebra. Indeed, any admirer of classics like the Auslander–Buchsbaum formula, the Auslander–Buchsbaum–Serre characterization of regular rings, and Bass' formula for injective dimension must be intrigued by the highlights of this monograph.

The reader is expected to be well-versed in commutative algebra and in the standard applications of homological methods within this realm. In chapters 2–6 we work consistently with complexes of modules, but for the benefit of those who prefer plain modules, all major results are restated for modules in traditional notation. The appendix offers a crash course in hyperhomological algebra, including homological dimensions. Hopefully, this easy reference will make the proofs accessible, also for casual users of hyperhomological methods. We work with categories because the language is convenient, but, apart from the basic definitions, no knowledge of category theory is required.

To the relief of some — and to the dismay of others — it should be emphasized that we do not use the derived category: we use equivalence of complexes, but we never formally invert the quasi-isomorphisms. This deficiency does not really give rise to problems, because we never need the deeper properties of the derived category, e.g., the triangulated structure. But we are prevented from using true derived functors of complexes, and, needless to say, this makes some proofs a little extra involved. The experienced user of derived categories is invited to redo these, somewhat clumsy, proofs and celebrate the power of derived functors.

While the form may have been changed and the proofs recast, most results in this book have appeared before in conference proceedings, research papers, etc. At the beginning of each chapter credit is given for the key ideas to be introduced, and further references are included in the notes found at the end of most sections. Any omission or inaccuracy in the references is unintended, and absence of references should not be interpreted as a claim for credit from the side of the author.

To set the record straight, once and for all, the author only wants to claim credit for Theorems (5.1.11), (5.3.8), (6.2.15), and (6.4.2). All other results can — if no one else is specifically credited — be ascribed to one or more of the gentlemen: Maurice Auslander, Mark Bridger, Edgar E. Enochs, Hans-Bjørn Foxby, Overtoun M. G. Jenda, Blas Torrecillas, Jinzhong Xu, and Siamak Yassemi.

Finally, one should be aware that the original papers by Auslander and Bridger [1, 2] have triggered work in other directions also. E.g., the study of maximal Cohen–Macaulay modules by Auslander, Buchweitz, and Reiten, to name a few, and studies of generalized G-dimensions by Golod, the author, and others. However, these aspects fall beyond the scope of this book.

# Synopsis

We are going to study refinements of some of the central notions in classical homological algebra: the projective, the flat, and the injective dimension for modules over commutative Noetherian rings. In the following $R$ denotes such a ring.

The projective dimension is a most important invariant for modules; this is illustrated by the next two classical and highly advertised results.

**Regularity Theorem.** *Let $R$ be a local ring with residue field $k$. The following are equivalent:*

(*i*) *$R$ is regular.*

(*ii*) $\mathrm{pd}_R\, k < \infty$.

(*iii*) $\mathrm{pd}_R\, M < \infty$ *for all finite (that is, finitely generated) $R$–modules $M$.*

(*iv*) $\mathrm{pd}_R\, M < \infty$ *for all $R$–modules $M$.*

**Auslander–Buchsbaum Formula.** *Let $R$ be a local ring, and let $M$ be a finite $R$–module. If $M$ is of finite projective dimension, then*

$$\mathrm{pd}_R\, M = \mathrm{depth}\, R - \mathrm{depth}_R\, M.$$

The Regularity Theorem ( [12, Theorem 2.2.7] and [49, Theorem 19.2]) is from the mid 1950s and due to Serre [54], and to Auslander and Buchsbaum [3]. Also Auslander and Buchsbaum's famous formula [12, Theorem 1.3.3] goes back to those heydays [4], when homological methods found their way into commutative algebra. We call attention to these classics because results of their kind will play a key role in our study.

The subject of the first chapter is Auslander's *G–dimension*, or *Gorenstein dimension*, for finite modules. Not only is it a *finer invariant* than projective dimension, i.e., there is always an inequality:

$$\mathrm{G\text{–}dim}_R\, M \leq \mathrm{pd}_R\, M,$$

but equality holds whenever the projective dimension is finite. We say that G–dimension is a *refinement* of projective dimension for finite modules. The G–dimension shares many of the nice properties of the projective dimension; there is for example an

**Auslander–Bridger Formula.** *Let $R$ be a local ring, and let $M$ be a finite $R$–module. If $M$ is of finite G–dimension, then*

$$\text{G–dim}_R\, M = \operatorname{depth} R - \operatorname{depth}_R M.$$

And the next result parallels the Regularity Theorem.

**Gorenstein Theorem, G–dimension Version.** *Let $R$ be a local ring with residue field $k$. The following are equivalent:*

(*i*) *$R$ is Gorenstein.*

(*ii*) $\text{G–dim}_R\, k < \infty$.

(*iii*) $\text{G–dim}_R\, M < \infty$ *for all finite $R$–modules $M$.*

However, part (*iv*) in the Regularity Theorem lacks a counterpart!

We make up for this in chapters 2 and 3, where the *Auslander class* $\mathcal{A}_0(R)$ is introduced for a local ring $R$ admitting a dualizing complex. The finite modules in $\mathcal{A}_0(R)$ are exactly those of finite G–dimension, but the class also contains non-finite modules, so the next theorem is an extension of the G–dimension version above.

**Gorenstein Theorem, $\mathcal{A}$ Version.** *Let $R$ be a local ring with residue field $k$. If $R$ admits a dualizing complex, then the following are equivalent:*

(*i*) *$R$ is Gorenstein.*

(*ii*) $k \in \mathcal{A}_0(R)$.

(*iii*) $M \in \mathcal{A}_0(R)$ *for all finite $R$–modules $M$.*

(*iv*) $M \in \mathcal{A}_0(R)$ *for all $R$–modules $M$.*

The next task is to establish a quantitative version of this Theorem, and to this end we extend the G–dimension to non-finite modules. In chapters 4, 5, and 6 we concentrate on Cohen–Macaulay local rings admitting a dualizing (canonical) module; for such rings two extensions of the G–dimension are possible.

In chapter 4 we introduce the *Gorenstein projective dimension* — a refinement of the projective dimension, also for non-finite modules — and we prove that a module has finite Gorenstein projective dimension if and only if it belongs to the Auslander class. In particular, we have the following special case of the $\mathcal{A}$ version:

**Gorenstein Theorem, GPD Version.** *Let $R$ be a Cohen–Macaulay local ring with residue field $k$. If $R$ admits a dualizing module, then the following are equivalent:*

(*i*) *$R$ is Gorenstein.*

(*ii*) $\mathrm{Gpd}_R\, k < \infty$.

(*iii*) $\mathrm{Gpd}_R\, M < \infty$ *for all finite $R$–modules $M$.*

(*iv*) $\mathrm{Gpd}_R\, M < \infty$ *for all $R$–modules $M$.*

In chapter 5 the *Gorenstein flat dimension* — a refinement of the usual flat dimension — is examined. The Gorenstein flat and projective dimensions behave much like the usual flat and projective dimensions. For every module $M$ there is an inequality:

$$\mathrm{Gfd}_R\, M \leq \mathrm{Gpd}_R\, M;$$

the two dimensions are simultaneously finite, that is,

$$\mathrm{Gfd}_R\, M < \infty \Leftrightarrow \mathrm{Gpd}_R\, M < \infty;$$

and for finite modules $M$ the two dimensions agree and coincide with the G–dimension:

$$\mathrm{Gfd}_R\, M = \mathrm{Gpd}_R\, M = \text{G–dim}_R\, M.$$

The Gorenstein flat dimension — just like the usual flat one — satisfies a formula of the Auslander–Buchsbaum type:

**AB Formula for Gorenstein Flat Dimension.** *Let $R$ be a Cohen–Macaulay local ring with a dualizing module. If $M$ is an $R$–module of finite Gorenstein flat dimension, i.e., $M \in \mathcal{A}_0(R)$, then*

$$\mathrm{Gfd}_R\, M = \sup\{\operatorname{depth} R_\mathfrak{p} - \operatorname{depth}_{R_\mathfrak{p}} M_\mathfrak{p} \mid \mathfrak{p} \in \operatorname{Spec} R\}.$$

(By definition $\operatorname{depth}_{R_\mathfrak{p}} M_\mathfrak{p}$ is the number of the first non-vanishing $\operatorname{Ext}_{R_\mathfrak{p}}(R_\mathfrak{p}/\mathfrak{p}_\mathfrak{p}, M_\mathfrak{p})$ module, see also page 183.)

We call it an *AB formula* because it is similar to not only the Auslander–Buchsbaum formula for flat dimension (proved by Chouinard [14]) but also the Auslander–Bridger formula for G–dimension. )

The classical characterization below of Gorenstein rings in terms of homological dimensions (see [12, 3.1.25]) is due to Bass [11] and Foxby [35].

**Gorenstein Theorem, PD/ID Version.** *Let $R$ be a local ring. The following are equivalent:*

(*i*) *$R$ is Gorenstein.*

(*ii*) $\operatorname{id}_R R < \infty$.

(*iii*) $\operatorname{id}_R M < \infty$ *and* $\operatorname{pd}_R M < \infty$ *for some finite $R$–module $M \neq 0$.*

(*iv*) *A finite $R$–module $M$ has finite projective dimension if and only if it has finite injective dimension; that is,* $\operatorname{pd}_R M < \infty \Leftrightarrow \operatorname{id}_R M < \infty$.

This can be improved. The dual notions of the Auslander class and the Gorenstein projective dimension are, respectively, the *Bass class* $\mathcal{B}_0(R)$ and the *Gorenstein injective dimension*; these are introduced and studied in chapters 3 and 6. The Gorenstein injective dimension is a refinement of the usual injective dimension; it is linked to the Bass class as one would expect,

$$\operatorname{Gid}_R M < \infty \quad \Longleftrightarrow \quad M \in \mathcal{B}_0(R),$$

and finiteness of Gorenstein injective dimensions characterizes Gorenstein rings:

**Gorenstein Theorem, GID Version.** *Let $R$ be a Cohen–Macaulay local ring with residue field $k$. If $R$ admits a dualizing module, then the following are equivalent:*

(*i*) *$R$ is Gorenstein.*

(*ii*) $\operatorname{Gid}_R k < \infty$.

(*iii*) $\operatorname{Gid}_R M < \infty$ *for all finite $R$–modules $M$.*

(*iv*) $\operatorname{Gid}_R M < \infty$ *for all $R$–modules $M$.*

A non-trivial finite $R$–module has finite depth, so the GFD/GID version below extends the PD/ID version in several ways.

**Gorenstein Theorem, GFD/GID Version.** *Let $R$ be a Cohen–Macaulay local ring. If $D$ is a dualizing module for $R$, then the following are equivalent:*

(*i*) *$R$ is Gorenstein.*

(*ii*) $\operatorname{Gid}_R R < \infty$.

(*ii'*) $\operatorname{Gfd}_R D < \infty$.

(*iii*) $\operatorname{Gid}_R M < \infty$ *and* $\operatorname{fd}_R M < \infty$ *for some $R$–module $M$ of finite depth.*

(*iii'*) $\operatorname{id}_R M < \infty$ *and* $\operatorname{Gfd}_R M < \infty$ *for some $R$–module $M$ of finite depth.*

(*iv*) *An $R$–module $M$ has finite Gorenstein flat dimension if and only if it has finite Gorenstein injective dimension; that is,* $\operatorname{Gfd}_R M < \infty \Leftrightarrow \operatorname{Gid}_R M < \infty$.

Non-trivial finite modules of finite injective dimension are only found over Cohen–Macaulay rings. This was conjectured by Bass [11] (and proved by Peskine and Szpiro [51] and Roberts [53]), and his celebrated formula,

$$\mathrm{id}_R\, M = \operatorname{depth} R,$$

for these modules tells us that, indeed, they are special. Also finite modules of finite Gorenstein injective dimension seem to be special, at least in the sense that there is a

**Bass Formula for Gorenstein Injective Dimension.** *Let $R$ be a Cohen–Macaulay local ring with a dualizing module. If $M \neq 0$ is a finite $R$–module of finite Gorenstein injective dimension, then*

$$\mathrm{Gid}_R\, M = \operatorname{depth} R.$$

The classical duality between flat and injective dimension — as captured by Ishikawa's formulas [42]:

$$\mathrm{id}_R(\mathrm{Hom}_R(M,E)) = \mathrm{fd}_R\, M \quad \text{and} \quad \mathrm{fd}_R(\mathrm{Hom}_R(M,E)) = \mathrm{id}_R\, M,$$

where $M$ is any $R$–module and $E$ is a faithfully injective one — extends, at least partially, to Gorenstein dimensions. In chapter 6 we prove the following:

**Theorem.** *Let $R$ be a Cohen–Macaulay local ring with a dualizing module, and let $E$ be an injective $R$–module. For every $R$–module $M$ there is an inequality:*

$$\mathrm{Gid}_R(\mathrm{Hom}_R(M,E)) \leq \mathrm{Gfd}_R\, M,$$

*and equality holds if $E$ is faithfully injective.*

**Proposition.** *Let $R$ be a Cohen–Macaulay local ring with a dualizing module, and let $E$ be an injective $R$–module. For every $R$–module $M$ there is an inequality:*

$$\mathrm{Gfd}_R(\mathrm{Hom}_R(M,E)) \leq \mathrm{Gid}_R\, M;$$

*and if $E$ is faithfully injective, then the two dimensions are simultaneously finite, that is,*

$$\mathrm{Gfd}_R(\mathrm{Hom}_R(M,E)) < \infty \iff \mathrm{Gid}_R\, M < \infty.$$

Ishikawa's formulas belong to a group of results, which we lump together under the label *stability*. A typical and well-known stability result says that the derived tensor product $M \otimes^{\mathbf{L}}_R M'$ of two modules of finite flat dimension has itself finite flat dimension; to be exact:

$$\mathrm{fd}_R(M \otimes^{\mathbf{L}}_R M') \leq \mathrm{fd}_R\, M + \mathrm{fd}_R\, M'.$$

There are also a number of stability results involving Gorenstein dimensions, but most of them — like the one above — are only interesting from a "derived category point of view". We prove a few stability results in chapter 6 and leave the rest of them as exercises for the interested reader.

In this book we consistently work with complexes of modules, and the results stated above are special cases of what we prove.

As indicated in this short outline, the monograph focuses on the Gorenstein dimensions' ability to characterize Gorenstein rings; and the coverage of interrelations between Gorenstein dimensions takes clues from the domestic triangle: projective–injective–flat dimension.

# Conventions and Prerequisites

This preliminary chapter records the blanket assumptions, some basic notation, and a few important results to be used throughout the book.

All rings are assumed to be **commutative and Noetherian with a unit** $1 \neq 0$; in particular, $R$ always denotes such a ring. All modules are assumed to be unitary.

We will often need to impose extra assumptions on the base ring $R$. Whenever such assumptions are needed throughout an entire section, they are stated at the beginning of the section in a separate paragraph labeled "setup".

## Notation and Basics

We use the standard notation of commutative ring theory and classical homological algebra, but to keep the record straight a few essentials are spelled out below. In the chapters to come we will, from time to time, recall definitions and results from the literature, just to make sure we speak the same language.

The vocabulary of hyperhomological algebra is explained in detail in the appendix. Modules are also complexes (concentrated in degree zero), and for modules the definitions given in the appendix agree with the usual ones, so it also provides a recap on classical homological algebra.

All notions and symbols defined within the text are listed in the index and the list of symbols; in the former bold face numbers point to definitions. Whenever the explanations given here turn out insufficient, please accept the author's apology and refer to the literature. Keeping to the author's personal favorites — [49] (Matsumura) and [12] (Bruns and Herzog) for commutative algebra, [13] (Cartan and Eilenberg) and [60] (Weibel) for homological algebra, and [47] (MacLane) for categories — should eliminate the risk of misunderstandings.

References to the literature are, as illustrated above, given in square brackets. References to paragraphs within the text usually include the paragraph label as well as the number, e.g., 'Lemma (1.2.6)', but sometimes the label is omitted.

Stand-alone labels also occur, and they always refer to the last paragraph with the label in question: e.g., 'the Lemma' would in paragraph (1.2.7) refer to Lemma (1.2.6).

As usual we denote the integers by $\mathbb{Z}$, the natural numbers by $\mathbb{N}$, and we set $\mathbb{N}_0 = \{0\} \cup \mathbb{N}$. When possible, we write down equalities an inequalities in such a way that they hold also when one or more of the quantities involved are infinite. The following rules are used for addition and subtraction in $\{-\infty\} \cup \mathbb{Z} \cup \{\infty\}$: $a + \infty = \infty + a = \infty$ and $a - \infty = -\infty$ for all $a \in \mathbb{Z}$, $\infty + \infty = \infty$ and $-\infty - \infty = -\infty$, while $\infty - \infty$ is undefined.

For supremum and infimum we use the conventions: $\sup \emptyset = -\infty$ and $\inf \emptyset = \infty$.

We use two-letter abbreviations (pd, fd, id) for the standard homological dimensions; and we write 'depth' and 'dim' for the depth and the Krull dimension.

For elements $x_1, \ldots, x_t$ in a ring $R$ we denote by $(x_1, \ldots, x_t)$ the ideal $Rx_1 + \cdots + Rx_t$ generated by the elements.

A ring is said to be *local* if it has a unique maximal ideal. The notation $(R, \mathfrak{m}, k)$ means that $R$ is local with maximal ideal $\mathfrak{m}$ and ***residue field*** $k = R/\mathfrak{m}$. In general, for a prime ideal $\mathfrak{p} \in \operatorname{Spec} R$ the residue field of the local ring $R_\mathfrak{p}$ is denoted by $k(\mathfrak{p})$, i.e., $k(\mathfrak{p}) = R_\mathfrak{p}/\mathfrak{p}_\mathfrak{p}$.

Finitely generated modules are, for short, called ***finite***. The ***injective hull*** of an $R$–module $M$ is denoted by $\mathrm{E}_R(M)$. In the literature it is also called the injective envelope of $M$.

When applied to maps the word ***natural*** is synonymous with 'functorial'.

A ***ladder*** is a commutative diagram with exactly two rows (or columns); it is said to be ***exact*** when the rows (columns) are so. In particular, a ***short exact ladder*** is a commutative diagram with two short exact rows (columns):

$$\begin{array}{ccccccccc} 0 & \longrightarrow & M' & \longrightarrow & M & \longrightarrow & M'' & \longrightarrow & 0 \\ & & \downarrow & & \downarrow & & \downarrow & & \\ 0 & \longrightarrow & N' & \longrightarrow & N & \longrightarrow & N'' & \longrightarrow & 0 \end{array}$$

For an $R$–module $M$ we denote by $\mathrm{z}_R\, M$ the set of ***zero-divisors*** for $M$:

$$\mathrm{z}_R\, M = \{r \in R \mid \exists x \in M - \{0\} : rx = 0\}.$$

In particular, $\mathrm{z}\, R = \mathrm{z}_R\, R$ is the set of zero-divisors in $R$:

$$\mathrm{z}\, R = \{r \in R \mid rx = 0 \text{ for some } x \neq 0 \text{ in } R\}.$$

The ***annihilator***, $\operatorname{Ann}_R M$, of an $R$–module $M$ is the set

$$\operatorname{Ann}_R M = \{r \in R \mid \forall x \in M : rx = 0\}.$$

The annihilator of an element $x \in M$ is the annihilator of the cyclic module $(x) = Rx$.

# Standard Tools

The results below are, by now, folklore.

***Nakayama's Lemma.*** *Let $(R, \mathfrak{m}, k)$ be local, and let $M$ be a finite $R$–module. If $M \neq 0$, then $\mathfrak{m}M \neq M$, so $M/\mathfrak{a}M \neq 0$ for every proper ideal $\mathfrak{a}$ in $R$; in particular, $M \otimes_R k \neq 0$.*

***Hom Vanishing Lemma.*** *Consider two $R$-modules $M$ and $N$. A necessary condition for the module $\operatorname{Hom}_R(M, N)$ to be non-zero is the existence of elements $m \in M$ and $n \neq 0$ in $N$ such that*

$$\operatorname{Ann}_R(m) \subseteq \operatorname{Ann}_R(n).$$

*This condition is also sufficient if $N$ is injective or if $M$ is finite.*

***Hom Vanishing Corollary.*** *If $M$ and $N$ are finite $R$–modules, then*

$$\operatorname{Hom}_R(M, N) \neq 0 \iff \operatorname{Ann}_R M \subseteq z_R N.$$

***Snake Lemma.*** *Consider a short exact ladder*

$$\begin{array}{ccccccccc} 0 & \longrightarrow & M' & \longrightarrow & M & \longrightarrow & M'' & \longrightarrow & 0 \\ & & \downarrow{\scriptstyle \psi'} & & \downarrow{\scriptstyle \psi} & & \downarrow{\scriptstyle \psi''} & & \\ 0 & \longrightarrow & N' & \longrightarrow & N & \longrightarrow & N'' & \longrightarrow & 0 \end{array}$$

*There is an exact sequence of kernels and cokernels:*

$$0 \to \operatorname{Ker} \psi' \to \operatorname{Ker} \psi \to \operatorname{Ker} \psi'' \to \operatorname{Coker} \psi' \to \operatorname{Coker} \psi \to \operatorname{Coker} \psi'' \to 0.$$

# Standard Homomorphisms

Let $S$ be an $R$–algebra. (In all our applications $S$ will be a homomorphic image of $R$, usually $R$ itself.) Let $M$ be an $R$–module and let $N$ and $P$ be $S$–modules (and thereby $R$–modules). Then $N \otimes_R M$, $\operatorname{Hom}_R(N, M)$, and $\operatorname{Hom}_R(M, N)$ are also $S$–modules, and there are five natural homomorphisms of $S$–modules:

The *associativity* homomorphism,

$$\sigma_{PNM} : (P \otimes_S N) \otimes_R M \longrightarrow P \otimes_S (N \otimes_R M)$$

given by

$$\sigma_{PNM}((p \otimes n) \otimes m) = p \otimes (n \otimes m),$$

is always invertible. The same holds for *adjointness*,

$$\rho_{PNM} : \operatorname{Hom}_R(P \otimes_S N, M) \xrightarrow{\cong} \operatorname{Hom}_S(P, \operatorname{Hom}_R(N, M))$$

given by

$$\rho_{PNM}(\psi)(p)(n) = \psi(p \otimes n),$$

and *swap*,

$$\varsigma_{PMN} : \operatorname{Hom}_S(P, \operatorname{Hom}_R(M, N)) \xrightarrow{\cong} \operatorname{Hom}_R(M, \operatorname{Hom}_S(P, N))$$

given by

$$\varsigma_{PMN}(\psi)(m)(p) = \psi(p)(m).$$

The *tensor evaluation* homomorphism,

$$\omega_{PNM} : \operatorname{Hom}_S(P, N) \otimes_R M \longrightarrow \operatorname{Hom}_S(P, N \otimes_R M)$$

given by

$$\omega_{PNM}(\psi \otimes m)(p) = \psi(p) \otimes m,$$

is invertible under each of the next two extra conditions:

- $P$ is finite and projective; or
- $P$ is finite and $M$ is flat.

The *Hom evaluation* homomorphism,

$$\theta_{PNM} : P \otimes_S \operatorname{Hom}_R(N, M) \longrightarrow \operatorname{Hom}_R(\operatorname{Hom}_S(P, N), M)$$

given by

$$\theta_{PNM}(p \otimes \psi)(\vartheta) = \psi\vartheta(p),$$

is invertible under each of the next two extra conditions:

- $P$ is finite and projective; or
- $P$ is finite and $M$ is injective.

These standard homomorphisms were used systematically in [13], but some of the criteria for invertibility of the evaluation homomorphisms were first described in [42].

## Homological Dimensions

The projective, flat, and injective dimension of $R$-modules are the standard (absolute) homological dimensions. They are defined in terms of resolutions, and they can be computed in terms of derived functors:

$$\begin{aligned}
\mathrm{pd}_R M &= \sup\{m \in \mathbb{N}_0 \mid \mathrm{Ext}_R^m(M,T) \neq 0 \text{ for some module } T\};\\
\mathrm{fd}_R M &= \sup\{m \in \mathbb{N}_0 \mid \mathrm{Tor}_m^R(T,M) \neq 0 \text{ for some module } T\}; \quad \text{and}\\
\mathrm{id}_R M &= \sup\{m \in \mathbb{N}_0 \mid \mathrm{Ext}_R^m(T,M) \neq 0 \text{ for some module } T\}.
\end{aligned}$$

For the flat and injective dimensions it is always sufficient to test by cyclic modules $T$, and if $M$ is finite, then the projective dimension can be tested by finite modules.

When they are finite, the homological dimensions of $R$-modules are bounded by the Krull dimension of the ring:

$$\begin{aligned}
\mathrm{pd}_R M < \infty &\implies \mathrm{pd}_R M \leq \dim R,\\
\mathrm{fd}_R M < \infty &\implies \mathrm{fd}_R M \leq \dim R; \quad \text{and}\\
\mathrm{id}_R M < \infty &\implies \mathrm{id}_R M \leq \dim R.
\end{aligned}$$

Actually, if $\dim R < \infty$, then the flat and injective dimensions are bounded by the number:

$$\sup\{\operatorname{depth} R_\mathfrak{p} \mid \mathfrak{p} \in \operatorname{Spec} R\} = \begin{cases} \dim R & \text{if } R \text{ is Cohen–Macaulay, and} \\ \dim R - 1 & \text{otherwise.} \end{cases}$$

These bounds go back to [5], [52], and [48].

The next two famous results [12, Theorems 1.3.3 and 3.1.17] first appeared as, respectively, [4, Theorem 3.7] and [11, Lemma (3.3)].

***Auslander–Buchsbaum Formula.*** *Let $R$ be a local ring, and let $M$ be a finite $R$-module. If $M$ is of finite projective dimension, then*

$$\mathrm{pd}_R M = \operatorname{depth} R - \operatorname{depth}_R M.$$

***Bass Formula.*** *Let $R$ be a local ring, and let $M \neq 0$ be a finite $R$-module. If $M$ is of finite injective dimension, then*

$$\mathrm{id}_R M = \operatorname{depth} R.$$

The fundamental stability results below follow straight from the definitions and the standard homomorphisms (e.g., see [42]).

***Stability.*** *Let $F, F'$ be flat, $I, I'$ be injective, and $P, P'$ be projective $R$–modules. Then the following hold:*

- *$F \otimes_R F'$ is flat;*
- *$\operatorname{Hom}_R(F, I)$ is injective;*
- *$I \otimes_R F$ is injective;*
- *$\operatorname{Hom}_R(I, I')$ is flat;*
- *$P \otimes_R P'$ is projective;*
- *$\operatorname{Hom}_R(P, P')$ is flat, and even projective if $P$ is finite; in particular*
- *$\operatorname{Hom}_R(P, P')$ is finite and projective if both $P$ and $P'$ are so.*

# A Hierarchy of Rings

A local ring $R$ is *Cohen–Macaulay* if some, equivalently every, system of parameters is an $R$–sequence. That is, $R$ is Cohen–Macaulay when the dimension equals the depth: $\dim R = \operatorname{depth} R$. The depth of $R$ cannot exceed the dimension, so the *Cohen–Macaulay defect*

$$\operatorname{cmd} R = \dim R - \operatorname{depth} R$$

is a non-negative integer, and $R$ is Cohen–Macaulay if and only if $\operatorname{cmd} R = 0$.

A local ring $R$ is *Gorenstein* if it has finite self-injective dimension, i.e., $\operatorname{id}_R R < \infty$. By [49, Theorem 18.1] a local ring is Gorenstein if and only if it is Cohen–Macaulay and some, equivalently every, system of parameters generates an irreducible ideal.

A local ring is *regular* if the maximal ideal is generated by a system of parameters. By [49, Theorem 19.2] a local ring $(R, \mathfrak{m}, k)$ is regular if and only if it has finite global dimension, i.e., $\operatorname{pd}_R M \leq \dim R$ for all $R$–modules $M$, and it is sufficient that $\operatorname{pd}_R k < \infty$.

A non-local ring $R$ is said to be, respectively, Cohen–Macaulay, Gorenstein, or regular if $R_\mathfrak{m}$ is Cohen–Macaulay, Gorenstein, or regular for every maximal ideal $\mathfrak{m}$ in $R$.

Every regular ring is Gorenstein, and every Gorenstein ring is Cohen–Macaulay [12, Proposition 3.1.20].

The Cohen–Macaulay defect of a non-local ring $R$ is the number

$$\begin{aligned}\operatorname{cmd} R &= \sup\{\operatorname{cmd} R_\mathfrak{m} \mid \mathfrak{m} \in \operatorname{Max} R\}\\ &= \sup\{\operatorname{cmd} R_\mathfrak{p} \mid \mathfrak{p} \in \operatorname{Spec} R\}.\end{aligned}$$

The first equality is the definition and the second follows by [11, Lemma (3.1)].

Prime examples of regular local rings are rings of formal power series with coefficients in a field, cf. [12, Theorem 2.2.13], and other rings from the hierarchy are easily constructed by factoring out elements in such rings. E.g., if $k$ is a field, then

- $R = k[\![X, Y]\!]$ is a 2-dimensional regular local ring;
- $R' = R/(XY)$ is a 1-dimensional Gorenstein ring but not a domain and, hence, not regular;
- $R'' = R/(X^2, XY)$ is not even Cohen–Macaulay: $\dim R'' = 1$ but $\operatorname{depth} R'' = 0$; and
- $R''' = R/(X^2, XY, Y^2)$ is 0-dimensional and, therefore, Cohen–Macaulay, but not Gorenstein: $(0) = (x) \cap (y)$, where $x$ and $y$ are the residue classes in $R'''$ of, respectively, $X$ and $Y$.

This follows easily by [49, Theorem 18.1] and [12, Propositions 2.2.3 and 3.1.19].

# Chapter 1

# The Classical Gorenstein Dimension

The G-dimension, short for Gorenstein dimension, of finite modules was introduced by Auslander in [1], and the finer details were developed in his joint paper [2] with Bridger. This chapter is based on [1], and most of the proofs, or at least the ideas behind them, can be found there. The chapter also includes some results from [2], while others have been deferred to the next chapter.

All results in this chapter will be reestablished in more general form (for complexes) in chapter 2.

## 1.1 The G–class

The G–dimension is a relative homological dimension, so the first step is to define the modules that will serve as "building blocks" in the resolutions. These modules will later — in chapter 4 — be given a proper name, but for the time being they will just be "modules in the G–class".

(1.1.1) ***Biduality.*** For an $R$–module $M$ the ***biduality map*** is the canonical map

$$\delta_M : M \longrightarrow \mathrm{Hom}_R(\mathrm{Hom}_R(M,R),R),$$

defined by: $\delta_M(x)(\psi) = \psi(x)$ for $\psi \in \mathrm{Hom}_R(M,R)$ and $x \in M$. It is a homomorphism of $R$–modules and natural in $M$.

The biduality map $\delta_M$ is closely related to the Hom evaluation homomorphism $\theta_{MRR}$ by the commutative diagram

(1.1.1.1)
$$\begin{array}{ccc} M & \xrightarrow{\delta_M} & \mathrm{Hom}_R(\mathrm{Hom}_R(M,R),R) \\ \Big\downarrow \cong & & \Big\uparrow \theta_{MRR} \\ M \otimes_R R & \xrightarrow{\cong} & M \otimes_R \mathrm{Hom}_R(R,R) \end{array}$$

(1.1.2) **Definition.** A finite $R$–module $M$ belongs to the *G–class* $\mathrm{G}(R)$ if and only if

(1) $\mathrm{Ext}^m_R(M,R) = 0$ for $m > 0$;
(2) $\mathrm{Ext}^m_R(\mathrm{Hom}_R(M,R),R) = 0$ for $m > 0$; and
(3) The biduality map $\delta_M : M \to \mathrm{Hom}_R(\mathrm{Hom}_R(M,R),R)$ is an isomorphism.

(1.1.3) **Remark.** A finite free $R$–module $L$ obviously satisfies the first two conditions in the Definition, and $\delta_L$ is clearly an isomorphism. Thus, finite free modules belong to the G–class.

It is, actually, not known if all three conditions in the Definition are necessary to characterize the G–class. The class of modules satisfying the first condition is studied in [50] and [58]. In (2.2.6) we give an example of a module which satisfies the third condition but is not in the G–class.

(1.1.4) ***Duality.*** For an $R$–module $M$ it is standard to set

$$M^* = \mathrm{Hom}_R(M,R) \quad \text{and} \quad M^{**} = (M^*)^*.$$

The modules $M^*$ and $M^{**}$ are called, respectively, the *dual* and *bidual* of $M$, and applying the (*algebraic*) *duality functor* $-^* = \mathrm{Hom}_R(-,R)$ is called *dualizing*.

(1.1.5) ***Torsion.*** For an $R$–module $M$ the *torsion submodule*, $M_{\mathrm{T}}$, is defined as

$$M_{\mathrm{T}} = \{x \in M \mid \exists r \in R - \mathrm{z}R \colon rx = 0\}.$$

The module is said to be *torsion* if $M_{\mathrm{T}} = M$ and *torsion-free* if the torsion submodule is the zero-module. Note that $M$ is a torsion-free $R$–module if and only if all zero-divisors for $M$ are also zero-divisors for $R$; that is,

$$M_{\mathrm{T}} = 0 \iff \mathrm{z}_R\, M \subseteq \mathrm{z}\, R. \tag{1.1.5.1}$$

(1.1.6) **Remark.** Free modules are torsion-free, and submodules of torsion-free modules are obviously torsion-free. Dualizing the sequence $R^\beta \to M \to 0$ we see that the dual of a finite $R$–module can be embedded in a finite free $R$–module and, consequently, it is torsion-free. In particular, all modules in $\mathrm{G}(R)$ are torsion-free.

(1.1.7) **Observation.** If an $R$–module $M$ belongs to the G–class, then so does its dual; that is,

$$M \in \mathrm{G}(R) \implies M^* \in \mathrm{G}(R).$$

This is evident from the definition; on the other hand, it is also clear that the reverse implication does not hold: suppose $G \in \mathrm{G}(R)$ and $M \neq 0$ is torsion, then $(G \oplus M)^* \cong G^* \in \mathrm{G}(R)$, but the module $G \oplus M$ does not belong to $\mathrm{G}(R)$ as, indeed, it is not torsion-free.

(1.1.8) **Lemma.** *Let $M$ be a finite $R$–module and consider the following three conditions:*

(*i*) *The biduality map $\delta_M$ is injective.*

(*ii*) *$M$ can be embedded in a finite free module.*

(*iii*) *$M$ is torsion-free.*

*Conditions (i) and (ii) are equivalent and imply (iii); furthermore, the three conditions are equivalent if $R$ is a domain.*

*Proof.* It is clear that (*ii*) implies (*iii*), and as remarked in (1.1.6) the dual, and hence the bidual, of a finite module can be embedded in a finite free module, so (*i*) implies (*ii*).

Suppose $M$ can be embedded in a finite free module: $M \xrightarrow{\iota} L$. It then follows immediately from the commutative diagram

$$\begin{array}{ccc} M & \xrightarrow{\iota} & L \\ \big\downarrow{\scriptstyle \delta_M} & & \cong\big\downarrow{\scriptstyle \delta_L} \\ M^{**} & \xrightarrow{\iota^{**}} & L^{**} \end{array}$$

that $\delta_M$ is injective, so (*i*) and (*ii*) are equivalent as claimed.

Suppose $R$ is a domain. Using the Hom vanishing lemma it is then straightforward to prove that $\operatorname{Ker}\delta_M = M_{\mathrm{T}}$; so if $M$ is torsion-free, then the biduality map is injective and, hence, all three conditions are equivalent. □

In the literature a finite $R$–module $M$ is often said to be *torsionless* if the biduality map $\delta_M$ is injective.

(1.1.9) **Proposition.** *The following hold for a finite $R$–module $M$:*

(a) *If $M$ is a dual, that is, $M \cong N^*$ for some finite $R$–module $N$, then the sequence*

$$0 \to M \xrightarrow{\delta_M} M^{**} \to \operatorname{Coker}\delta_M \to 0$$

*splits.*

(b) *If $M$ is isomorphic to $M^{**}$, then the biduality map $\delta_M$ is an isomorphism.*

*Proof.* (a): We consider the sequence

$$(\dagger) \qquad 0 \to M \xrightarrow{\delta_M} M^{**} \to \operatorname{Coker}\delta_M \to 0.$$

Exactness in $M^{**}$ and $\operatorname{Coker}\delta_M$ is implicit, so to prove that (†) splits it is sufficient to prove that $\delta_M$ has a section. Assume that there is an isomorphism $\varphi\colon M \xrightarrow{\cong} N^*$, for some finite $R$–module $N$. It is straightforward to check that $(\delta_N)^*\delta_{N^*} = 1_{N^*}$, and it follows that also $\delta_M$ has a section, namely $\varphi^{-1}(\delta_N)^*\varphi^{**}$.

(b): Suppose $M \cong M^{**}$, and set $C = \operatorname{Coker}\delta_M$; then it follows by (a) that the sequence $0 \to M \xrightarrow{\delta_M} M^{**} \to C \to 0$ splits. For each maximal ideal $\mathfrak{m}$ we then have $M_\mathfrak{m} \cong M_\mathfrak{m} \oplus C_\mathfrak{m}$, in particular $M_\mathfrak{m}$ and $M_\mathfrak{m} \oplus C_\mathfrak{m}$ have the same minimal number of generators, so $C_\mathfrak{m} = 0$ and therefore $C = 0$. □

(1.1.10) **Lemma.** *Let $0 \to K \to N \to M \to 0$ be an exact sequence of finite $R$–modules. The following hold:*

(a) *If $M \in \mathrm{G}(R)$, then the sequences*

$$0 \to M^* \to N^* \to K^* \to 0$$

*and*

$$0 \to K^{**} \to N^{**} \to M^{**} \to 0$$

*are exact, and $K$ belongs to the G–class if and only if $N$ does; that is,*

$$K \in \mathrm{G}(R) \iff N \in \mathrm{G}(R).$$

(b) *If $N \in \mathrm{G}(R)$, then there are isomorphisms:*

$$\mathrm{Ext}^m_R(K,R) \cong \mathrm{Ext}^{m+1}_R(M,R)$$

*for $m > 0$.*

(c) *If the sequence $0 \to K \to N \to M \to 0$ splits, then $N$ belongs to the G–class if and only if both $K$ and $M$ do so; that is,*

$$N \in \mathrm{G}(R) \iff K \in \mathrm{G}(R) \wedge M \in \mathrm{G}(R).$$

*Proof.* Dualizing the short exact sequence $0 \to K \to N \to M \to 0$ yields a long exact sequence

$$(\dagger) \qquad \begin{aligned} &0 \to M^* \to N^* \to K^* \to \mathrm{Ext}^1_R(M,R) \to \cdots \\ &\cdots \to \mathrm{Ext}^m_R(M,R) \to \mathrm{Ext}^m_R(N,R) \to \mathrm{Ext}^m_R(K,R) \to \cdots . \end{aligned}$$

(a): Assume that $M \in \mathrm{G}(R)$, then, in particular, $\mathrm{Ext}^1_R(M,R) = 0$, and exactness of the first sequence in (a) is obvious from (†). The biduality maps are natural, so we have the following commutative diagram with exact rows:

$$\begin{array}{ccccccccc} 0 & \longrightarrow & K & \longrightarrow & N & \longrightarrow & M & \longrightarrow & 0 \\ & & \downarrow{\scriptstyle \delta_K} & & \downarrow{\scriptstyle \delta_N} & & \cong\downarrow{\scriptstyle \delta_M} & & \\ 0 & \longrightarrow & K^{**} & \longrightarrow & N^{**} & \longrightarrow & M^{**} & & \end{array}$$

The second row is obtained by dualizing the short exact sequence just established. The diagram shows that also the sequence $0 \to K^{**} \to N^{**} \to M^{**} \to 0$ is exact, and applying the snake lemma we see that $\delta_N$ is an isomorphism if and only if $\delta_K$ is so. Since $M$ is in the G–class, $\mathrm{Ext}^m_R(M,R) = 0$ for $m > 0$, so by exactness of (†) we have isomorphisms $\text{Ext}_R^m(N,R) \cong \text{Ext}_R^m(K,R)$ for $m > 0$. Similarly, it follows from the long exact sequence

$$\begin{aligned} 0 \to K^{**} \to N^{**} \to M^{**} \xrightarrow{0} \text{Ext}_R^1(K^*,R) \to \cdots \\ \cdots \to \text{Ext}_R^m(K^*,R) \to \text{Ext}_R^m(N^*,R) \to \text{Ext}_R^m(M^*,R) \to \cdots, \end{aligned}$$

that $\text{Ext}_R^m(N^*,R) \cong \text{Ext}_R^m(K^*,R)$ for $m > 0$. Thus, $K$ belongs to the G–class if and only if $N$ does so.

(b): If $N \in \text{G}(R)$, then $\text{Ext}_R^m(N,R) = 0$ for $m > 0$, and the desired isomorphisms are evident from (†).

(c): Suppose $0 \to K \to N \to M \to 0$ splits, then so do the dualized sequences $0 \to M^* \to N^* \to K^* \to 0$ and $0 \to K^{**} \to N^{**} \to M^{**} \to 0$. The rows in the short exact ladder

$$(\ddagger) \qquad \begin{array}{ccccccccc} 0 & \longrightarrow & K & \longrightarrow & N & \longrightarrow & M & \longrightarrow & 0 \\ & & \downarrow{\scriptstyle \delta_K} & & \downarrow{\scriptstyle \delta_N} & & \downarrow{\scriptstyle \delta_M} & & \\ 0 & \longrightarrow & K^{**} & \longrightarrow & N^{**} & \longrightarrow & M^{**} & \longrightarrow & 0 \end{array}$$

split, so $\delta_N$ is an isomorphism if and only if $\delta_M$ and $\delta_K$ are isomorphisms. The functors $\text{Ext}_R^m(-,R)$ are additive, so for $m > 0$ we have isomorphisms

$$\text{Ext}_R^m(N,R) \cong \text{Ext}_R^m(K,R) \oplus \text{Ext}_R^m(M,R)$$

and

$$\text{Ext}_R^m(N^*,R) \cong \text{Ext}_R^m(K^*,R) \oplus \text{Ext}_R^m(M^*,R).$$

This proves (c), and the proof is complete. □

Every finite projective module is a direct summand in a finite free module, so the next result is an immediate consequence of part (c) in the Lemma, cf. Remark (1.1.3). However, to stress the kinship with Proposition (2.1.9), we rephrase the proof in terms of Hom evaluation.

(1.1.11) **Proposition.** *Every finite projective R–module belongs to* $\text{G}(R)$.

*Proof.* If $P$ is a finite projective module, then also $\text{Hom}_R(P,R) = P^*$ is projective, so it is immediate that $\text{Ext}_R^m(P,R) = 0$ and $\text{Ext}_R^m(P^*,R) = 0$ for $m > 0$. Furthermore, the Hom evaluation homomorphism $\theta_{PRR}$ is invertible and, hence, so is the biduality map $\delta_P$, cf. diagram (1.1.1.1). □

We now have a fair collection of modules in the G–class, and we are ready to define the G–dimension. However, before doing so we should give an example of a ring with non-projective modules in $\text{G}(R)$.

(1.1.12) **Observation.** Let $(R, \mathfrak{m}, k)$ be a local ring. The dual of the residue field $k^* = \text{Hom}_R(k,R)$ is a $k$–vector space of dimension $\mu_R^0$. ($k^*$ is isomorphic to the annihilator of $\mathfrak{m}$ and also called the socle of $R$.) Hence, the bidual $k^{**}$ is a vector space of dimension $(\mu_R^0)^2$, and it follows by Proposition (1.1.9)(b) that $\delta_k$ is an isomorphism if and only if $\mu_R^0 = 1$.

(1.1.13) **Example.** Let $k$ be a field. The ring $R = k[\![X]\!]/(X^2)$ is self-injective (by Baer's criterion for example), so it is immediate that $\mathrm{Ext}_R^m(k, R) = 0$ and $\mathrm{Ext}_R^m(k^*, R) = 0$ for $m > 0$. The annihilator of the maximal ideal in $R$ is generated by the residue class of $X$, so $\mu_R^0 = 1$ and hence $\delta_k$ is invertible, cf. the Observation. Thus, the $R$–module $k$ belongs $\mathrm{G}(R)$, but $k$ is not projective ($\mathrm{pd}_R\, k = \infty$) as $R$ is not regular.

In a sense — to be made clear by Theorem (1.4.9) — this example is both canonical ($R$ is Gorenstein) and minimal ($\mathrm{id}_R\, R = \dim R = 0$). For more general examples (of non-projective modules in the G–class of a non-Gorenstein ring) see (4.1.5).

## 1.2 G–dimension of Finite Modules

We will first define the G–dimension of finite modules in terms of resolutions by modules from the G–class; and then we will show that (when finite) it can be computed in terms of the derived functors $\mathrm{Ext}_R^m(-, R)$. These two descriptions are merged in, what we call, the GD Theorem (1.2.7). It is modeled on the characterizations of homological dimensions in Cartan and Eilenberg's book — e.g., [13, Proposition VI.2.1] on projective dimension — and this practise is continued through the following chapters.

(1.2.1) **Definition.** A *G-resolution* of a finite $R$–module $M$ is a sequence of modules in $\mathrm{G}(R)$,

$$\cdots \to G_\ell \to G_{\ell-1} \to \cdots \to G_1 \to G_0 \to 0,$$

which is exact at $G_\ell$ for $\ell > 0$ and has $G_0/\mathrm{Im}(G_1 \to G_0) \cong M$. That is, there is an exact sequence

$$\cdots \to G_\ell \to G_{\ell-1} \to \cdots \to G_1 \to G_0 \to M \to 0.$$

The resolution is said to be of *length* $n$ if $G_n \neq 0$ and $G_\ell = 0$ for $\ell > n$.

(1.2.2) **Remark.** Every finite $R$–module has a resolution by finite free modules and, thereby, a G–resolution.

(1.2.3) **Definition.** A finite $R$–module $M$ is said to have *finite G–dimension*, and we write $\mathrm{G\text{-}dim}_R\, M < \infty$ for short, if it has a G–resolution of finite length. We set $\mathrm{G\text{-}dim}_R\, 0 = -\infty$, and for $M \neq 0$ we define the G–dimension of $M$ as follows: For $n \in \mathbb{N}_0$ we say that $M$ has *G–dimension at most n*, and write $\mathrm{G\text{-}dim}_R\, M \leq n$ for short, if and only if $M$ has a G–resolution of length $n$. If $M$

has no G–resolution of finite length, then we say that it has *infinite G–dimension* and write $\text{G-dim}_R M = \infty$.

**(1.2.4) Remark.** Note that also the zero–module is said to have finite G–dimension, and that $\text{G-dim}_R M \in \{-\infty\} \cup \mathbb{N}_0 \cup \{\infty\}$ for any finite $R$–module $M$. If $\text{G-dim}_R M \leq n$, then $M$ has a G–resolution of length $m$ for all $m \geq n$; this follows by adding free summands to the resolution of length $n$. If $M$ is non-zero, then the G–dimension of $M$ is the length of the shortest possible G–resolution of $M$; in particular,

(1.2.4.1) $$M \in \text{G}(R) \iff \text{G-dim}_R M = 0 \vee M = 0.$$

**(1.2.5) Observation.** Let $M$ be a finite $R$–module and consider an exact sequence

$$\cdots \to G_\ell \to G_{\ell-1} \to \cdots \to G_1 \to G_0 \to M \to 0,$$

where the modules $G_\ell$ belong to $\text{G}(R)$. We set

(1.2.5.1) $$\begin{aligned} &K_0 = M, \quad K_1 = \text{Ker}(G_0 \to M), \quad \text{and} \\ &K_\ell = \text{Ker}(G_{\ell-1} \to G_{\ell-2}) \quad \text{for } \ell \geq 2. \end{aligned}$$

For each $\ell \in \mathbb{N}$ we then have a short exact sequence

(†) $$0 \to K_\ell \to G_{\ell-1} \to K_{\ell-1} \to 0.$$

Applying Lemma (1.1.10)(b) to (†) we get isomorphisms

$$\text{Ext}_R^m(K_\ell, R) \cong \text{Ext}_R^{m+1}(K_{\ell-1}, R),$$

which piece together to give isomorphisms

$$\text{Ext}_R^m(K_\ell, R) \cong \text{Ext}_R^{m+\ell}(M, R)$$

for $m > 0$.

Suppose $K_n \in \text{G}(R)$, that is, $\text{G-dim}_R M \leq n$. For $\ell < n$ we then have an exact sequence

$$0 \to K_n \to G_{n-1} \to \cdots \to G_\ell \to K_\ell \to 0,$$

showing that $\text{G-dim}_R K_\ell \leq n - \ell$, and we note that equality holds if $\text{G-dim}_R M = n$.

**(1.2.6) Lemma.** *Let $M$ be a finite $R$–module of finite G–dimension. If $\text{Ext}_R^m(M, R) = 0$ for all $m > 0$, then $M \in \text{G}(R)$.*

*Proof.* First we assume that $\text{G-dim}_R M \leq 1$, then we have an exact sequence

$$0 \to G_1 \to G_0 \to M \to 0,$$

where the modules $G_1$ and $G_0$ belong to $\mathrm{G}(R)$. As $\mathrm{Ext}_R^1(M,R) = 0$ this sequence dualizes to give a short exact sequence

$$0 \to M^* \to {G_0}^* \to {G_1}^* \to 0,$$

and it follows by Observation (1.1.7) and Lemma (1.1.10)(a) that $M^* \in \mathrm{G}(R)$; in particular, $\mathrm{Ext}_R^m(M^*,R) = 0$ for $m > 0$. Dualizing once more we get the second row in the short exact ladder

$$\begin{array}{ccccccccc} 0 & \longrightarrow & G_1 & \longrightarrow & G_0 & \longrightarrow & M & \longrightarrow & 0 \\ & & \cong \big\downarrow \delta_{G_1} & & \cong \big\downarrow \delta_{G_0} & & \big\downarrow \delta_M & & \\ 0 & \longrightarrow & {G_1}^{**} & \longrightarrow & {G_0}^{**} & \longrightarrow & M^{**} & \longrightarrow & 0 \end{array}$$

which, by the snake lemma, allows us to conclude that $\delta_M$ is an isomorphism and, hence, that $M \in \mathrm{G}(R)$.

Now, let $n > 1$ and assume that $\text{G-dim}_R M \leq n-1$ implies $M \in \mathrm{G}(R)$. If $\text{G-dim}_R M \leq n$, then $M$ has a G–resolution of length $n$:

$$0 \to G_n \to \cdots \to G_1 \to G_0 \to M \to 0.$$

We define $K_{n-1}$ as in the Observation; then $\text{G-dim}_R K_{n-1} \leq 1$ and $\mathrm{Ext}_R^m(K_{n-1},R) \cong \mathrm{Ext}_R^{m+n-1}(M,R) = 0$ for $m > 0$, so $K_{n-1} \in \mathrm{G}(R)$ by the above. Now, the exact sequence

$$0 \to K_{n-1} \to G_{n-2} \to \cdots \to G_0 \to M \to 0$$

shows that $\text{G-dim}_R M \leq n-1$, so by the induction hypothesis $M \in \mathrm{G}(R)$. □

(1.2.7) **GD Theorem for Finite Modules.** *Let $M$ be a finite $R$–module and $n \in \mathbb{N}_0$. The following are equivalent:*

(*i*) $\text{G-dim}_R M \leq n$.

(*ii*) $\text{G-dim}_R M < \infty$ *and* $\mathrm{Ext}_R^m(M,R) = 0$ *for* $m > n$.

(*iii*) *In any G–resolution of $M$,*

$$\cdots \to G_\ell \to G_{\ell-1} \to \cdots \to G_0 \to M \to 0,$$

*the kernel*[1] $K_n = \mathrm{Ker}(G_{n-1} \to G_{n-2})$ *belongs to* $\mathrm{G}(R)$.

*Furthermore, if* $\text{G-dim}_R M < \infty$, *then*

$$\text{G-dim}_R M = \sup\{m \in \mathbb{N}_0 \mid \mathrm{Ext}_R^m(M,R) \neq 0\}.$$

[1] Appropriately interpreted for small $n$ as $K_0 = M$ and $K_1 = \mathrm{Ker}(G_0 \to M)$, cf. (1.2.5.1).

*Proof.* Note, right away, that once the equivalence of (*i*) and (*ii*) is established, then the equality $\operatorname{G-dim}_R M = \sup\{m \in \mathbb{N}_0 \mid \operatorname{Ext}_R^m(M,R) \neq 0\}$ for modules of finite G–dimension is immediate.

For $n = 0$ the three conditions are equivalent by the Lemma and (1.2.4.1); they all say that $M \in \mathrm{G}(R)$. We may now assume that $n$ is positive.

(*i*) ⇒ (*ii*): If $\operatorname{G-dim}_R M \leq n$, then $M$ has a G–resolution of length $n$:

$$0 \to G_n \to \cdots \to G_1 \to G_0 \to M \to 0.$$

It follows from Observation (1.2.5) that $\operatorname{Ext}_R^{m+n}(M,R) \cong \operatorname{Ext}_R^m(G_n,R) = 0$ for $m > 0$, that is, $\operatorname{Ext}_R^m(M,R) = 0$ for $m > n$.

(*ii*) ⇒ (*i*): By assumption $M$ has a G–resolution of finite length, say $p$:

$$0 \to G_p \to \cdots \to G_1 \to G_0 \to M \to 0.$$

If $p \leq n$ there is nothing to prove, so we assume that $p > n$. Defining $K_n$ as in (1.2.5.1) we get an exact sequence

$$0 \to K_n \to G_{n-1} \to \cdots \to G_0 \to M \to 0,$$

where $K_n$ has finite G–dimension, at most $p-n$. We assume that $\operatorname{Ext}_R^m(M,R) = 0$ for $m > n$; by the Observation $\operatorname{Ext}_R^m(K_n,R) \cong \operatorname{Ext}_R^{m+n}(M,R) = 0$ for $m > 0$, and then $K_n \in \mathrm{G}(R)$ by the Lemma. Thus, $M$ has a G–resolution of length $n$ as desired.

(*i*)⇔(*iii*): It is clear that (*iii*) implies (*i*), so we assume that $\operatorname{G-dim}_R M \leq n$, i.e., there is an exact sequence

$$0 \to G_n \to \cdots \to G_1 \to G_0 \to M \to 0,$$

where the modules $G_\ell$ belong to $\mathrm{G}(R)$. To prove the assertion it is now sufficient to see that: if

$$0 \to H_n \to P_{n-1} \to \cdots \to P_0 \to M \to 0$$

and

$$0 \to K_n \to G_{n-1} \to \cdots \to G_0 \to M \to 0$$

are exact sequences, $P_0, \ldots, P_{n-1}$ are finite projective modules, and $G_0, \ldots, G_{n-1}$ belong to $\mathrm{G}(R)$, then the kernel $H_n$ belongs to $\mathrm{G}(R)$ if and only if $K_n$ does so.

Since the modules $P_0, \ldots, P_{n-1}$ are projective, there exist homomorphisms $\gamma_0, \ldots, \gamma_n$ making the diagram

$$\begin{array}{ccccccccccccc}
0 & \longrightarrow & H_n & \xrightarrow{\pi_n} & P_{n-1} & \xrightarrow{\pi_{n-1}} & \cdots & \xrightarrow{\pi_1} & P_0 & \xrightarrow{\pi_0} & M & \longrightarrow & 0 \\
 & & \downarrow{\scriptstyle \gamma_n} & & \downarrow{\scriptstyle \gamma_{n-1}} & & & & \downarrow{\scriptstyle \gamma_0} & & \downarrow{\scriptstyle 1_M} & & \\
0 & \longrightarrow & K_n & \xrightarrow{\pi'_n} & G_{n-1} & \xrightarrow{\pi'_{n-1}} & \cdots & \xrightarrow{\pi'_1} & G_0 & \xrightarrow{\pi'_0} & M & \longrightarrow & 0
\end{array}$$

commutative. This diagram gives rise to a sequence

$$0 \to H_n \to K_n \oplus P_{n-1} \to G_{n-1} \oplus P_{n-2} \to \cdots \to G_1 \oplus P_0 \to G_0 \to 0,$$

which we now[2] show is exact.

- The map $H_n \to K_n \oplus P_{n-1}$ is given by $h \mapsto (\gamma_n(h), -\pi_n(h))$, and this map is injective because $\pi_n$ is so. This proves exactness in $H_n$.
- The map $K_n \oplus P_{n-1} \to G_{n-1} \oplus P_{n-2}$ is given by

$$(k,p) \longmapsto (\pi'_n(k) + \gamma_{n-1}(p), -\pi_{n-1}(p)),$$

so an element $(\gamma_n(h), -\pi_n(h))$ is mapped to $(\pi'_n\gamma_n(h) - \gamma_{n-1}\pi_n(h), \pi_{n-1}\pi_n(h)) = (0,0)$. On the other hand, if $(k,p)$ is mapped to $(0,0)$, then $p = -\pi_n(h)$ for some $h \in H_n$, and from the computation

$$\begin{aligned} 0 &= \pi'_n(k) + \gamma_{n-1}(p) \\ &= \pi'_n(k) - \gamma_{n-1}\pi_n(h) \\ &= \pi'_n(k) - \pi'_n\gamma_n(h) \\ &= \pi'_n(k - \gamma_n(h)) \end{aligned}$$

we conclude, by injectivity of $\pi'_n$, that $k = \gamma_n(h)$. This proves exactness in $K_n \oplus P_{n-1}$.
- The general map $G_\ell \oplus P_{\ell-1} \to G_{\ell-1} \oplus P_{\ell-2}$ is given by

$$(g,p) \longmapsto (\pi'_\ell(g) + \gamma_{\ell-1}(p), -\pi_{\ell-1}(p)).$$

As above it is easy to see that an element on the form $(g,p) = (\pi'_{\ell+1}(\tilde{g}) + \gamma_\ell(\tilde{p}), -\pi_\ell(\tilde{p}))$, for some $(\tilde{g},\tilde{p}) \in G_{\ell+1} \oplus P_\ell$, is mapped to $(0,0)$. And on the other hand, if $(\pi'_\ell(g) + \gamma_{\ell-1}(p), -\pi_{\ell-1}(p)) = (0,0)$, then $p = -\pi_\ell(\tilde{p})$ for some $\tilde{p} \in P_\ell$, and from the computation

$$\begin{aligned} 0 &= \pi'_\ell(g) + \gamma_{\ell-1}(p) \\ &= \pi'_\ell(g) - \gamma_{\ell-1}\pi_\ell(\tilde{p}) \\ &= \pi'_\ell(g - \gamma_\ell(\tilde{p})) \end{aligned}$$

it follows that $g - \gamma_\ell(\tilde{p}) = \pi'_{\ell+1}(\tilde{g})$ for some $\tilde{g} \in G_{\ell+1}$. Thus, $(g,p) = (\pi'_{\ell+1}(\tilde{g}) + \gamma_\ell(\tilde{p}), -\pi_\ell(\tilde{p}))$, and this proves exactness in $G_\ell \oplus P_{\ell-1}$.
- The map $G_1 \oplus P_0 \to G_0$, given by $(g,p) \mapsto \pi'_1(g) + \gamma_0(p)$, is surjective: let an element $x \in G_0$ be given and choose, by surjectivity of $\pi_0$, an element $p \in P_0$ such that $\pi_0(p) = \pi'_0(x)$, then

$$\begin{aligned} \pi'_0(x - \gamma_0(p)) &= \pi'_0(x) - \pi'_0\gamma_0(p) \\ &= \pi'_0(x) - 1_M\pi_0(p) \\ &= 0. \end{aligned}$$

Hence, $x - \gamma_0(p) = \pi'_1(g)$ for some $g \in G_1$, and $x = \pi'_1(g) + \gamma_0(p)$ as desired.

---

[2] What follows is actually a mapping cone argument.

The finite projective modules belong to G($R$) by Proposition (1.1.11), so the modules $G_\ell \oplus P_{\ell-1}$ belong to G($R$) by Lemma (1.1.10)(c). In the exact sequence (†) all the modules $G_0, G_1 \oplus P_0, G_2 \oplus P_1, \ldots, G_{n-1} \oplus P_{n-2}$ now belong to the G–class, and it follows by repeated applications of Lemma (1.1.10)(a) that $H_n$ belongs to G($R$) if and only if $K_n \oplus P_{n-1} \in \mathrm{G}(R)$. That is, $H_n \in \mathrm{G}(R) \Leftrightarrow K_n \in \mathrm{G}(R)$ as desired, again by (1.1.10)(c). □

(1.2.8) **Remark.** Let $M$ be a finite $R$-module of G–dimension at most $n$, take a G–resolution of $M$, and break it off in degree $n$ to get an exact sequence

$$0 \to K_n \to G_{n-1} \to \cdots \to G_1 \to G_0 \to M \to 0.$$

If we could only know in advance that the module $K_n$ is of finite G–dimension, then — as in the proof of "(*ii*) ⇒ (*i*)" above — it would follow from Observation (1.2.5) and Lemma (1.2.6) that $K_n$ is in the G–class. Alas, it is not immediate that the kernel $K_n$ has finite G–dimension, and that is why we have to work a little to establish the equivalence of (*i*) and (*iii*) in the Theorem. In chapter 2 a different proof becomes available, see Observation (2.2.4).

(1.2.9) **Corollary.** *Let $0 \to M' \to M \to M'' \to 0$ be an exact sequence of finite $R$–modules. The following hold:*

(a) *If $n \in \mathbb{N}_0$ and* $\text{G–dim}_R\, M'' \leq n$, *then*

$$\text{G–dim}_R\, M' \leq n \quad \Longleftrightarrow \quad \text{G–dim}_R\, M \leq n;$$

*and there are inequalities:*

$$\text{G–dim}_R\, M' \leq \max\{\text{G–dim}_R\, M, \text{G–dim}_R\, M''\}$$

*and*

$$\text{G–dim}_R\, M \leq \max\{\text{G–dim}_R\, M', \text{G–dim}_R\, M''\}.$$

(b) *If* $\text{G–dim}_R\, M' > \text{G–dim}_R\, M''$ *or* $\text{G–dim}_R\, M > \text{G–dim}_R\, M''$, *then*

$$\text{G–dim}_R\, M' = \text{G–dim}_R\, M.$$

(c) *If* $\text{G–dim}_R\, M'' > 0$ *and* $M \in \mathrm{G}(R)$, *then*

$$\text{G–dim}_R\, M' = \text{G–dim}_R\, M'' - 1.$$

*In particular: if two modules in the sequence have finite G–dimension, then so has the third.*

*Proof.* The last assertion is immediate by (a), (b), and (c).

(a): First note that if $\text{G–dim}_R\, M'' \leq 0$, that is, $M'' \in \mathrm{G}(R)$, then the biconditional is known from Lemma (1.1.10)(a). We now assume that $\text{G–dim}_R\, M'' \leq n$ and $n \in \mathbb{N}$. Let $\cdots \to P'_\ell \to P'_{\ell-1} \to \cdots \to P'_0 \to 0$ and

$\cdots \to P''_\ell \to P''_{\ell-1} \to \cdots \to P''_0 \to 0$ be resolutions by finite projective modules of, respectively, $M'$ and $M''$, then we have a commutative diagram

$$\begin{array}{ccccccccc}
 & & 0 & & 0 & & 0 & & \\
 & & \uparrow & & \uparrow & & \uparrow & & \\
0 & \longrightarrow & M' & \longrightarrow & M & \longrightarrow & M'' & \longrightarrow & 0 \\
 & & \uparrow & & \uparrow & & \uparrow & & \\
0 & \longrightarrow & P'_0 & \longrightarrow & P'_0 \oplus P''_0 & \longrightarrow & P''_0 & \longrightarrow & 0 \\
 & & \uparrow & & \uparrow & & \uparrow & & \\
 & & \vdots & & \vdots & & \vdots & & \\
 & & \uparrow & & \uparrow & & \uparrow & & \\
0 & \longrightarrow & P'_{n-1} & \longrightarrow & P'_{n-1} \oplus P''_{n-1} & \longrightarrow & P''_{n-1} & \longrightarrow & 0 \\
 & & \uparrow & & \uparrow & & \uparrow & & \\
0 & \longrightarrow & K'_n & \longrightarrow & K_n & \longrightarrow & K''_n & \longrightarrow & 0 \\
 & & \uparrow & & \uparrow & & \uparrow & & \\
 & & 0 & & 0 & & 0 & &
\end{array}$$

with exact rows and columns. By GD Theorem (1.2.7) we have $K''_n \in \mathrm{G}(R)$, so by Lemma (1.1.10)(a) $K'_n$ is in $\mathrm{G}(R)$ if and only if $K_n$ is so. This proves that $\mathrm{G\text{-}dim}_R\, M' \leq n$ if and only if $\mathrm{G\text{-}dim}_R\, M \leq n$, as claimed, and the inequalities are obvious.

(b): Assume that $\mathrm{G\text{-}dim}_R\, M' > \mathrm{G\text{-}dim}_R\, M''$. By the second inequality in (a) we then have $\mathrm{G\text{-}dim}_R\, M \leq \mathrm{G\text{-}dim}_R\, M'$, and since $\mathrm{G\text{-}dim}_R\, M < \mathrm{G\text{-}dim}_R\, M'$ leads to a contradiction by the first inequality, we conclude that $\mathrm{G\text{-}dim}_R\, M = \mathrm{G\text{-}dim}_R\, M'$. A parallel argument applies if we assume that $\mathrm{G\text{-}dim}_R\, M > \mathrm{G\text{-}dim}_R\, M''$.

(c): A G–resolution of $M'$ of finite length, say $n$, gives a G–resolution of $M''$ of length $n+1$, so if $\mathrm{G\text{-}dim}_R\, M'' = \infty$, then also $M'$ has infinite G–dimension. If $M''$ has finite G–dimension, then so has $M'$, cf. (a), and since $\mathrm{Ext}^m_R(M', R) \cong \mathrm{Ext}^{m+1}_R(M'', R)$ for $m > 0$, cf. Lemma (1.1.10)(b), it follows from the GD Theorem (1.2.7) that $\mathrm{G\text{-}dim}_R\, M' = \mathrm{G\text{-}dim}_R\, M'' - 1$. □

The last result of this section shows that G–dimension is a refinement of projective dimension for finite modules.

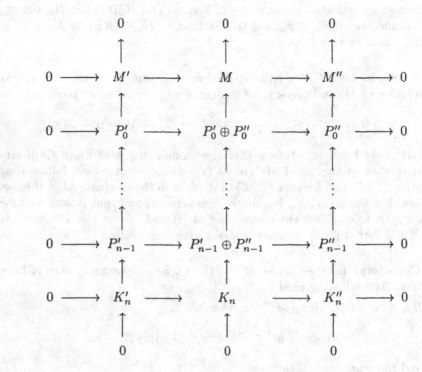

(1.2.10) **Proposition (GD–PD Inequality for Finite Modules).** *For every finite $R$–module $M$ there is an inequality:*

$$\operatorname{G-dim}_R M \leq \operatorname{pd}_R M,$$

*and equality holds if* $\operatorname{pd}_R M < \infty$.

*Proof.* The inequality certainly holds if $M$ has infinite projective dimension, and equality holds for the zero-module. Assume that $M$ is non-zero of finite projective dimension, say $p$. Since $M$ is finite it has a resolution of minimal length by finite projective modules: $0 \to P_p \to \cdots \to P_1 \to P_0 \to M \to 0$. This resolution is, in particular, a G–resolution, so $\operatorname{G-dim}_R M \leq p$. By assumption $\operatorname{Ext}_R^{p+1}(M,-) = 0$, and a finite $R$–module $T$ exists, such that $\operatorname{Ext}_R^p(M,T) \neq 0$, cf. (A.5.4.1). Applying the functor $\operatorname{Hom}_R(M,-)$ to the short exact sequence $0 \to K \to R^\beta \to T \to 0$, we get an exact sequence

$$\cdots \to \operatorname{Ext}_R^p(M,K) \to \operatorname{Ext}_R^p(M,R^\beta) \to \operatorname{Ext}_R^p(M,T) \to 0,$$

showing that $\operatorname{Ext}_R^p(M,R^\beta) \neq 0$ and, therefore, $\operatorname{Ext}_R^p(M,R) \neq 0$. By the GD Theorem (1.2.7) it now follows that $\operatorname{G-dim}_R M = p$, as wanted. □

**Notes**

The proof of "$(iii) \Rightarrow (i)$" in GD Theorem (1.2.7) is, actually, a mapping cone argument in disguise: the lifting $\gamma$ of $1_M$ in the diagram on page 25 is a morphism of complexes from $0 \to H_n \to P_{n-1} \to \cdots \to P_0 \to 0$ to $0 \to K_n \to G_{n-1} \to \cdots \to G_0 \to 0$. It is a quasi-isomorphism, because the two complexes are resolutions of the same module, and the exact sequence below the diagram is the mapping cone. This argument is considerably shorter than the original proof [1, pp. 56–60] which is, so to speak, an approximation to Schanuel's lemma. The author was advised of the mapping cone argument by Iyengar.

As remarked already in (1.2.8), everything would be so much easier if one could always compute the G–dimension of a module $M$ by non-vanishing of the modules $\operatorname{Ext}_R^m(M,R)$. But it is not known if this is actually the case. That is to say, no example is known (to the author) of a finite module $M$ with $\sup\{m \in \mathbb{N}_0 \mid \operatorname{Ext}_R^m(M,R) \neq 0\} < \infty$ but $\operatorname{G-dim}_R M = \infty$. We return to this point in (2.3.9). It should be mentioned that a couple of attempts have been made — in [50] and [58] — to circumvent this problem altogether by defining a weak G–dimension in terms of non-vanishing of Ext modules.

## 1.3 Standard Operating Procedures

The formation of fraction and residue class modules are standard procedures in commutative algebra, and we now study the behavior of G–dimension under these constructions. The study of residue class modules continues in the next section, where we turn to local rings.

**(1.3.1) Lemma.** *A finite $R$–module $M$ belongs to* $\mathrm{G}(R)$ *if and only if* $M_\mathfrak{p} \in \mathrm{G}(R_\mathfrak{p})$ *for all prime ideals* $\mathfrak{p} \in \operatorname{Spec} R$.

*Proof.* The higher modules $\mathrm{Ext}^m_R(M,R)$ and $\mathrm{Ext}^m_R(M^*,R)$ vanish if and only if their localizations at all prime ideals do so, that is, if and only if $\mathrm{Ext}^m_R(M,R)_\mathfrak{p} \cong \mathrm{Ext}^m_{R_\mathfrak{p}}(M_\mathfrak{p},R_\mathfrak{p}) = 0$ and $\mathrm{Ext}^m_R(M^*,R)_\mathfrak{p} \cong \mathrm{Ext}^m_{R_\mathfrak{p}}(\mathrm{Hom}_{R_\mathfrak{p}}(M_\mathfrak{p},R_\mathfrak{p}),R_\mathfrak{p}) = 0$ for all $m > 0$. Localization is an exact process, so the biduality map $\delta_M$ is an isomorphism if and only if all the localized maps $(\delta_M)_\mathfrak{p}$ are isomorphisms. For each $\mathfrak{p} \in \operatorname{Spec} R$ we have a commutative diagram

$$\begin{array}{ccc} M_\mathfrak{p} & \xrightarrow{(\delta_M)_\mathfrak{p}} & \mathrm{Hom}_R(\mathrm{Hom}_R(M,R),R)_\mathfrak{p} \\ \big\downarrow = & & \big\downarrow \cong \\ M_\mathfrak{p} & \xrightarrow{\delta_{M_\mathfrak{p}}} & \mathrm{Hom}_{R_\mathfrak{p}}(\mathrm{Hom}_R(M_\mathfrak{p},R_\mathfrak{p}),R_\mathfrak{p}) \end{array}$$

showing that $(\delta_M)_\mathfrak{p}$ is an isomorphism if and only if $\delta_{M_\mathfrak{p}}$ is so, and it follows that $\delta_M$ is invertible if and only if all the biduality maps $\delta_{M_\mathfrak{p}}$ are so. □

**(1.3.2) Proposition.** *Let $M$ be a finite $R$–module and $n \in \mathbb{N}_0$; then* $\mathrm{G\text{–}dim}_R\, M \le n$ *if and only if* $\mathrm{G\text{–}dim}_{R_\mathfrak{p}}\, M_\mathfrak{p} \le n$ *for all prime ideals* $\mathfrak{p} \in \operatorname{Spec} R$.

*Proof.* We have already dealt with the case $n = 0$ in the Lemma, so we fix an $n \in \mathbb{N}$ and consider an exact sequence of $R$–modules

$$0 \to K_n \to G_{n-1} \to \cdots \to G_1 \to G_0 \to M \to 0,$$

where $G_0, \ldots, G_{n-1} \in \mathrm{G}(R)$. Let $\mathfrak{p} \in \operatorname{Spec} R$, localization at $\mathfrak{p}$ is exact, and by the Lemma the modules $G_\ell$ localize to give modules in $\mathrm{G}(R_\mathfrak{p})$; now apply the Lemma to the module $K_n$, and we are done. □

**(1.3.3)** ***Regular Elements.*** Let $M$ be an $R$–module; an element $x \in R$ is said to be *$M$–regular* if and only if $x \notin \mathrm{z}_R\, M$. A sequence $x_1, \ldots, x_t$ of elements in $R$ is called an *$M$–sequence* if and only if

(1) $x_m \notin \mathrm{z}_R\, M/(x_1, \ldots, x_{m-1})M$;[3] and
(2) $M/(x_1, \ldots, x_t)M \neq 0$ or $M = 0$.

**(1.3.4) Lemma.** *Let $M$ be a finite $R$–module. The following hold if $x \in R$ is $M$– and $R$–regular.*

(a) $\mathrm{Tor}^R_m(M, R/(x)) = 0$ *for* $m > 0$.
(b) *If* $\mathrm{Ext}^1_R(M,R) = 0$, *then* $\mathrm{Hom}_{R/(x)}(M/xM, R/(x)) \cong M^*/xM^*$.
(c) *If* $\mathrm{Ext}^1_R(M,R) = 0 = \mathrm{Ext}^1_R(M^*,R)$, *then*

$$\mathrm{Hom}_{R/(x)}(\mathrm{Hom}_{R/(x)}(M/xM, R/(x)), R/(x)) \cong M^{**}/xM^{**}.$$

[3] For $m = 1$ this means $x_1 \notin \mathrm{z}_R\, M$.

*Proof.* Set $\bar{R} = R/(x)$ and $\overline{M} = M/xM$. Assuming that $x$ is $M$– and $R$–regular we have two exact sequences

$$(\dagger) \qquad 0 \to R \xrightarrow{x} R \to \bar{R} \to 0 \quad \text{and}$$

$$(\ddagger) \qquad 0 \to M \xrightarrow{x} M \to \overline{M} \to 0.$$

(a): Tensoring (†) by $M$ we get the long exact sequence

$$\cdots \to \operatorname{Tor}_m^R(M,R) \xrightarrow{x} \operatorname{Tor}_m^R(M,R) \to \operatorname{Tor}_m^R(M,\bar{R}) \to \cdots$$
$$\cdots \to \operatorname{Tor}_1^R(M,\bar{R}) \to M \xrightarrow{x} M \to \overline{M} \to 0.$$

Obviously, the modules $\operatorname{Tor}_m^R(M,\bar{R})$ vanish for $m > 1$, as $\operatorname{Tor}_m^R(M,R) = 0$ for $m > 0$, and by exactness of (‡) also $\operatorname{Tor}_1^R(M,\bar{R}) = 0$.

(b): We assume that $\operatorname{Ext}_R^1(M,R) = 0$. Applying $\operatorname{Hom}_R(M,-)$ to (†) we get the short exact sequence

$$0 \to M^* \xrightarrow{x} M^* \to \operatorname{Hom}_R(M,\bar{R}) \to 0,$$

showing that $M^*/xM^* \cong \operatorname{Hom}_R(M,\bar{R})$, and it follows by adjointness that $\operatorname{Hom}_{\bar{R}}(\overline{M},\bar{R}) \cong \operatorname{Hom}_R(M,\bar{R})$, so $M^*/xM^* \cong \operatorname{Hom}_{\bar{R}}(\overline{M},\bar{R})$ as wanted.

(c): The dual of $M$ is torsion-free, so $x$ is $M^*$–regular by (1.1.5.1). Applying (b) twice, first to $M$ and next to $M^*$, we establish the desired isomorphism:

$$\operatorname{Hom}_{\bar{R}}(\operatorname{Hom}_{\bar{R}}(\overline{M},\bar{R}),\bar{R}) \cong \operatorname{Hom}_{\bar{R}}(M^*/xM^*,\bar{R}) \cong M^{**}/xM^{**}.$$

This completes the proof. □

(1.3.5) **Lemma.** *Let $M$ be a finite $R$–module, and let $x \in R$ be $R$–regular. If $M \in \mathrm{G}(R)$, then $M/xM \in \mathrm{G}(R/(x))$.*

*Proof.* Set $\bar{R} = R/(x)$ and $\overline{M} = M/xM$. We assume that $M \in \mathrm{G}(R)$, that is, $\operatorname{Ext}_R^m(M,R) = 0 = \operatorname{Ext}_R^m(M^*,R)$ for $m > 0$ and $\delta_M$ is an isomorphism. Furthermore, $M$ is torsion-free as remarked in (1.1.6), so $x$ is also $M$–regular, cf. (1.1.5.1). Applying $\operatorname{Hom}_R(M,-)$ to the short exact sequence $0 \to R \xrightarrow{x} R \to \bar{R} \to 0$ we get the long exact sequence

$$(\dagger) \qquad \begin{aligned} \cdots \to \operatorname{Ext}_R^m(M,R) \xrightarrow{x} \operatorname{Ext}_R^m(M,R) \to \\ \operatorname{Ext}_R^m(M,\bar{R}) \to \operatorname{Ext}_R^{m+1}(M,R) \to \cdots. \end{aligned}$$

It is evident from (†) that $\operatorname{Ext}_R^m(M,\bar{R}) = 0$ for $m > 0$, so $\operatorname{Ext}_{\bar{R}}^m(\overline{M},\bar{R}) \cong \operatorname{Ext}_R^m(M,\bar{R}) = 0$ for $m > 0$. The element $x$ is not a zero-divisor for the torsion-free module $M^*$, so similarly we see that $\operatorname{Ext}_{\bar{R}}^m(M^*/xM^*,\bar{R}) = 0$ for $m > 0$, and by Lemma (1.3.4)(b) this means that $\operatorname{Ext}_{\bar{R}}^m(\operatorname{Hom}_{\bar{R}}(\overline{M},\bar{R}),\bar{R}) = 0$ for $m > 0$. The biduality map $\delta_M$ is an isomorphism, and so is $\delta_M \otimes_R \bar{R}$. By

Lemma (1.3.4)(c) we have the commutative diagram

$$\begin{array}{ccc} M \otimes_R \bar{R} & \xrightarrow[\cong]{\delta_M \otimes_R \bar{R}} & M^{**} \otimes_R \bar{R} \\ \downarrow \cong & & \downarrow \cong \\ \overline{M} & \xrightarrow{\delta_{\overline{M}}} & \mathrm{Hom}_{\bar{R}}(\mathrm{Hom}_{\bar{R}}(\overline{M}, \bar{R}), \bar{R}) \end{array}$$

showing that also $\delta_{\overline{M}} \colon \overline{M} \to \mathrm{Hom}_{\bar{R}}(\mathrm{Hom}_{\bar{R}}(\overline{M}, \bar{R}), \bar{R})$ is an isomorphism. □

(1.3.6) **Proposition.** *Let $M$ be a finite $R$-module. If $x \in R$ is $M$- and $R$-regular, then*

$$\text{G-dim}_{R/(x)} M/xM \leq \text{G-dim}_R M.$$

*Proof.* If $\text{G-dim}_R M = \infty$ the inequality obviously holds, so we assume that $M$ has finite G-dimension, say $n$, and consider a G-resolution of $M$ of minimal length:

$$(\dagger) \qquad 0 \to G_n \to \cdots \to G_1 \to G_0 \to M \to 0.$$

As remarked in (1.1.6) the modules $G_0, \ldots, G_n$ are torsion-free, so $x$ is not a zero-divisor for these modules; in particular, $x$ is not a zero-divisor for any of the kernels $K_n$, defined as in (1.2.5.1). By Lemma (1.3.4)(a) it then follows that tensoring by $R/(x)$ leaves all the short exact sequences $0 \to K_\ell \to G_{\ell-1} \to K_{\ell-1} \to 0$ exact, so from (†) we get an exact sequence

$$0 \to G_n/xG_n \to \cdots \to G_1/xG_1 \to G_0/xG_0 \to M/xM \to 0.$$

Now, $G_\ell/xG_\ell \in \mathrm{G}(R/(x))$ by the Lemma, so $\text{G-dim}_{R/(x)} M/xM \leq n$ as desired. □

## 1.4 Local Rings

The highlights of this section — and this chapter — are the Auslander–Bridger formula (1.4.8) and the characterization in Theorem (1.4.9) of Gorenstein rings in terms of finiteness of G-dimensions.

(1.4.1) **Setup.** In this section $(R, \mathfrak{m}, k)$ is a **local ring**.

(1.4.2) ***Depth.*** The *depth* of an $R$-module $M$ can be defined as

$$\mathrm{depth}_R M = \inf\{m \in \mathbb{N}_0 \mid \mathrm{Ext}_R^m(k, M) \neq 0\}.$$

If $M$ is finite, then $\mathrm{depth}_R M < \infty$ and all maximal $M$-sequences have length $\mathrm{depth}_R M$.

Since $R$ is Noetherian, we can, if $M$ is finite and both $M$ and $R$ have positive depth, choose an element $x \in R$ which is both $M$- and $R$-regular.

(1.4.3) **Observation.** If $M$ is torsion-free, e.g., $M \in \mathrm{G}(R)$, then all $R$-regular elements are also $M$-regular, cf. (1.1.5.1). From Nakayama's lemma and Lemma (1.3.5) it follows, by a simple induction argument, that if $M \in \mathrm{G}(R)$, then every $R$-sequence is also an $M$-sequence, in particular, $\operatorname{depth}_R M \geq \operatorname{depth} R$ (and we shall soon see that equality holds).

Consider a short exact sequence of non-zero $R$-modules,

$$0 \to K \to N \to M \to 0,$$

where $\operatorname{depth}_R M = d$ and $\operatorname{depth}_R N > d$. Inspection of the long exact sequence

$$\cdots \to \operatorname{Ext}_R^m(k,K) \to \operatorname{Ext}_R^m(k,N) \to \operatorname{Ext}_R^m(k,M) \to \operatorname{Ext}_R^{m+1}(k,K) \to \cdots$$

shows that $\operatorname{depth}_R K = d+1$.

(1.4.4) **Lemma.** *Let $M$ be a finite $R$-module. If $x \in R$ is $M$- and $R$-regular, then $M \in \mathrm{G}(R)$ if and only if $M/xM \in \mathrm{G}(R/(x))$.*

*Proof.* The "only if" part was proved in Lemma (1.3.5). Set $\bar{R} = R/(x)$ and $\overline{M} = M/xM$. We assume that $\overline{M} \in \mathrm{G}(\bar{R})$, that is, $\delta_{\overline{M}} : \overline{M} \to \operatorname{Hom}_{\bar{R}}(\operatorname{Hom}_{\bar{R}}(\overline{M},\bar{R}),\bar{R})$ is an isomorphism and $\operatorname{Ext}_{\bar{R}}^m(\overline{M},\bar{R}) = 0 = \operatorname{Ext}_{\bar{R}}^m(\operatorname{Hom}_{\bar{R}}(\overline{M},\bar{R}),\bar{R})$ for $m > 0$. Now, $\operatorname{Ext}_{\bar{R}}^m(\overline{M},\bar{R}) \cong \operatorname{Ext}_R^m(M,\bar{R})$, so $\operatorname{Ext}_R^m(M,\bar{R}) = 0$ for $m > 0$. Applying $\operatorname{Hom}_R(M,-)$ to the short exact sequence $0 \to R \xrightarrow{x} R \to \bar{R} \to 0$ we get an exact sequence

$$\cdots \to \operatorname{Ext}_R^m(M,R) \xrightarrow{x} \operatorname{Ext}_R^m(M,R) \to \operatorname{Ext}_R^m(M,\bar{R}) \to \cdots,$$

from which it is then evident, by Nakayama's lemma, that $\operatorname{Ext}_R^m(M,R) = 0$ for $m > 0$. Now, $\operatorname{Hom}_{\bar{R}}(\overline{M},\bar{R}) \cong M^*/xM^*$ by Lemma (1.3.4)(b), and since $x$ is also $M^*$-regular we can analogously conclude that $\operatorname{Ext}_R^m(M^*,R) = 0$ for $m > 0$. It follows by Lemma (1.3.4)(c) that
$tp\delta_M \bar{R}$ is an isomorphism, cf. the commutative diagram

$$\begin{array}{ccc} M \otimes_R \bar{R} & \xrightarrow{\delta_M \otimes_R \bar{R}} & M^{**} \otimes_R \bar{R} \\ \downarrow \cong & & \downarrow \cong \\ \overline{M} & \underset{\cong}{\xrightarrow{\delta_{\overline{M}}}} & \operatorname{Hom}_{\bar{R}}(\operatorname{Hom}_{\bar{R}}(\overline{M},\bar{R}),\bar{R}) \end{array}$$

In order to show that $\delta_M$ is invertible we consider the exact sequence

$$0 \to K \to M \xrightarrow{\delta_M} M^{**} \to C \to 0.$$

Tensoring by $\bar{R}$ yields the exact sequence

$$\overline{M} \xrightarrow{\delta_M \otimes_R \bar{R}} M^{**}/xM^{**} \to C/xC \to 0,$$

which shows that $C/xC = 0$, as $\delta_M \otimes_R \bar{R}$ is surjective, and hence $C = 0$ by Nakayama's lemma. Tensoring $0 \to K \to M \xrightarrow{\delta_M} M^{**} \to 0$ by $\bar{R}$ gives a short exact sequence

$$0 \to K/xK \to \bar{M} \xrightarrow{\delta_M \otimes_R \bar{R}} M^{**}/xM^{**} \to 0,$$

as $\mathrm{Tor}_1^R(M^{**}, \bar{R}) = 0$ by Lemma (1.3.4)(a). This sequence shows that $K/xK = 0$ and, hence, $K = 0$. Thus, $\delta_M$ is an isomorphism and we have, hereby, proved the "if" part. □

(1.4.5) **Proposition.** *Let $M$ be a finite $R$–module. If $x \in R$ is $M$– and $R$–regular, then*

$$\text{G–dim}_{R/(x)} M/xM = \text{G–dim}_R M.$$

*Proof.* We recall that by Proposition (1.3.6) we have $\text{G–dim}_{R/(x)} M/xM \leq \text{G–dim}_R M$, it is therefore sufficient to prove the inequality $\text{G–dim}_R M \leq \text{G–dim}_{R/(x)} M/xM$. This inequality obviously holds if $M/xM$ has infinite G–dimension over $R/(x)$, so we assume that $\text{G–dim}_{R/(x)} M/xM = n < \infty$ and proceed by induction on $n$. The Lemma furnishes the induction base, so we let $n > 0$ and assume that the inequality holds for $R/(x)$–modules of G–dimension at most $n - 1$. By (1.3.6) the G–dimension of $M$ over $R$ is at least $n$, so we can consider a short exact sequence of $R$–modules

(†) $$0 \to K \to G \to M \to 0,$$

where $G \in \mathrm{G}(R)$ and $K \neq 0$. The modules $G$ and $K$ are torsion-free, so by Lemma (1.3.4)(a) also the sequence

$$0 \to K/xK \to G/xG \to M/xM \to 0,$$

obtained by tensoring (†) by $R/(x)$, is exact, and $G/xG \in \mathrm{G}(R/(x))$, cf. Lemma (1.4.4). By Corollary (1.2.9)(c) we have $\text{G–dim}_{R/(x)} K/xK = n - 1$, and since the element $x$ is also $K$–regular, we have $\text{G–dim}_R K \leq n - 1$ by the induction hypothesis. From (†) we now conclude that $\text{G–dim}_R M \leq n$. □

(1.4.6) **Corollary.** *Let $M$ be a finite $R$–module. If $\boldsymbol{x} = x_1, \dots, x_t$ is an $M$– and $R$–sequence, then*

$$\text{G–dim}_R M = \text{G–dim}_{R/(\boldsymbol{x})} M/(\boldsymbol{x})M.$$

*Proof.* Immediate from the Proposition by induction on the length, $t$, of the sequence. □

(1.4.7) **Lemma.** *If* $\operatorname{depth} R = 0$, *then all finite $R$–modules of finite G–dimension belong to* $\mathrm{G}(R)$.

*Proof.* Let $n \in \mathbb{N}$, and let $M \neq 0$ be a finite $R$–module with $\text{G-dim}_R\, M \leq n$. We proceed by induction on $n$.

First we assume that $\text{G-dim}_R\, M \leq 1$, then $\text{Ext}_R^m(M,R) = 0$ for $m > 1$ by GD Theorem (1.2.7), and it is sufficient to prove that $\text{Ext}_R^1(M,R) = 0$. By assumption $M$ has a G–resolution of length 1:

$$(\dagger) \qquad 0 \to G_1 \to G_0 \to M \to 0.$$

Dualizing twice gives an exact sequence

$$0 \to \text{Ext}_R^1(M,R)^* \to G_1^{**} \to G_0^{**},$$

which, when compared to (†), shows that $\text{Ext}_R^1(M,R)^* = 0$. Now we have

$$\emptyset = \text{Ass}_R(\text{Ext}_R^1(M,R)^*) = \text{Ass}\, R \cap \text{Supp}_R(\text{Ext}_R^1(M,R)),$$

and since $\mathfrak{m} \in \text{Ass}\, R$, as $\text{depth}\, R = 0$, we conclude that $\text{Supp}_R(\text{Ext}_R^1(M,R)) = \emptyset$ and, therefore, $\text{Ext}_R^1(M,R) = 0$.

Next, let $n > 1$ and assume that all modules of G–dimension at most $n-1$ belong to $\text{G}(R)$. By assumption $M$ has a G–resolution of length $n$:

$$0 \to G_n \to \cdots \to G_1 \to G_0 \to M \to 0.$$

We define $K_{n-1}$ as in (1.2.5.1); as noted there $\text{G-dim}_R\, K_{n-1} \leq 1$, so $K_{n-1} \in \text{G}(R)$ by the induction base. The exact sequence

$$0 \to K_{n-1} \to G_{n-2} \to \cdots \to G_0 \to M \to 0$$

now shows that $\text{G-dim}_R\, M \leq n-1$, whence $M \in \text{G}(R)$ by the induction hypothesis. □

By the GD–PD inequality (1.2.10) the next result extends the classical Auslander–Buchsbaum formula.

**(1.4.8) Theorem (Auslander–Bridger Formula).** *Let $R$ be a local ring. If $M$ is a finite $R$–module of finite G–dimension, then*

$$\text{G-dim}_R\, M = \text{depth}\, R - \text{depth}_R\, M.$$

*Proof.* Note that $\text{G-dim}_R\, 0 = \text{depth}\, R - \text{depth}_R\, 0 = -\infty$. We now assume that $M \neq 0$ and proceed by induction on the depth of $R$.

First assume that $\text{depth}\, R = 0$, then $M \in \text{G}(R)$ by the Lemma, i.e., $\text{G-dim}_R\, M = 0$, so we want to prove that also $\text{depth}_R\, M = 0$. Because $M$ is isomorphic to its bidual we have

$$\text{Ass}_R\, M = \text{Ass}_R\, M^{**} = \text{Ass}\, R \cap \text{Supp}_R\, M^*,$$

and since $\text{depth}\, R = 0$ and $M^* \neq 0$, the maximal ideal $\mathfrak{m}$ is contained in both sets on the right hand side. Therefore $\mathfrak{m} \in \text{Ass}_R\, M$, and this gives the desideratum.

Next, let $\operatorname{depth} R = n > 0$ and assume that the desired equality holds for finite modules over rings of depth $n-1$. There are two cases to consider: $\operatorname{depth}_R M > 0$ and $\operatorname{depth}_R M = 0$. In the first case we choose an element $x \in R$ which is both $M$– and $R$–regular, then $\operatorname{depth} R/(x) = \operatorname{depth} R - 1$ and $\operatorname{depth}_{R/(x)} M/xM = \operatorname{depth}_R M - 1$, so by Proposition (1.4.5) and the induction hypothesis we have

$$\begin{aligned} \text{G-dim}_R M &= \text{G-dim}_{R/(x)} M/xM \\ &= \operatorname{depth} R/(x) - \operatorname{depth}_{R/(x)} M/xM \\ &= \operatorname{depth} R - \operatorname{depth}_R M. \end{aligned}$$

Finally we consider the case $\operatorname{depth}_R M = 0$; here $M$ cannot belong to $\mathrm{G}(R)$, cf. Observation (1.4.3), so we can consider an exact sequence

$$0 \to K \to G \to M \to 0,$$

where $G \in \mathrm{G}(R)$ and $\text{G-dim}_R K = \text{G-dim}_R M - 1$, cf. Corollary (1.2.9)(c). As $\operatorname{depth} R > 0$ implies $\operatorname{depth}_R G > 0$, we have $\operatorname{depth}_R K = 1$ by Observation (1.4.3). Now, from what we have already proved it follows that

$$\text{G-dim}_R K = \operatorname{depth} R - \operatorname{depth}_R K = \operatorname{depth}_R R - 1,$$

and, therefore, $\text{G-dim}_R M = \operatorname{depth} R$ as wanted. □

Recall that a local ring $R$ is Gorenstein if and only if it has finite injective dimension as module over itself, and in the affirmative case $\operatorname{id}_R R = \operatorname{depth}_R R$ by the Bass formula.

(1.4.9) **Gorenstein Theorem, GD Version.** *Let $R$ be a local ring with residue field $k$. The following are equivalent:*

(*i*) *$R$ is Gorenstein.*
(*ii*) $\text{G-dim}_R k < \infty$.
(*iii*) $\text{G-dim}_R M < \infty$ *for all finite $R$–modules $M$.*

*Proof.* Evidently, (*iii*) is stronger than (*ii*), so it is sufficient to prove that (*i*) implies (*iii*) and (*ii*) implies (*i*).

(*i*) ⇒ (*iii*): Assume that $R$ is Gorenstein with $\operatorname{id}_R R = \operatorname{depth} R = d$, and let $M \neq 0$ be a finite $R$–module; we proceed by induction on $d$ to prove that $\text{G-dim}_R M \leq d$. If $d = 0$, then $R$ is an injective $R$–module, and hence $\operatorname{Ext}_R^m(M,R) = 0$ and $\operatorname{Ext}_R^m(M^*,R) = 0$ for all $m > 0$. Furthermore, the Hom evaluation homomorphism $\theta_{MRR}$ is an isomorphism, and the commutative diagram

$$\begin{array}{ccc} M & \xrightarrow{\delta_M} & M^{**} \\ \downarrow \cong & & \cong \uparrow \theta_{MRR} \\ M \otimes_R R & \xrightarrow{\cong} & M \otimes_R \operatorname{Hom}_R(R,R) \end{array}$$

shows that also $\delta_M$ is an isomorphism.

Now, let $d > 0$ and assume that the claim holds for rings of depth $d-1$. If $\operatorname{depth}_R M > 0$ we choose an element $x \in R$ both $M$– and $R$–regular, then $\operatorname{depth} R/(x) = d-1$, and by Proposition (1.4.5) and the induction hypothesis we have

$$\text{G-dim}_R M = \text{G-dim}_{R/(x)} M/xM \leq d-1.$$

If $\operatorname{depth}_R M = 0$ the module $M$ cannot be in the G–class, cf. Observation (1.4.3), so we can consider an exact sequence

$$0 \to K \to G \to M \to 0,$$

where $K \neq 0$ and $G \in \mathrm{G}(R)$. By the same Observation $\operatorname{depth}_R K = 1$, so it follows from what we have already proved that $\text{G-dim}_R K \leq d-1$, and therefore $\text{G-dim}_R M \leq d$, cf. Corollary (1.2.9)(c).

*(ii)* ⇒ *(i)*: We assume that $\text{G-dim}_R k < \infty$, then $\text{G-dim}_R k = \operatorname{depth} R = d$ by the Auslander–Bridger formula (1.4.8). It follows by the GD Theorem (1.2.7) that the Bass numbers $\mu_R^m$ vanish for $m > d$ while $\mu_R^d \neq 0$, so $\operatorname{id}_R R = d$. □

A local ring $(R, \mathfrak{m}, k)$ is regular if and only if $\operatorname{pd}_R k < \infty$, so in view of the Theorem the canonical example of a module of finite G–dimension but infinite projective dimension is the residue field of a non-regular Gorenstein ring, cf. Example (1.1.13).

(1.4.10) **Remark.** We have actually proved a little more than stated in the Theorem, namely that the G–dimension of all finite modules over a Gorenstein local ring is limited by the number $d = \operatorname{depth} R = \dim R$ and, in particular, $\text{G-dim}_R k = d$. However, this is not surprising: the Auslander–Bridger formula shows (for any local ring $R$) that the G–dimension of a finite $R$–module cannot exceed the depth of the ring (unless, of course, it is infinite), and $\operatorname{depth} R \leq \dim R$ with equality when $R$ is Cohen–Macaulay. We return to this question of bounds in Observation (2.4.9).

## 1.5 G–dimension versus Projective Dimension

We have already seen how the G–dimension shares many of the nice properties of the projective dimension. The Auslander–Buchsbaum formula is just one example of a result for projective dimension that can be extended to G–dimension. In this section we give a couple of contrasting examples. First we prove a result for G–dimension which does not hold for projective dimension; and next we give examples of results for modules of finite projective dimension which do not hold (in general) for modules of finite G–dimension.

(1.5.1) **Setup.** In this section $R$ is a **local ring**.

(1.5.2) **Lemma.** *Let $M$ be a finite $R$–module, and assume that $x \in \operatorname{Ann}_R M$ is $R$–regular. If $n \in \mathbb{N}_0$ and* $\text{G-dim}_{R/(x)} M = n$, *then* $\text{G-dim}_R M = n+1$.

*Proof.* Tensoring the exact sequence $0 \to R \xrightarrow{x} R \to R/(x) \to 0$ by $M$ we get an exact sequence

$$0 \to \operatorname{Tor}_1^R(M, R/(x)) \to M \xrightarrow{x} M.$$

Since $x \in \operatorname{Ann}_R M$ the homothety $x_M$ is the zero-map, so $\operatorname{Tor}_1^R(M, R/(x)) \cong M$. Because the element $x$ is not $M$–regular, the module $M$ cannot belong to $\mathrm{G}(R)$; we can therefore consider an exact sequence of $R$–modules

$$(\dagger) \qquad 0 \to K \to G \to M \to 0,$$

where $G \in \mathrm{G}(R)$ and $K \neq 0$. We now set out to prove that $\text{G-dim}_R K = n$. Since $\operatorname{Tor}_1^R(M, R/(x)) \cong M$ and $\operatorname{Tor}_1^R(G, R/(x)) = 0$ by Lemma (1.3.4)(a), tensoring $(\dagger)$ by $R/(x)$ yields an exact sequence

$$0 \to M \to K/xK \to G/xG \to M \to 0,$$

where $G/xG \in \mathrm{G}(R/(x))$ by Lemma (1.3.5). Setting $N = \operatorname{Ker}(G/xG \to M)$ we have two exact sequences of $R/(x)$–modules:

$$(\ddagger) \qquad 0 \to M \to K/xK \to N \to 0 \quad \text{and}$$

$$(\star) \qquad 0 \to N \to G/xG \to M \to 0.$$

If $n = \text{G-dim}_{R/(x)} M = 0$, it follows from $(\star)$ and Lemma (1.1.10)(a) that $\text{G-dim}_{R/(x)} N = 0$, and then $(\ddagger)$ shows that also $\text{G-dim}_{R/(x)} K/xK = 0 = n$. If $n > 0$ we use Corollary (1.2.9)(c) to conclude from $(\star)$ that $\text{G-dim}_{R/(x)} N = n-1$, and then it follows from $(\ddagger)$ and (b) in the same Corollary that $\text{G-dim}_{R/(x)} K/xK = n$. Being $R$–regular $x$ is also $G$–regular and, in particular, $K$–regular, so $\text{G-dim}_R K = n$ by Proposition (1.4.5), and from $(\dagger)$ it then follows that $\text{G-dim}_R M = n+1$, again by (1.2.9)(c). □

(1.5.3) **Proposition.** *Let $M$ be a finite $R$–module. If $\boldsymbol{x} = x_1, \ldots, x_t$ is an $R$–sequence in $\operatorname{Ann}_R M$ and* $\text{G-dim}_{R/(\boldsymbol{x})} M < \infty$, *then*

$$\text{G-dim}_R M = \text{G-dim}_{R/(\boldsymbol{x})} M + t.$$

*Proof.* If $M = 0$ the equality is trivial. If $n \in \mathbb{N}_0$ and $\text{G-dim}_{R/(x)} M = n$, then the equality follows from the Lemma by induction on the length, $t$, of the $R$–sequence. □

(1.5.4) **Remarks (Change of Rings).** It is well-known — or can easily be seen as the same proof applies — that the result above also holds for projective dimension; that is, $\operatorname{pd}_{R/(x)} M < \infty$ implies $\operatorname{pd}_R M < \infty$, when $x \in \operatorname{Ann}_R M$ is $R$–regular. Weibel calls this "The First Change of Rings Theorem" [60, 4.3.3].

It is also well-known that the reverse implication does not hold: for example, let $k$ be a field and set $R = k[\![X]\!]$, then $\mathrm{pd}_R\, k < \infty$ as $R$ is regular, but $R/(X^2)$ is not regular, so $\mathrm{pd}_{R/(X^2)}\, k = \infty$ even though $X^2 \in \mathrm{Ann}_R\, k - \mathfrak{z}\, R$. It is therefore interesting that the following holds:

**Change of Rings Theorem for G–dimension.** *Let $M$ be a finite $R$–module. If $\boldsymbol{x} = x_1, \dots, x_t$ is an $R$–sequence in $\mathrm{Ann}_R\, M$, then*

$$\text{G-dim}_R\, M = \text{G-dim}_{R/(\boldsymbol{x})}\, M + t.$$

*In particular, the two dimensions are simultaneously finite.*

It is easy to see that Proposition (1.5.3) holds also over non-local rings, and in the next chapter we prove the Change of Rings Theorem — it is (2.2.8) — for Noetherian rings in general.

(1.5.5) **Remark (Zero-divisors).** Let $M \neq 0$ be a finite $R$–module of finite projective dimension, then any $M$–regular element $x \in R$ is also $R$-regular. This result is known as *Auslander's zero-divisor conjecture*, and one could ask if it also holds for modules of finite G–dimension; the next example shows that the answer is negative.

(1.5.6) **Example.** Let $k$ be a field, let $R$ be the local ring $R = k[\![X,Y]\!]/(XY)$, and denote by $x$ and $y$ the residue classes in $R$ of, respectively, $X$ and $Y$. The ring $R$ is Gorenstein (see page 14), so all finite $R$–modules have finite G–dimension by Theorem (1.4.9). In particular, the module $(x)$ has finite G–dimension, and the element $x$ is $(x)$–regular but, certainly, not $R$–regular.

This last example and the argument given above in (1.5.4) reflect the same fact: Gorensteinness of a local ring is preserved when a regular element is factored out [12, Proposition 3.1.19(b)], but regularity is not (indeed, regular local rings are domains [12, Proposition 2.2.3]).

(1.5.7) **Remarks (Intersections).** If $M$ and $N$ are finite $R$–modules, and $M$ has finite projective dimension, then

$$(\dagger) \qquad \dim_R N \leq \mathrm{pd}_R\, M + \dim_R(M \otimes_R N),$$

cf. [12, Corollary 9.4.6]. This result belongs to the family of, so-called, intersection theorems, and it does not extend to G–dimension: let $R$ be as in the Example, and set $M = R/(x)$ and $N = R/(y)$. It is immediate by the Auslander–Bridger formula (1.4.8) that $\text{G-dim}_R\, M = 0 = \text{G-dim}_R\, N$; and $M \otimes_R N \cong k$, so we have

$$\dim_R N = 1 \; > \; 0 = \text{G-dim}_R\, M + \dim_R(M \otimes_R N).$$

If there exists a finite $R$–module $M$ which is Cohen–Macaulay ($\dim_R M = \operatorname{depth}_R M$) and of finite projective dimension, then $R$ is itself Cohen–Macaulay. This follows, as demonstrated in [9, (2.6.2)], by (†) and the Auslander–Buchsbaum formula. It is, however, not known if the existence of a Cohen–Macaulay module of finite G–dimension implies that $R$ is Cohen–Macaulay.

## Notes

The Change of Rings Theorem advertised in (1.5.4) can be derived from a result [2, Proposition (4.35)] due to Peskine and Szpiro, and a generalized version has been proved by Golod [40, Proposition 5]. The related issue of the behavior of G–dimension under flat base change will not be treated in this book. The interested reader is referred to [8, Section 4] or [15, Section 5].

A couple of noteworthy results from [1] and [2] have, so far, been omitted, because the proofs tend to get (unnecessarily) intricate. We make up for this in the next chapter, where we extend the G–dimension to complexes; using hyperhomological techniques we can then — with relative ease — prove even stronger results. From a "module point of view" the following results in chapter 2 are of particular interest:

- Theorem (2.2.8);
- Corollary (2.4.2);
- Corollary (2.4.4);
- Corollary (2.4.6);
- Corollary (2.4.8); and
- Observation (2.4.9).

Also Proposition (4.1.3) deserves mention: it characterizes modules in the G–class in a way that is quite different from anything we have considered here.

# Chapter 2

# G–dimension and Reflexive Complexes

In chapter 1 we studied Auslander's Gorenstein dimension for finite modules, now we extend it to complexes with finite homology.

G–dimension for complexes was first studied by Yassemi, and most of the results in this chapter can be traced back to [62]. Most of the proofs, however, are new, because we define the G–dimension in terms of resolutions, while Yassemi went straight for the throat and gave the definition in terms of derived functors. In this presentation of the theory it becomes a theorem that the G–dimension can be characterized in terms of derived functors, and the characterization we end up with is, of course, Yassemi's definition. Thus, the two definitions are equivalent, and they are both rooted in a result — due to Foxby — saying that a finite module has finite G–dimension in the sense of chapter 1 if and only if it is reflexive as a complex in the sense defined below.

## 2.1 Reflexive Complexes

We establish the basic properties of a full subcategory, $\mathcal{R}(R)$, of the category $\mathcal{C}(R)$ of all $R$–complexes and all morphisms of $R$–complexes. The objects in $\mathcal{R}(R)$ are so-called reflexive complexes.

(2.1.1) ***Biduality and Homothety Morphisms.*** Let $X$ and $Y$ be $R$–complexes. In degree $\ell$ the homomorphism complex $\mathrm{Hom}_R(\mathrm{Hom}_R(X,Y),Y)$ has the module

$$\mathrm{Hom}_R(\mathrm{Hom}_R(X,Y),Y)_\ell = \prod_{p\in\mathbb{Z}} \mathrm{Hom}_R(\mathrm{Hom}_R(X,Y)_p, Y_{p+\ell}).$$

An element $x \in X_\ell$ determines an element in $\mathrm{Hom}_R(\mathrm{Hom}_R(X,Y),Y)_\ell$, namely the family $(\delta_\ell(x)_p)_{p\in\mathbb{Z}}$, where $\delta_\ell(x)_p$ maps a family $(\psi_q)_{q\in\mathbb{Z}}$ in $\mathrm{Hom}_R(X,Y)_p =$

$\prod_{q\in\mathbb{Z}} \mathrm{Hom}_R(X_q, Y_{p+q})$ to $(-1)^{\ell p}\psi_\ell(x) \in Y_{p+\ell}$. It is easy check that this map from $X$ to $\mathrm{Hom}_R(\mathrm{Hom}_R(X,Y),Y)$ commutes with the differentials, and the *biduality morphism*,

$$\delta_X^Y : X \longrightarrow \mathrm{Hom}_R(\mathrm{Hom}_R(X,Y),Y),$$

is the one with $p$-th component of the map in degree $\ell$ given on a family $\psi = (\psi_q)_{q\in\mathbb{Z}}$ by[1]

$$((\delta_X^Y)_\ell(x))_p(\psi) = (-1)^{\ell p}\psi_\ell(x).$$

The morphism is natural in $X$ and $Y$; the action is, perhaps, better visualized from the graphical definition:

$$x \longmapsto [\psi \mapsto (-1)^{|x||\psi|}\psi(x)].$$

For $R$–modules $M$ and $E$ we note that $\delta_M^E$ is the homomorphism given by $\delta_M^E(m)(\psi) = \psi(m)$; in particular, $\delta_M^R$ is just the biduality map $\delta_M$, cf. (1.1.1).

For an $R$–complex $Y$ the *homothety morphism*,

$$\chi_Y^R : R \longrightarrow \mathrm{Hom}_R(Y,Y),$$

is the natural map given by

$$r \longmapsto [y \mapsto ry].$$

That is, $\chi_Y^R(r)$ is the family of homotheties

$$(r_{Y_p})_p \in \mathrm{Hom}_R(Y,Y)_0 = \prod_{p\in\mathbb{Z}} \mathrm{Hom}_R(Y_p, Y_p).$$

Biduality is linked to Hom evaluation: the diagram

$$\begin{array}{ccc} X & \xrightarrow{\delta_X^Y} & \mathrm{Hom}_R(\mathrm{Hom}_R(X,Y),Y) \\ \downarrow{\cong} & & \uparrow{\theta_{XYY}} \\ X\otimes_R R & \xrightarrow{X\otimes_R \chi_Y^R} & X\otimes_R \mathrm{Hom}_R(Y,Y) \end{array} \tag{2.1.1.1}$$

is commutative for all $R$–complexes $X$ and $Y$.

(2.1.2) **Observation.** Let $\iota\colon R \xrightarrow{\simeq} I$ be an injective resolution of $R$, then $\mathrm{Hom}_R(I,I)$ represents $\mathbf{R}\mathrm{Hom}_R(R,R) = R$. The functor $\mathrm{Hom}_R(-,I)$ preserves quasi-isomorphisms, and the commutative diagram

$$\begin{array}{ccc} R & \xrightarrow{\chi_I^R} & \mathrm{Hom}_R(I,I) \\ \simeq\downarrow{\iota} & & \simeq\downarrow{\mathrm{Hom}_R(\iota,I)} \\ I & \xrightarrow{\cong} & \mathrm{Hom}_R(R,I) \end{array}$$

shows that the homothety morphism $\chi_I^R$ is a quasi-isomorphism.

[1] The sign is not required to make $\delta$ a morphism, but it is introduced in accordance with the "universal sign rule", cf. (A.2.12), and without it, e.g., the diagram (2.1.1.1) would not be commutative.

(2.1.3) **Definition.** Let $X$ be an $R$–complex and let $I \in \mathcal{C}_{\sqsubset}^{\mathrm{I}}(R)$ be an injective resolution of $R$, then $\mathbf{R}\mathrm{Hom}_R(X,R)$ is represented by $\mathrm{Hom}_R(X,I)$ and $\mathbf{R}\mathrm{Hom}_R(\mathbf{R}\mathrm{Hom}_R(X,R),R)$ by $\mathrm{Hom}_R(\mathrm{Hom}_R(X,I),I)$. We say that $X$ *represents* $\mathbf{R}\mathrm{Hom}_R(\mathbf{R}\mathrm{Hom}_R(X,R),R)$ *canonically* if and only if the biduality morphism

$$\delta_X^I : X \longrightarrow \mathrm{Hom}_R(\mathrm{Hom}_R(X,I),I)$$

is a quasi-isomorphism.

(2.1.4) **Remark.** To see that this definition of canonical representation makes sense, take two injective resolutions $I, I' \in \mathcal{C}_{\sqsubset}^{\mathrm{I}}(R)$ of $R$. There is then by (A.3.5) a quasi-isomorphism $\iota\colon I' \xrightarrow{\simeq} I$, and by the quasi-isomorphism preserving properties of the various functors we get the commutative diagram

$$\begin{array}{ccc}
X & \xrightarrow{\delta_X^I} & \mathrm{Hom}_R(\mathrm{Hom}_R(X,I),I) \\
\Big\downarrow \delta_X^{I'} & & \simeq \Big\downarrow \mathrm{Hom}_R(\mathrm{Hom}_R(X,\iota),I) \\
\mathrm{Hom}_R(\mathrm{Hom}_R(X,I'),I') & \xrightarrow[\simeq]{\mathrm{Hom}_R(\mathrm{Hom}_R(X,I'),\iota)} & \mathrm{Hom}_R(\mathrm{Hom}_R(X,I'),I)
\end{array}$$

from which it is obvious that $\delta_X^I$ is a quasi-isomorphism if and only if $\delta_X^{I'}$ is so.

(2.1.5) **Lemma.** *Let $I \in \mathcal{C}_{\sqsubset}^{\mathrm{I}}(R)$ be an injective resolution of $R$, and let $X \in \mathcal{C}_{(\sqsupset)}(R)$. If $P \in \mathcal{C}_{\sqsupset}^{\mathrm{P}}(R)$ is a projective resolution of $X$, then the following are equivalent:*

(*i*) *$X$ represents $\mathbf{R}\mathrm{Hom}_R(\mathbf{R}\mathrm{Hom}_R(X,R),R)$ canonically.*
(*ii*) *The biduality morphism $\delta_P^I$ is a quasi-isomorphism.*
(*iii*) *The Hom evaluation morphism $\theta_{PRI}$ is a quasi-isomorphism.*

*Proof.* Let the resolutions $\pi\colon P \xrightarrow{\simeq} X$ and $\iota\colon R \xrightarrow{\simeq} I$ be given. By the Observation and the quasi-isomorphism preserving properties of the various functors, cf. (A.4.1), we get the following commutative diagram

$$\begin{array}{ccccc}
X & \xrightarrow{\delta_X^I} & \mathrm{Hom}_R(\mathrm{Hom}_R(X,I),I) & & \\
\simeq \Big\uparrow \pi & & \simeq \Big\uparrow \mathrm{Hom}_R(\mathrm{Hom}_R(\pi,I),I) & & \\
P & \xrightarrow{\delta_P^I} & \mathrm{Hom}_R(\mathrm{Hom}_R(P,I),I) & \xrightarrow[\simeq]{\mathrm{Hom}_R(\mathrm{Hom}_R(P,\iota),I)} & \mathrm{Hom}_R(\mathrm{Hom}_R(P,R),I) \\
\cong \Big\downarrow & & \Big\uparrow \theta_{PII} & & \Big\uparrow \theta_{PRI} \\
P \otimes_R R & \xrightarrow[\simeq]{P \otimes_R \chi_I^R} & P \otimes_R \mathrm{Hom}_R(I,I) & \xrightarrow[\simeq]{P \otimes_R \mathrm{Hom}_R(\iota,I)} & P \otimes_R \mathrm{Hom}_R(R,I)
\end{array}$$

The equivalence of the three conditions is now clear. □

(2.1.6) **Definition.** An $R$–complex $X$ is said to be *reflexive* if and only if

(1) $X \in \mathcal{C}_{(\square)}^{(\mathrm{f})}(R)$;

(2) $\mathbf{R}\mathrm{Hom}_R(X,R) \in \mathcal{C}_{(\square)}^{(\mathrm{f})}(R)$; and

(3) $X$ represents $\mathbf{R}\mathrm{Hom}_R(\mathbf{R}\mathrm{Hom}_R(X,R),R)$ canonically.

By $\mathcal{R}(R)$ we denote the full subcategory of $\mathcal{C}(R)$, actually of $\mathcal{C}_{(\square)}^{(\mathrm{f})}(R)$, whose objects are the reflexive complexes. We also use the notation $\mathcal{R}_0(R)$ with the usual definition:

$$\mathcal{R}_0(R) = \mathcal{R}(R) \cap \mathcal{C}_0(R).$$

If $X$ is reflexive, then so is every equivalent complex $X' \simeq X$; this follows by the Lemma as a projective resolution $P$ of $X$ is also a resolution of $X'$, cf. (A.3.6). We say that an equivalence class $\boldsymbol{X}$ of $R$–complexes is reflexive, and we write $\boldsymbol{X} \in \mathcal{R}(R)$, if some, equivalently every, representative $X$ of $\boldsymbol{X}$ is reflexive.

(2.1.7) **Remarks.** Note that the full subcategory $\mathcal{R}(R)$ is closed under shifts and finite direct sums.

A word of caution: in the literature, in [12] for example, a module is often said to be reflexive when it is isomorphic to its bidual; that is, a finite $R$–module $M$ is reflexive if and only if the biduality map $\delta_M \colon M \to M^{**}$ is an isomorphism, cf. Proposition (1.1.9). This definition, however, does not agree with the one above — see (2.2.6) for examples — and we have reserved the term 'reflexive' for complexes.

(2.1.8) ***Duality.*** The module functor $-^*$ induces a functor on complexes, which we also call the (algebraic) duality functor and denote by $-^*$; that is, $X^* = \mathrm{Hom}_R(X,R)$ for $X \in \mathcal{C}(R)$.

The next result is the natural extension to complexes of Proposition (1.1.11); this will be clarified in the next section.

(2.1.9) **Proposition.** *Every $R$–complex with finite homology and finite projective dimension is reflexive. That is, there is a full embedding:*

$$\mathcal{P}^{(\mathrm{f})}(R) \subseteq \mathcal{R}(R).$$

*Proof.* If $X \in \mathcal{P}^{(\mathrm{f})}(R)$, then $X$ is equivalent to a complex $P \in \mathcal{C}_{\square}^{\mathrm{fP}}(R)$, and it follows that $\mathbf{R}\mathrm{Hom}_R(X,R)$, represented by $P^*$, belongs to $\mathcal{C}_{(\square)}^{(\mathrm{f})}(R)$. Whenever $I \in \mathcal{C}_{\sqsubset}^{\mathrm{I}}(R)$ is an injective resolution of $R$ the Hom evaluation morphism $\theta_{PRI}$ is an isomorphism, cf. (A.2.11), in particular a quasi-isomorphism, so $X$ is reflexive by Lemma (2.1.5). $\square$

(2.1.10) **Theorem.** *An $R$–complex $X$ with finite homology is reflexive if and only if* $\mathbf{R}\mathrm{Hom}_R(X,R)$ *is so and of finite projective dimension if and only if* $\mathbf{R}\mathrm{Hom}_R(X,R)$ *is so. That is, the following hold for* $X \in \mathcal{C}^{(\mathrm{f})}_{(\square)}(R)$:

(a) $$X \in \mathcal{R}(R) \iff \mathbf{R}\mathrm{Hom}_R(X,R) \in \mathcal{R}(R); \quad \textit{and}$$

(b) $$X \in \mathcal{P}^{(\mathrm{f})}(R) \iff \mathbf{R}\mathrm{Hom}_R(X,R) \in \mathcal{P}^{(\mathrm{f})}(R).$$

*Proof.* (a): If $X$ or $\mathbf{R}\mathrm{Hom}_R(X,R)$ is reflexive, then both $\mathbf{R}\mathrm{Hom}_R(X,R)$ and $\mathbf{R}\mathrm{Hom}_R(\mathbf{R}\mathrm{Hom}_R(X,R),R)$ belong to $\mathcal{C}^{(\mathrm{f})}_{(\square)}(R)$. Let $I \in \mathcal{C}^{\mathrm{I}}_{\sqsubset}(R)$ be an injective resolution of $R$; it is easy to see that

(†) $$\mathrm{Hom}_R(\delta^I_X, I)\delta^I_{\mathrm{Hom}_R(X,I)} = 1_{\mathrm{Hom}_R(X,I)}.$$

If $X$ is reflexive, then $\delta^I_X$ is a quasi-isomorphism and, hence, so is $\mathrm{Hom}_R(\delta^I_X, I)$ because $\mathrm{Hom}_R(-,I)$ preserves quasi-isomorphisms. From (†) it then follows that $\delta^I_{\mathrm{Hom}_R(X,I)}$ is a quasi-isomorphism, and $\mathrm{Hom}_R(X,I)$ represents $\mathbf{R}\mathrm{Hom}_R(X,R)$, so $\mathbf{R}\mathrm{Hom}_R(X,R)$ belongs to $\mathcal{R}(R)$ as wanted. On the other hand, if $\mathbf{R}\mathrm{Hom}_R(X,R) \in \mathcal{R}(R)$, then the representative $\mathrm{Hom}_R(X,I)$ is reflexive, so $\delta^I_{\mathrm{Hom}_R(X,I)}$ is a quasi-isomorphism, and then, by (†), so is $\mathrm{Hom}_R(\delta^I_X, I)$. It now follows by (A.8.11) that $\delta^I_X$ is a quasi-isomorphism; and $X$ is, therefore, reflexive as wanted.

(b): Since $\mathbf{R}\mathrm{Hom}_R(\mathbf{R}\mathrm{Hom}_R(X,R),R)$ is represented by $X$ when $X \in \mathcal{R}(R)$, and $\mathcal{P}^{(\mathrm{f})}(R) \subseteq \mathcal{R}(R)$, by Proposition (2.1.9), it is, in view of (a), sufficient to prove that $X \in \mathcal{P}^{(\mathrm{f})}(R)$ implies $\mathbf{R}\mathrm{Hom}_R(X,R) \in \mathcal{P}^{(\mathrm{f})}(R)$. But this is easy: if $X \in \mathcal{P}^{(\mathrm{f})}(R)$, then $X$ is equivalent to a complex $P \in \mathcal{C}^{\mathrm{fP}}_{\square}(R)$, cf. (A.5.4.2), and $\mathbf{R}\mathrm{Hom}_R(X,R)$ is represented by $\mathrm{Hom}_R(P,R)$, which is a bounded complex of finite projective modules, so $\mathbf{R}\mathrm{Hom}_R(X,R) \in \mathcal{P}^{(\mathrm{f})}(R)$. □

Part (b) in the Theorem is a stability result and, actually, so is (a), but this will only be clear from (2.3.8).

(2.1.11) **Observation.** Let $\mathfrak{p} \in \operatorname{Spec} R$ and $X \in \mathcal{C}^{(\mathrm{f})}_{(\sqsupset)}(R)$. Let $I \in \mathcal{C}^{\mathrm{I}}_{\sqsubset}(R)$ and $L \in \mathcal{C}^{\mathrm{L}}_{\sqsupset}(R)$ be resolutions of, respectively, $R$ and $X$, then $I_\mathfrak{p} \in \mathcal{C}^{\mathrm{I}}_{\sqsubset}(R_\mathfrak{p})$ and $L_\mathfrak{p} \in \mathcal{C}^{\mathrm{L}}_{\sqsupset}(R_\mathfrak{p})$ are resolutions of $R_\mathfrak{p}$ and $X_\mathfrak{p}$. The commutative diagram

$$\begin{array}{ccc}
\mathrm{Hom}_R(\mathrm{Hom}_R(L,I),I)_\mathfrak{p} & \xrightarrow{\cong} & \mathrm{Hom}_{R_\mathfrak{p}}(\mathrm{Hom}_{R_\mathfrak{p}}(L_\mathfrak{p},I_\mathfrak{p}),I_\mathfrak{p}) \\
\uparrow (\delta^I_L)_\mathfrak{p} & & \uparrow \delta^{I_\mathfrak{p}}_{L_\mathfrak{p}} \\
L_\mathfrak{p} & =\!=\!= & L_\mathfrak{p}
\end{array}$$

shows that $\delta^{I_\mathfrak{p}}_{L_\mathfrak{p}}$ is a quasi-isomorphism if and only if $(\delta^I_L)_\mathfrak{p}$ is so. In particular, $\delta^{I_\mathfrak{p}}_{L_\mathfrak{p}}$ is a quasi-isomorphism if $\delta^I_L$ is so.

If $X$ belongs to $\mathcal{R}(R)$, then the homological boundedness of $X$ and $\mathbf{R}\mathrm{Hom}_R(X,R)$ implies that of $X_\mathfrak{p}$ and $\mathbf{R}\mathrm{Hom}_{R_\mathfrak{p}}(X_\mathfrak{p},R_\mathfrak{p}) = \mathbf{R}\mathrm{Hom}_R(X,R)_\mathfrak{p}$, so

$$(2.1.11.1) \qquad X \in \mathcal{R}(R) \implies X_\mathfrak{p} \in \mathcal{R}(R_\mathfrak{p}).$$

(2.1.12) **Lemma.** *Let $0 \to X' \to X \to X'' \to 0$ be a short exact sequence in $\mathcal{C}^{(\mathrm{f})}_{(\sqsupset)}(R)$. If two of the complexes are reflexive, then so is the third.*

*Proof.* By (A.3.4) we can choose a short exact sequence

$$(\dagger) \qquad 0 \to P' \to P \to P'' \to 0$$

in $\mathcal{C}^{\mathrm{P}}_{\sqsupset}(R)$, such that $P'$, $P$, and $P''$ are projective resolutions of, respectively, $X'$, $X$, and $X''$. If two of the complexes $X$, $X'$, and $X''$ are reflexive, then two of the complexes in (†) are homologically bounded, and by inspection of the associated long exact sequence,

$$\cdots \to \mathrm{H}_{\ell+1}(P'') \to \mathrm{H}_\ell(P') \to \mathrm{H}_\ell(P) \to \mathrm{H}_\ell(P'') \to \cdots,$$

we see that also the third is homologically bounded. That is, if two of the complexes in the original short exact sequence belong to $\mathcal{C}^{(\mathrm{f})}_{(\square)}(R)$, then so does the third.

Let $I \in \mathcal{C}^{\mathrm{I}}_{\sqsubset}(R)$ be an injective resolution of $R$. From (†) we get another short exact sequence of complexes

$$(\ddagger) \qquad 0 \to \mathrm{Hom}_R(P'',I) \to \mathrm{Hom}_R(P,I) \to \mathrm{Hom}_R(P',I) \to 0,$$

which represent $\mathbf{R}\mathrm{Hom}_R(X'',R)$, $\mathbf{R}\mathrm{Hom}_R(X,R)$, and $\mathbf{R}\mathrm{Hom}_R(X',R)$. As above it follows that if two of the complexes in the original short exact sequence are reflexive, then two of the complexes in (‡) belong to $\mathcal{C}^{(\mathrm{f})}_{(\square)}(R)$, and hence so does the third.

Using the abbreviated notation $[[-,I]I] = \mathrm{Hom}_R(\mathrm{Hom}_R(-,I),I)$ we have the following commutative diagram:

$$\begin{array}{ccccccccc}
0 & \longrightarrow & P' & \longrightarrow & P & \longrightarrow & P'' & \longrightarrow & 0 \\
 & & \downarrow{\scriptstyle \delta^I_{P'}} & & \downarrow{\scriptstyle \delta^I_P} & & \downarrow{\scriptstyle \delta^I_{P''}} & & \\
0 & \longrightarrow & [[P',I]I] & \longrightarrow & [[P,I]I] & \longrightarrow & [[P'',I]I] & \longrightarrow & 0
\end{array}$$

The top row is the short exact sequence (†), and the bottom row is also exact (apply $\mathrm{Hom}_R(-,I)$ to (‡)). When we pass to homology, this diagram yields a long exact ladder

$$\begin{array}{ccccccc}
\cdots \to & \mathrm{H}_{\ell+1}(P'') & \longrightarrow & \mathrm{H}_\ell(P') & \longrightarrow & \mathrm{H}_\ell(P) & \to \cdots \\
 & \downarrow{\scriptstyle \mathrm{H}_{\ell+1}(\delta^I_{P''})} & & \downarrow{\scriptstyle \mathrm{H}_\ell(\delta^I_{P'})} & & \downarrow{\scriptstyle \mathrm{H}_\ell(\delta^I_P)} & \\
\cdots \to & \mathrm{H}_{\ell+1}([[P'',I]I]) & \longrightarrow & \mathrm{H}_\ell([[P',I]I]) & \longrightarrow & \mathrm{H}_{\ell+1}([[P,I]I]) & \to \cdots
\end{array}$$

showing that if two of the morphisms $\delta^I_{P'}$, $\delta^I_P$, and $\delta^I_{P''}$ are quasi-isomorphisms, then so is the third. □

(2.1.13) **Proposition.** *A bounded complex of modules from $\mathcal{R}_0(R)$ is reflexive.*

*Proof.* Let $X \neq 0$ be a bounded complex of modules from $\mathcal{R}_0(R)$. We can, without loss of generality, assume that $X$ is concentrated in non-negative degrees and set $u = \sup\{\ell \in \mathbb{Z} \mid X_\ell \neq 0\}$. If $u = 0$ then $X \in \mathcal{R}_0(R)$. If $u > 0$ we consider the short exact sequence of complexes $0 \to \sqsubset_{u-1} X \to X \to \Sigma^u X_u \to 0$, where $\Sigma^u X_u \in \mathcal{R}(R)$ as $X_u \in \mathcal{R}_0(R)$, and $\sqsubset_{u-1} X$ is concentrated in degrees at most $u-1$. In view of the Lemma, the claim is now obvious by induction on $u$. □

## 2.2 The Module Case

The reflexive complexes defined in the previous section will play a key role in the following. We start by investigating what it means for a module to be a reflexive complex. The answer — provided by Theorem (2.2.3) — is that a module is reflexive as a complex if and only if it has finite G–dimension in the sense of chapter 1. This result enables us to prove the Change of Rings Theorem advertised in (1.5.4) and give the examples promised in (2.1.7).

(2.2.1) **Lemma.** *Let $M$ be a finite $R$–module. The following hold:*

(a) *$\mathbf{R}\mathrm{Hom}_R(M,R)$ has homology concentrated in degree zero if and only if $\mathrm{Ext}^m_R(M,R) = 0$ for $m > 0$, i.e., if and only if $-\inf(\mathbf{R}\mathrm{Hom}_R(M,R)) \leq 0$.*

(b) *If $\mathbf{R}\mathrm{Hom}_R(M,R) \in \mathcal{C}_{(0)}(R)$, then $M^*$ represents $\mathbf{R}\mathrm{Hom}_R(M,R)$, and $\mathbf{R}\mathrm{Hom}_R(\mathbf{R}\mathrm{Hom}_R(M,R),R)$ belongs to $\mathcal{C}_{(0)}(R)$ if and only if $\mathrm{Ext}^m_R(M^*,R) = 0$ for $m > 0$.*

(c) *If both $\mathbf{R}\mathrm{Hom}_R(M,R)$ and $\mathbf{R}\mathrm{Hom}_R(\mathbf{R}\mathrm{Hom}_R(M,R),R)$ have homology concentrated in degree zero, then the biduality map $\delta_M$ is an isomorphism if and only if $M$ represents $\mathbf{R}\mathrm{Hom}_R(\mathbf{R}\mathrm{Hom}_R(M,R),R)$ canonically.*

*Proof.* (a) is immediate from (A.4.3) and (A.4.6.1).

We take resolutions $\pi\colon P \xrightarrow{\simeq} M$ and $\iota\colon R \xrightarrow{\simeq} I$, where $P \in \mathcal{C}^{\mathrm{P}}_{\sqsupset}(R)$ and $I \in \mathcal{C}^{\mathrm{I}}_{\sqsubset}(R)$ have $P_\ell = 0$ for $\ell < 0$ and $I_\ell = 0$ for $\ell > 0$.

(b): The complex $P^*$ represents $\mathbf{R}\mathrm{Hom}_R(M,R)$, and the induced map $\mathrm{H}_0(\pi^*)\colon M^* \to \mathrm{H}_0(P^*)$ is an isomorphism, cf. (A.4.6), so when $\mathbf{R}\mathrm{Hom}_R(M,R)$ belongs to $\mathcal{C}_{(0)}(R)$, the induced morphism $\pi^*$ is a quasi-isomorphism. In particular, $M^*$ represents $\mathbf{R}\mathrm{Hom}_R(M,R)$. Now it follows by (A.4.6.1) that $\mathbf{R}\mathrm{Hom}_R(\mathbf{R}\mathrm{Hom}_R(M,R),R)$ has homology concentrated in non-positive degrees, and for $m \geq 0$ we have

$$\mathrm{H}_{-m}(\mathbf{R}\mathrm{Hom}_R(\mathbf{R}\mathrm{Hom}_R(M,R),R)) = \mathrm{H}_{-m}(\mathbf{R}\mathrm{Hom}_R(M^*,R)) = \mathrm{Ext}^m_R(M^*,R).$$

Hence, $\mathbf{R}\mathrm{Hom}_R(\mathbf{R}\mathrm{Hom}_R(X,R),R) \in \mathcal{C}_{(0)}(R)$ if and only if $\mathrm{Ext}^m_R(M^*,R) = 0$ for $m > 0$.

(c): From the above it follows that the complex $\mathrm{Hom}_R(M^*, I)$ represents $\mathbf{R}\mathrm{Hom}_R(\mathbf{R}\mathrm{Hom}_R(M, R), R)$. The induced map

$$\mathrm{H}_0(\mathrm{Hom}_R(M^*, \iota)) : M^{**} \longrightarrow \mathrm{H}_0(\mathrm{Hom}_R(M^*, I))$$

is an isomorphism, so $\mathrm{Hom}_R(M^*, \iota)$ is a quasi-isomorphism because $\mathbf{R}\mathrm{Hom}_R(\mathbf{R}\mathrm{Hom}_R(M, R), R)$ has homology concentrated in degree zero. Using that the functors $\mathrm{Hom}_R(P, -)$ and $\mathrm{Hom}_R(-, I)$ preserve quasi-isomorphisms, we establish the following diagram:

$$\begin{array}{ccc}
P & \xrightarrow{\ \delta_P^I\ } & \mathrm{Hom}_R(\mathrm{Hom}_R(P, I), I) \\
\simeq \big\downarrow \pi & & \simeq \big\downarrow \mathrm{Hom}_R(\mathrm{Hom}_R(P,\iota),I) \\
M & & \mathrm{Hom}_R(\mathrm{Hom}_R(P, R), I) \\
\big\downarrow \delta_M & & \simeq \big\downarrow \mathrm{Hom}_R(\pi^*,I) \\
M^{**} & \xrightarrow[\simeq]{\ \mathrm{Hom}_R(M^*,\iota)\ } & \mathrm{Hom}_R(M^*, I)
\end{array}$$

It is easy to check that the diagram is commutative. It shows that $\delta_M$ is a quasi-isomorphism if and only if $\delta_P^I$ is so; that is, $\delta_M$ is an isomorphism of modules if and only if $M$ represents $\mathbf{R}\mathrm{Hom}_R(\mathbf{R}\mathrm{Hom}_R(M, R), R)$ canonically, cf. Lemma (2.1.5). □

(2.2.2) **Proposition.** *A finite $R$–module $M$ belongs to the G–class if and only if it is reflexive as a complex and has* $-\inf(\mathbf{R}\mathrm{Hom}_R(M, R)) \leq 0$; *that is,*

$$M \in \mathrm{G}(R) \iff M \in \mathcal{R}_0(R) \wedge -\inf(\mathbf{R}\mathrm{Hom}_R(M, R)) \leq 0.$$

*Proof.* "⇒": Suppose $M \in \mathrm{G}(R)$, then $\delta_M$ is an isomorphism,

$$-\inf(\mathbf{R}\mathrm{Hom}_R(M, R)) = \sup\{m \in \mathbb{Z} \mid \mathrm{Ext}_R^m(M, R) \neq 0\} \leq 0,$$

and $\mathrm{Ext}_R^m(M^*, R) = 0$ for $m > 0$. From the Lemma it now follows that both $\mathbf{R}\mathrm{Hom}_R(M, R)$ and $\mathbf{R}\mathrm{Hom}_R(\mathbf{R}\mathrm{Hom}_R(M, R), R)$ have homology concentrated in degree zero, and that $M$ represents $\mathbf{R}\mathrm{Hom}_R(\mathbf{R}\mathrm{Hom}_R(M, R), R)$ canonically, so $M \in \mathcal{R}_0(R)$.

"⇐": If $M$ belongs to $\mathcal{R}_0(R)$ and has $-\inf(\mathbf{R}\mathrm{Hom}_R(M, R)) \leq 0$, then both $\mathbf{R}\mathrm{Hom}_R(M, R)$ and $\mathbf{R}\mathrm{Hom}_R(\mathbf{R}\mathrm{Hom}_R(M, R), R) = M$ have homology concentrated in degree zero, so it follows by the Lemma that $\mathrm{Ext}_R^m(M, R) = 0 = \mathrm{Ext}_R^m(M^*, R)$ for $m > 0$, and that the biduality map $\delta_M$ is an isomorphism. □

(2.2.3) **Theorem.** *A finite $R$–module $M$ has finite G–dimension if and only if it is reflexive as a complex; that is,*

$$\mathrm{G\text{–}dim}_R\, M < \infty \iff M \in \mathcal{R}_0(R).$$

*Furthermore, if $M \in \mathcal{R}_0(R)$, then*

$$\mathrm{G\text{–}dim}_R\, M = -\inf(\mathbf{R}\mathrm{Hom}_R(M, R)).$$

*Proof.* First note that $0 \in \mathcal{R}_0(R)$ and $\text{G-dim}_R\, 0 = -\infty = -\inf 0$. We can now assume that $M$ is non-zero.

"$\Rightarrow$": If $M$ has finite G–dimension, then $M$ has a G–resolution of finite length, say $n$:

$$0 \to G_n \to \cdots \to G_1 \to G_0 \to M \to 0.$$

The module $M$ is therefore equivalent to the complex $G = 0 \to G_n \to \cdots \to G_1 \to G_0 \to 0$, and since $G$ belongs to $\mathcal{R}(R)$, cf. Proposition (2.1.13), so does $M$.

"$\Leftarrow$": When $M \in \mathcal{R}_0(R)$ the number $g = -\inf(\mathbf{R}\mathrm{Hom}_R(M,R))$ belongs to $\mathbb{N}_0$, as $\mathbf{R}\mathrm{Hom}_R(M,R)$ is homologically non-trivial; we proceed by induction on $g$. If $g = 0$ then $M \in \mathrm{G}(R)$ by the Proposition, so $\text{G-dim}_R\, M = 0$. Let $g > 0$ and assume that modules $K \in \mathcal{R}_0(R)$ with $-\inf(\mathbf{R}\mathrm{Hom}_R(K,R)) = g-1$ have finite G–dimension. We can now consider a short exact sequence of modules

$$0 \to K \to L \to M \to 0,$$

where $L$ is a finite free $R$–module and $K \neq 0$. Inspecting the associated long exact sequence,

$$\cdots \to \mathrm{Ext}_R^{m-1}(K,R) \to \mathrm{Ext}_R^{m}(M,R) \to \mathrm{Ext}_R^{m}(L,R) \to \mathrm{Ext}_R^{m}(K,R) \to \cdots,$$

we see that $\mathrm{Ext}_R^m(K,R) = 0$ for $m \geq g$ and $\mathrm{Ext}_R^{g-1}(K,R) \neq 0$, that is, $-\inf(\mathbf{R}\mathrm{Hom}_R(K,R)) = g-1$. By assumption $M \in \mathcal{R}_0(R)$, and $L \in \mathcal{R}_0(R)$ by Proposition (2.1.9), so it follows by Lemma (2.1.12) that $K$ belongs to $\mathcal{R}_0(R)$. By the induction hypothesis $K$ has finite G–dimension, and hence so has $M$, cf. Corollary (1.2.9)(c).

Now it follows from GD Theorem (1.2.7) that

$$\begin{aligned} \text{G-dim}_R\, M &= \sup\{m \in \mathbb{Z} \mid \mathrm{Ext}_R^m(M,R) \neq 0\} \\ &= -\inf(\mathbf{R}\mathrm{Hom}_R(M,R)) \end{aligned}$$

for $M \in \mathcal{R}_0(R)$. □

(2.2.4) **Observation.** Let $M$ be a finite $R$–module, and let

$$G = \cdots \to G_\ell \to G_{\ell-1} \to \cdots \to G_1 \to G_0 \to 0$$

be a G–resolution of $M$. For $n \in \mathbb{N}$ the diagram

$$\begin{array}{ccccccccccccc} 0 & \longrightarrow & K_n & \longrightarrow & 0 & & & & & & & & \\ \downarrow & & \downarrow & & \downarrow & & & & & & & & \\ 0 & \longrightarrow & G_{n-1} & \longrightarrow & G_{n-2} & \longrightarrow & \cdots & \longrightarrow & G_1 & \longrightarrow & G_0 & \longrightarrow & 0 \\ & & & & & & & & \downarrow & & \downarrow & & \downarrow \\ & & & & & & & & 0 & \longrightarrow & M & \longrightarrow & 0 \end{array}$$

where $K_n$ is as defined in (1.2.5.1), is commutative. It follows that we have a short exact sequence of complexes

$$0 \to \Sigma^{n-1} K_n \to \sqsubset_{n-1} G \to M \to 0;$$

and since $\sqsubset_{n-1} G$ belongs to $\mathcal{R}(R)$ by Propositions (2.2.2) and (2.1.13), it follows by Lemma (2.1.12) and the Theorem that $M$ is of finite G–dimension if and only if the same holds for $K_n$. This provides a new proof of "(*iii*) $\Rightarrow$ (*i*)" in Theorem (1.2.7) because, as we noted in (1.2.8), it is now easy to see that $K_n \in \mathrm{G}(R)$ if $\mathrm{G\text{–}dim}_R M \leq n$.

(2.2.5) **Corollary.** *Let $0 \to M' \to M \to M'' \to 0$ be a short exact sequence of finite $R$–modules. If two of the modules have finite G–dimension, then so has the third.*

*Proof.* Immediate by Theorem (2.2.3) and Lemma (2.1.12). □

Now that we understand the modules in $\mathcal{R}(R)$, we are ready to give the promised examples, showing that a module may be reflexive — in the sense that the biduality map $\delta_M : M \to M^{**}$ is an isomorphism — without being reflexive as a complex, and vice versa.

(2.2.6) **Examples.** Let $(R, \mathfrak{m}, k)$ be a local ring with $\operatorname{depth} R > 0$, and let $x$ be an $R$–regular element. Set $S = R/(x)$, then $\mathrm{Hom}_R(S, R) = 0$, cf. the Hom vanishing corollary, so the biduality map $\delta_S$ for the $R$–module $S$ is not an isomorphism. But $\mathrm{G\text{–}dim}_R S = \mathrm{pd}_R S = 1$, cf. Proposition (1.2.10), so by Theorem (2.2.3) $S$ is a reflexive $R$–complex.

Let $k$ be a field and consider the local ring $R = k[\![X, Y]\!]/(X^2, XY)$. This ring is not Gorenstein (it is not even Cohen–Macaulay: $\operatorname{depth} R = 0$ but $\dim R = 1$), so by Theorems (1.4.9) and (2.2.3) the $R$–complex $k$ is not reflexive. It is, however, easy to see that $\mathrm{Hom}_R(k, R) \cong (x)$, where $x$ is the residue class of $X$, so $\mu_R^0 = 1$ and $\delta_k$ is invertible, cf. Observation (1.1.12).

(2.2.7) **Lemma.** *Let $x_1, \ldots, x_t$ be an $R$–sequence and $S = R/(x_1, \ldots, x_t)$. If $X \in \mathcal{C}_{(\sqsupset)}^{(\mathrm{f})}(S)$, then $X$ is reflexive as $S$–complex if and only if it is so as $R$–complex; that is,*

$$X \in \mathcal{R}(S) \iff X \in \mathcal{R}(R).$$

*Furthermore, there is an equality:*

$$-\inf\left(\mathbf{R}\mathrm{Hom}_S(X, S)\right) = -\inf\left(\mathbf{R}\mathrm{Hom}_R(X, R)\right) - t.$$

*Proof.* First note that $X$ belongs to $\mathcal{C}_{(\sqsupset)}^{(\mathrm{f})}(R)$, because $S$ is a finite $R$–module. In particular, we have $X \in \mathcal{C}_{(\square)}^{(\mathrm{f})}(S)$ if and only if $X \in \mathcal{C}_{(\square)}^{(\mathrm{f})}(R)$. Let $L$ be the Koszul complex on the $R$–sequence $x_1, \ldots, x_t$, then $L$ is a projective resolution of $S$ and concentrated in degrees $t, t-1, \ldots, 0$. Using induction on $t$, it is easy to verify that $\mathrm{Hom}_R(L,R) \cong \Sigma^{-t}L$, and since $\mathrm{Hom}_R(L,R)$ represents $\mathbf{R}\mathrm{Hom}_R(S,R)$, the module $S$ represents $\Sigma^t\mathbf{R}\mathrm{Hom}_R(S,R)$. The calculation

$$\begin{aligned}\mathbf{R}\mathrm{Hom}_S(X,S) &= \mathbf{R}\mathrm{Hom}_S(X,\Sigma^t\mathbf{R}\mathrm{Hom}_R(S,R))\\ &= \Sigma^t\mathbf{R}\mathrm{Hom}_S(X,\mathbf{R}\mathrm{Hom}_R(S,R))\\ &= \Sigma^t\mathbf{R}\mathrm{Hom}_R(X\otimes^{\mathbf{L}}_S S,R)\\ &= \Sigma^t\mathbf{R}\mathrm{Hom}_R(X,R),\end{aligned}$$

where the third identity is adjointness (A.4.21), shows that

$$\inf\left(\mathbf{R}\mathrm{Hom}_S(X,S)\right) = \inf\left(\mathbf{R}\mathrm{Hom}_R(X,R)\right) + t$$

as wanted. In particular, $\mathbf{R}\mathrm{Hom}_S(X,S)$ is homologically bounded if and only if $\mathbf{R}\mathrm{Hom}_R(X,R)$ is so.

What remains to be proved is that the $S$–complex $X$ represents the equivalence class $\mathbf{R}\mathrm{Hom}_S(\mathbf{R}\mathrm{Hom}_S(X,S),S)$ canonically if and only if the $R$–complex $X$ represents $\mathbf{R}\mathrm{Hom}_R(\mathbf{R}\mathrm{Hom}_R(X,R),R)$ canonically. For $W \in \mathcal{C}(R)$ we denote the $S$–complex $\mathrm{Hom}_R(S,W)$ by $\underline{W}$. Let $I \in \mathcal{C}^{\mathrm{I}}_{\sqsubset}(R)$ be an injective resolution of $R$. Since $S$ is an $R$–algebra, the modules $\underline{I}_\ell = \mathrm{Hom}_R(S,I_\ell) = \underline{I_\ell}$ are injective over $S$, so $\underline{I} \in \mathcal{C}^{\mathrm{I}}_{\sqsubset}(S)$. The complex $\underline{I}$ represents $\mathbf{R}\mathrm{Hom}_R(S,R)$ and is, therefore, equivalent with $\Sigma^{-t}S$, so $\Sigma^t\underline{I}$ is an injective resolution of the $S$–module $S$. The aim is now to show that the biduality morphism $\delta^I_X$ is a quasi-isomorphism if and only if $\delta^{\Sigma^t \underline{I}}_X$ is so. By (A.2.1.1) and (A.2.1.3) there is a natural isomorphism

$$\mathrm{Hom}_S(\mathrm{Hom}_S(X,\Sigma^t\underline{I}),\Sigma^t\underline{I}) \cong \mathrm{Hom}_S(\mathrm{Hom}_S(X,\underline{I}),\underline{I}),$$

so it is sufficient to prove that $\delta^I_X$ is a quasi-isomorphism if and only if $\delta^{\underline{I}}_X$ is so. For an $S$–complex $V$ and an $R$–complex $W$ there are natural isomorphisms

$$\mathrm{Hom}_R(V,W) \cong \mathrm{Hom}_R(V\otimes_S S,W) \cong \mathrm{Hom}_S(V,\mathrm{Hom}_R(S,W)) = \mathrm{Hom}_S(V,\underline{W}),$$

cf. (A.2.8), and this accounts for the unlabeled isomorphisms in the diagram

$$\begin{array}{ccc} X & \xrightarrow{\ \delta^{\underline{I}}_X\ } & \mathrm{Hom}_S(\mathrm{Hom}_S(X,\underline{I}),\underline{I}) \\ \big\downarrow{\scriptstyle \delta^I_X} & & \big\uparrow{\scriptstyle \cong} \\ \mathrm{Hom}_R(\mathrm{Hom}_R(X,I),I) & \xrightarrow{\ \cong\ } & \mathrm{Hom}_R(\mathrm{Hom}_S(X,\underline{I}),I) \end{array}$$

The diagram is commutative and it follows, as wanted, that $\delta^I_X$ is a quasi-isomorphism if and only if $\delta^{\underline{I}}_X$ is so. □

(2.2.8) **Theorem (Change of Rings).** *Let $M$ be a finite $R$–module. If $\boldsymbol{x} = x_1, \ldots, x_t$ is an $R$–sequence in $\mathrm{Ann}_R\, M$, then*

$$\text{G-dim}_R\, M = \text{G-dim}_{R/(\boldsymbol{x})}\, M + t.$$

*In particular, the two dimensions are simultaneously finite.*

*Proof.* Set $S = R/(\boldsymbol{x})$, then $M$ is an $S$–module, and it follows by the Lemma and Theorem (2.2.3) that $\text{G–dim}_R\, M < \infty$ if and only if $\text{G–dim}_S\, M < \infty$. The equality also follows from these two results. □

### Notes

The characterization of modules in $\mathcal{R}(R)$, Theorem (2.2.3), is due to Foxby and appeared in [62].

The Change of Rings Theorem (2.2.8) can be derived from a result [2, Proposition (4.35)] due to Peskine and Szpiro. It has been generalized in different directions by Golod [40, Proposition 5], Avramov and Foxby [8, Theorem (7.11)], and the author [15, Theorem (6.5)].

## 2.3 G–dimension of Complexes with Finite Homology

We define G–dimension for complexes with finite homology, and we show how the principal results from chapter 1 can be extended to complexes.

(2.3.1) **Definition.** We use the notation $\mathcal{C}^{\mathrm{G}}(R)$ for the full subcategory (of $\mathcal{C}(R)$) of complexes of modules from $\mathrm{G}(R)$, and we use it with subscripts $\square$ and $\sqsupset$ with the usual definitions:

$$\mathcal{C}^{\mathrm{G}}_{\square}(R) = \mathcal{C}^{\mathrm{G}}(R) \cap \mathcal{C}_{\square}(R) \quad \text{and} \quad \mathcal{C}^{\mathrm{G}}_{\sqsupset}(R) = \mathcal{C}^{\mathrm{G}}(R) \cap \mathcal{C}_{\sqsupset}(R).$$

(2.3.2) **Definition.** The *G–dimension*, $\text{G–dim}_R\, X$, of $X \in \mathcal{C}^{(\mathrm{f})}_{(\sqsupset)}(R)$ is defined as

$$\text{G–dim}_R\, X = \inf \{\sup \{\ell \in \mathbb{Z} \mid G_\ell \neq 0\} \mid X \simeq G \in \mathcal{C}^{\mathrm{G}}_{\sqsupset}(R)\}.$$

Note that the set over which infimum is taken is non-empty: any complex $X \in \mathcal{C}^{(\mathrm{f})}_{(\sqsupset)}(R)$ has a resolution by finite free modules $X \xleftarrow{\simeq} L \in \mathcal{C}^{\mathrm{L}}_{\sqsupset}(R)$, and $\mathcal{C}^{\mathrm{L}}_{\sqsupset}(R) \subseteq \mathcal{C}^{\mathrm{G}}_{\sqsupset}(R)$.

(2.3.3) **Observation.** We note the following facts about the G–dimension of $X \in \mathcal{C}^{(\mathrm{f})}_{(\sqsupset)}(R)$:

$$\begin{aligned} &\text{G–dim}_R\, X \in \{-\infty\} \cup \mathbb{Z} \cup \{\infty\}; \\ &\text{G–dim}_R\, X \geq \sup X; \quad \text{and} \\ &\text{G–dim}_R\, X = -\infty \Leftrightarrow X \simeq 0. \end{aligned}$$

The next three results are auxiliaries needed in the proof of Theorem (2.3.7).

(2.3.4) **Lemma.** *Let $G \in \mathcal{C}^{\mathrm{G}}_{\sqsupset}(R)$. If $G$ is homologically trivial, then so is the dual complex $G^*$.*

*Proof.* For $G = 0$ there is nothing to prove, so we assume that $G$ is non-zero and set $v = \inf\{\ell \in \mathbb{Z} \mid G_\ell \neq 0\}$. To see that the complex

$$G^* = 0 \to G_v{}^* \to G_{v+1}{}^* \to G_{v+2}{}^* \to \cdots$$

is homologically trivial, it is sufficient to prove that the short exact sequences $0 \to \mathrm{Z}_\ell^G \to G_\ell \to \mathrm{Z}_{\ell-1}^G \to 0$, $\ell > v$, stay exact under dualization. But this follows immediately from Lemma (1.1.10)(a), since the kernels $\mathrm{Z}_\ell^G$ belong to $\mathrm{G}(R)$ by GD Theorem (1.2.7) (or by Lemma (1.1.10)(a) and induction). □

**(2.3.5) Proposition.** *If* $X \simeq G \in \mathcal{C}_\sqsupset^{\mathrm{G}}(R)$, *then* $G^*$ *represents* $\mathbf{R}\mathrm{Hom}_R(X,R)$.

*Proof.* Take a resolution $X \xleftarrow{\simeq} P \in \mathcal{C}_\sqsupset^{\mathrm{P}}(R)$, then $P^*$ represents $\mathbf{R}\mathrm{Hom}_R(X,R)$. Since $P \simeq X \simeq G$ there is by (A.3.6) a quasi-isomorphism $\pi\colon P \xrightarrow{\simeq} G$, and hence a morphism $\pi^*\colon G^* \to P^*$. If $\pi^*$ is a quasi-isomorphism, then $G^*$ represents $\mathbf{R}\mathrm{Hom}_R(X,R)$ as desired, so it is sufficient to prove that the mapping cone $\mathcal{M}(\pi^*)$ is homologically trivial. By (A.2.1.4) we have $\mathcal{M}(\pi^*) \simeq \Sigma^1 \mathcal{M}(\pi)^*$ The mapping cone $\mathcal{M}(\pi)$ is bounded to the right and consists of direct sums of modules in the G–class, so by Lemma (1.1.10)(c) we have $\mathcal{M}(\pi) \in \mathcal{C}_\sqsupset^{\mathrm{G}}(R)$. Furthermore, $\mathcal{M}(\pi)$ is homologically trivial as $\pi$ is a quasi-isomorphism, so by the Lemma we have $\mathcal{M}(\pi)^* \simeq 0$, and hence $\mathcal{M}(\pi^*)$ is homologically trivial as wanted. □

**(2.3.6) Lemma.** *If* $X \in \mathcal{C}_{(\square)}^{(\mathrm{f})}(R)$ *is equivalent to* $G \in \mathcal{C}_\sqsupset^{\mathrm{G}}(R)$ *and* $n \geq \sup X$, *then*

$$\mathrm{Ext}_R^m(\mathrm{C}_n^G, R) = \mathrm{H}_{-(m+n)}(\mathbf{R}\mathrm{Hom}_R(X,R))$$

*for* $m > 0$. *In particular, there is an inequality:*

$$\inf(\mathbf{R}\mathrm{Hom}_R(\mathrm{C}_n^G, R)) \geq \inf(\mathbf{R}\mathrm{Hom}_R(X,R)) + n.$$

*Proof.* Since $n \geq \sup X = \sup G$ we have $G_n\sqsupset \simeq \Sigma^n \mathrm{C}_n^G$, cf. (A.1.14.3), and it follows by the Proposition that $\mathbf{R}\mathrm{Hom}_R(\mathrm{C}_n^G, R)$ is represented by $\mathrm{Hom}_R(\Sigma^{-n}(G_n\sqsupset), R)$. For $m > 0$ the isomorphism class $\mathrm{Ext}_R^m(\mathrm{C}_n^G, R)$ is then represented by

$$\begin{aligned}
\mathrm{H}_{-m}(\mathrm{Hom}_R(\Sigma^{-n}(G_n\sqsupset), R)) &= \mathrm{H}_{-m}(\Sigma^n \mathrm{Hom}_R(G_n\sqsupset, R)) \\
&= \mathrm{H}_{-(m+n)}(\mathrm{Hom}_R(G_n\sqsupset, R)) \\
&= \mathrm{H}_{-(m+n)}(\sqsubset_{-n}\mathrm{Hom}_R(G,R)) \\
&= \mathrm{H}_{-(m+n)}(G^*),
\end{aligned}$$

cf. (A.2.1.3), (A.1.3.1), and (A.1.20.2). It also follows from the Proposition that the complex $G^*$ represents $\mathbf{R}\mathrm{Hom}_R(X,R)$, so $\mathrm{Ext}_R^m(\mathrm{C}_n^G, R) = \mathrm{H}_{-(m+n)}(\mathbf{R}\mathrm{Hom}_R(X,R))$ as wanted, and the inequality of infima follows. □

(2.3.7) **GD Theorem.** *Let $X \in \mathcal{C}_{(\sqsupset)}^{(\mathrm{f})}(R)$ and $n \in \mathbb{Z}$. The following are equivalent:*

(*i*) *$X$ is equivalent to a complex $G \in \mathcal{C}_{\square}^{\mathrm{G}}(R)$ concentrated in degrees at most $n$; and $G$ can be chosen with $G_\ell = 0$ for $\ell < \inf X$.*

(*ii*) $\mathrm{G\text{-}dim}_R\, X \leq n$.

(*iii*) *$X \in \mathcal{R}(R)$ and $n \geq -\inf(\mathbf{R}\mathrm{Hom}_R(X,R))$.*

(*iv*) *$n \geq \sup X$ and the module $\mathrm{C}_n^G$ belongs to $\mathrm{G}(R)$ whenever $G \in \mathcal{C}_{\sqsupset}^{\mathrm{G}}(R)$ is equivalent to $X$.*

*Proof.* It is immediate by Definition (2.3.2) that (*i*) implies (*ii*).

(*ii*) $\Rightarrow$ (*iii*): Choose a complex $G \in \mathcal{C}_{\square}^{\mathrm{G}}(R)$ concentrated in degrees at most $n$ and equivalent to $X$. It follows by Proposition (2.1.13) that $G$, and thereby $X$, belongs to $\mathcal{R}(R)$. By Proposition (2.3.5) the dual complex $G^*$ represents $\mathbf{R}\mathrm{Hom}_R(X,R)$, and $(G^*)_\ell = \mathrm{Hom}_R(G_{-\ell},R) = 0$ for $-\ell > n$. In particular, $\mathrm{H}_\ell(G^*) = 0$ for $\ell < -n$, so $\inf(\mathbf{R}\mathrm{Hom}_R(X,R)) = \inf G^* \geq -n$, as desired.

(*iii*) $\Rightarrow$ (*iv*): First note that since $X$ is reflexive we have

$$\sup X = \sup(\mathbf{R}\mathrm{Hom}_R(\mathbf{R}\mathrm{Hom}_R(X,R),R)) \leq -\inf(\mathbf{R}\mathrm{Hom}_R(X,R)) \leq n,$$

by (A.4.6.1). Suppose $G \in \mathcal{C}_{\sqsupset}^{\mathrm{G}}(R)$ is equivalent to $X$, and consider the short exact sequence of complexes $0 \to \sqsubset_{n-1}G \to \subset_n G \to \Sigma^n \mathrm{C}_n^G \to 0$. As $\sup G = \sup X \leq n$ it follows that $\subset_n G \simeq G \simeq X \in \mathcal{R}(R)$, cf. (A.1.14.2), and since $\sqsubset_{n-1}G \in \mathcal{R}(R)$ by Proposition (2.1.13) it follows from Lemma (2.1.12) that $\mathrm{C}_n^G \in \mathcal{R}_0(R)$. Furthermore, by Lemma (2.3.6) we have $-\inf(\mathbf{R}\mathrm{Hom}_R(\mathrm{C}_n^G,R)) \leq -\inf(\mathbf{R}\mathrm{Hom}_R(X,R)) - n \leq 0$, so $\mathrm{C}_n^G \in \mathrm{G}(R)$ by Proposition (2.2.2).

(*iv*)$\Rightarrow$(*i*): Choose by (A.3.2) a resolution by finite free modules $G \in \mathcal{C}_{\sqsupset}^{\mathrm{L}}(R) \subseteq \mathcal{C}_{\sqsupset}^{\mathrm{G}}(R)$ of $X$ with $G_\ell = 0$ for $\ell < \inf X$. Since $n \geq \sup X = \sup G$ it follows by (A.1.14.2) that $G \simeq \subset_n G$, so $X \simeq \subset_n G$ and $\subset_n G \in \mathcal{C}_{\square}^{\mathrm{G}}(R)$ as $\mathrm{C}_n^G \in \mathrm{G}(R)$. $\square$

(2.3.8) **GD Corollary.** *A complex $X \in \mathcal{C}_{(\sqsupset)}^{(\mathrm{f})}(R)$ has finite G–dimension if and only if it is reflexive; that is,*

$$\mathrm{G\text{-}dim}_R\, X < \infty \iff X \in \mathcal{R}(R).$$

*Furthermore, if $X \in \mathcal{R}(R)$, then*

$$\mathrm{G\text{-}dim}_R\, X = -\inf(\mathbf{R}\mathrm{Hom}_R(X,R)).$$

*Proof.* The biconditional as well as the equality follows by the equivalence of (*ii*) and (*iii*) in the Theorem. $\square$

(2.3.9) **Remarks.** It follows from Theorem (2.2.3) and the Corollary that the G–dimension defined in this section extends the G–dimension for finite modules introduced in chapter 1. For modules the contents of (2.3.7) are covered by GD Theorem (1.2.7), but, admittedly, things look a little different here, in as much as the word 'resolution' is missing. As we shall see — in Lemma (2.3.15) — usual G–resolutions of a finite module $M$ are, however, essentially the same as complexes $G \in \mathcal{C}^{\mathrm{G}}_{\sqsupset}(R)$ equivalent to $M$. This allows us — in (2.3.16) — to spell out for modules a more familiar looking version of Theorem (2.3.7).

We can now see that also part (a) in Theorem (2.1.10) is a stability result. It tells us, in particular, that a finite module $M$ has finite G–dimension, i.e., $M \in \mathcal{R}_0(R)$ if (and only if) $\text{G-dim}_R(\mathbf{R}\mathrm{Hom}_R(M,R)) < \infty$; but we do still not have an example of a finite module $M \notin \mathcal{R}_0(R)$ with $\mathbf{R}\mathrm{Hom}_R(M,R)$ homologically bounded.

The next proposition shows that G–dimension is a refinement of projective dimension for complexes with finite homology.

**(2.3.10) Proposition (GD–PD Inequality).** *For every complex $X \in \mathcal{C}^{(\mathrm{f})}_{(\sqsupset)}(R)$ there is an inequality:*

$$\text{G-dim}_R X \leq \mathrm{pd}_R X,$$

*and equality holds if* $\mathrm{pd}_R X < \infty$.

*Proof.* The inequality is trivial if $X$ is of infinite projective dimension, and equality holds if $X$ is homologically trivial. We now assume that $X$ is homologically non-trivial and of finite projective dimension, i.e., $X \in \mathcal{P}^{(\mathrm{f})}(R)$. By Proposition (2.1.9) $X$ is then reflexive, that is, of finite G–dimension, and by (A.5.4.3) we can choose a finite $R$–module $T$ such that $\mathrm{pd}_R X = -\inf(\mathbf{R}\mathrm{Hom}_R(X,T))$. The desired equality now follows from the next calculation, in which the third equality follows by tensor evaluation (A.4.23), the fourth by (A.4.16), and the last by GD Corollary (2.3.8).

$$\begin{aligned}
\mathrm{pd}_R X &= -\inf(\mathbf{R}\mathrm{Hom}_R(X,T)) \\
&= -\inf(\mathbf{R}\mathrm{Hom}_R(X, R \otimes^{\mathbf{L}}_R T)) \\
&= -\inf(\mathbf{R}\mathrm{Hom}_R(X,R) \otimes^{\mathbf{L}}_R T) \\
&= -\inf(\mathbf{R}\mathrm{Hom}_R(X,R)) \\
&= \text{G-dim}_R X. \quad \square
\end{aligned}$$

The proof presented above is certainly not the canonical one; actually, an argument quite parallel to the proof of the GD–PD inequality for finite modules (1.2.10) applies. However, the proof above has an interesting feature: note that even if $X$ does not have finite projective dimension, the identity (A.4.23) still applies if $T$ has finite flat dimension. We shall follow this track in the next section.

**(2.3.11) Proposition.** *Let $X \in \mathcal{C}^{(\mathrm{f})}_{(\sqsupset)}(R)$. For every $\mathfrak{p} \in \operatorname{Spec} R$ there is an inequality:*

$$\text{G-dim}_{R_\mathfrak{p}} X_\mathfrak{p} \leq \text{G-dim}_R X.$$

*Proof.* Suppose $\text{G-dim}_R X < \infty$, then $X \in \mathcal{R}(R)$ by GD Corollary (2.3.8), so $X_\mathfrak{p} \in \mathcal{R}(R_\mathfrak{p})$, cf. (2.1.11.1), and we have

$$\begin{aligned} \text{G-dim}_{R_\mathfrak{p}} X_\mathfrak{p} &= -\inf(\mathbf{R}\text{Hom}_{R_\mathfrak{p}}(X_\mathfrak{p}, R_\mathfrak{p})) \\ &= -\inf(\mathbf{R}\text{Hom}_R(X,R)_\mathfrak{p}) \\ &\leq -\inf(\mathbf{R}\text{Hom}_R(X,R)) \\ &= \text{G-dim}_R X. \quad \square \end{aligned}$$

(2.3.12) **Change of Rings Theorem for GD.** *Let $x_1, \ldots, x_t$ be an $R$–sequence and $S = R/(x_1, \ldots, x_t)$. For $X \in \mathcal{C}^{(\text{f})}_{(\sqsupset)}(S)$ there is an equality:*

$$\text{G-dim}_R X = \text{G-dim}_S X + t.$$

*In particular, the two dimensions are simultaneously finite.*

*Proof.* Immediate by Lemma (2.2.7) and GD Corollary (2.3.8). □

(2.3.13) **Theorem (AB Formula for GD).** *If $R$ is local, and $X$ is a complex of finite G–dimension, i.e., $X \in \mathcal{R}(R)$, then*

$$\text{G-dim}_R X = \operatorname{depth} R - \operatorname{depth}_R X.$$

*Proof.* Let $X \in \mathcal{R}(R)$; by (A.7.5.1), (A.7.7), and (A.7.4.1) we then have

$$\begin{aligned} \operatorname{depth}_R X &= \operatorname{depth}_R(\mathbf{R}\text{Hom}_R(\mathbf{R}\text{Hom}_R(X,R),R)) \\ &= \operatorname{ord} \mathrm{I}_R^{\mathbf{R}\text{Hom}_R(\mathbf{R}\text{Hom}_R(X,R),R)}(t) \\ &= \operatorname{ord}(\mathrm{P}^R_{\mathbf{R}\text{Hom}_R(X,R)}(t)\, \mathrm{I}_R(t)) \\ &= \inf(\mathbf{R}\text{Hom}_R(X,R)) + \operatorname{depth} R. \end{aligned}$$

The desired equality now follows as $\text{G-dim}_R X = -\inf(\mathbf{R}\text{Hom}_R(X,R))$ by GD Corollary (2.3.8). □

(2.3.14) **Gorenstein Theorem, $\mathcal{R}$ Version.** *Let $R$ be a local ring with residue field $k$. The following are equivalent:*

(*i*) *$R$ is Gorenstein.*

(*ii*) *$k \in \mathcal{R}_0(R)$.*

(*iii*) *$\mathcal{R}_0(R) = \mathcal{C}^{\text{f}}_0(R)$.*

(*iii'*) *$\mathcal{R}(R) = \mathcal{C}^{(\text{f})}_{(\square)}(R)$.*

*Proof.* First note that conditions (*i*)–(*iii*) are merely the equivalent conditions of the GD version (1.4.9) rewritten in agreement with Theorem (2.2.3). Also note that (*iii'*) implies (*iii*); it is then sufficient to prove that (*i*) implies (*iii'*):

Suppose $R$ is Gorenstein, and let $X \in \mathcal{C}^{(\text{f})}_{(\square)}(R)$. The ring $R$ has finite injective dimension as module over itself, so $\mathbf{R}\text{Hom}_R(X,R) \in \mathcal{C}^{(\text{f})}_{(\square)}(R)$, cf. (A.5.2). Since $X$ belongs to $\mathcal{C}^{(\mathrm{f})}_{(\square)}(R)$ it has a resolution by finite free modules: $X \xleftarrow{\simeq} L \in \mathcal{C}^{\mathrm{L}}_{\sqsupset}(R)$; now take a bounded injective resolution $I \in \mathcal{C}^{\mathrm{I}}_{\square}(R)$ of $R$, then the Hom evaluation morphism $\theta_{LRI}$ is an isomorphism, cf. (A.2.11), and in particular a quasi-isomorphism, so $X$ is reflexive by Lemma (2.1.5). ☐

(2.3.15) **Lemma.** *Let $M$ be a finite $R$–module. If $M \simeq G \in \mathcal{C}^{\mathrm{G}}_{\sqsupset}(R)$, then the truncated complex*

$$G_0\supset \; = \; \cdots \to G_\ell \to \cdots \to G_2 \to G_1 \to \mathrm{Z}^G_0 \to 0$$

*is a usual G–resolution of $M$.*

*Proof.* Suppose $M$ is equivalent to $G \in \mathcal{C}^{\mathrm{G}}_{\sqsupset}(R)$, then $\inf G = 0$, so $G_0\supset \simeq G \simeq M$ by (A.1.14.4), and we have an exact sequence of modules:

$$(\dagger) \qquad \cdots \to G_\ell \to \cdots \to G_2 \to G_1 \to \mathrm{Z}^G_0 \to M \to 0.$$

Set $v = \inf\{\ell \in \mathbb{Z} \mid G_\ell \neq 0\}$, then also the sequence

$$0 \to \mathrm{Z}^G_0 \to G_0 \to \cdots \to G_{v+1} \to G_v \to 0$$

is exact. All the modules $G_0, \ldots, G_v$ belong to $\mathrm{G}(R)$, so it follows by GD Theorem (1.2.7) (or by repeated applications of Lemma (1.1.10)(a)) that $\mathrm{Z}^G_0 \in \mathrm{G}(R)$ and, therefore, $G_0\supset$ is a usual G–resolution of $M$, cf. (†). ☐

(2.3.16) **GD Theorem for Finite Modules, $\mathcal{R}$ Version.** *Let $M$ be a finite $R$–module and $n \in \mathbb{N}_0$. The following are equivalent:*

- (*i*) *$M$ has a G–resolution of length at most $n$. That is, there is an exact sequence of modules $0 \to G_n \to \cdots \to G_1 \to G_0 \to M \to 0$, where $G_0, \ldots, G_n$ belong to $\mathrm{G}(R)$.*
- (*ii*) $\mathrm{G\text{–}dim}_R\, M \leq n$.
- (*iii*) *$M \in \mathcal{R}_0(R)$ and $\mathrm{Ext}^m_R(M,R) = 0$ for $m > n$.*
- (*iv*) *In any G–resolution of $M$,*

$$\cdots \to G_\ell \to G_{\ell-1} \to \cdots \to G_0 \to M \to 0,$$

*the kernel*[2] *$K_n = \mathrm{Ker}(G_{n-1} \to G_{n-2})$ belongs to $\mathrm{G}(R)$.*

*Proof.* If the sequence $\cdots \to G_\ell \to G_{\ell-1} \to \cdots \to G_0 \to M \to 0$ is exact, then $M$ is equivalent to $G = \cdots \to G_\ell \to G_{\ell-1} \to \cdots \to G_0 \to 0$. The complex $G$ belongs to $\mathcal{C}^{\mathrm{G}}_{\sqsupset}(R)$, and it has $\mathrm{C}^G_0 \cong M$, $\mathrm{C}^G_1 \cong \mathrm{Ker}(G_0 \to M)$, and $\mathrm{C}^G_\ell \cong \mathrm{Z}^G_{\ell-1} = \mathrm{Ker}(G_{\ell-1} \to G_{\ell-2})$ for $\ell \geq 2$. In view of the Lemma the equivalence of the four conditions now follows from Theorem (2.3.7). ☐

---

[2] Appropriately interpreted for small $n$ as $K_0 = M$ and $K_1 = \mathrm{Ker}(G_0 \to M)$, cf. (1.2.5.1).

(2.3.17) **Exercise (Stability).** Assume that $X \in \mathcal{R}(R)$ and $U \in \mathcal{P}^{(\mathrm{f})}(R)$. Prove that

(a) $\mathbf{R}\mathrm{Hom}_R(U,X)$ is reflexive and
$$\text{G-dim}_R(\mathbf{R}\mathrm{Hom}_R(U,X)) = \text{G-dim}_R X - \inf U.$$

(b) $X \otimes^{\mathbf{L}}_R U$ is reflexive and
$$\text{G-dim}_R(X \otimes^{\mathbf{L}}_R U) = \text{G-dim}_R X + \mathrm{pd}_R U.$$

**Notes**

Comparing GD Corollary (2.3.8) to [62, Definition 2.8] we see that the G–dimension, as defined in this chapter, coincides with the one studied by Yassemi.

The stability results in Exercise (2.3.17) we proved by Yassemi in [62] and generalized in [15].

A generalized version of the Change of Rings Theorem (2.3.12), dealing with so-called quasi-Gorenstein homomorphisms, has been established by Avramov and Foxby in [8, Section 7]. An even more general version can be found in [15, Section 6].

## 2.4 Testing G–dimension

We study the interaction between reflexive complexes and complexes of, respectively, finite flat and finite injective dimension. We arrive at a series of test expressions for G–dimension, and we shall later — in chapters 4 and 5 — recognize these as tests for the Gorenstein projective and the Gorenstein flat dimension.

(2.4.1) **Proposition.** *The following hold for $X \in \mathcal{R}(R)$ and $U \in \mathcal{F}(R)$:*

(a) $$-\inf(\mathbf{R}\mathrm{Hom}_R(X,U)) \le \text{G-dim}_R X - \inf U; \quad \textit{and}$$

(b) $$\sup(U \otimes^{\mathbf{L}}_R X) \le \sup U + \text{G-dim}_R X.$$

*Furthermore, suppose $X$ and $U$ are homologically non-trivial and set $s = \sup U$, $i = \inf U$, and $g = \inf(\mathbf{R}\mathrm{Hom}_R(X,R))$. Then equality holds in* (a) *if and only if $\mathrm{H}_g(\mathbf{R}\mathrm{Hom}_R(X,R)) \otimes_R \mathrm{H}_i(U) \neq 0$, and equality holds in* (b) *if and only if $\mathrm{Hom}_R(\mathrm{H}_g(\mathbf{R}\mathrm{Hom}_R(X,R)),\mathrm{H}_s(U)) \neq 0$.*

*Proof.* First note that equality holds in both (a) and (b) if $X \simeq 0$ or $U \simeq 0$. In the following we assume that $X$ and $U$ are homologically non-trivial.

(a): Since $U \in \mathcal{F}(R)$ tensor evaluation (A.4.23) accounts for the second equality in the computation:

$$\begin{aligned} -\inf(\mathbf{R}\mathrm{Hom}_R(X,U)) &= -\inf(\mathbf{R}\mathrm{Hom}_R(X, R \otimes^{\mathbf{L}}_R U)) \\ &= -\inf(\mathbf{R}\mathrm{Hom}_R(X,R) \otimes^{\mathbf{L}}_R U) \\ &\le -\inf(\mathbf{R}\mathrm{Hom}_R(X,R)) - \inf U \\ &= \text{G-dim}_R X - \inf U. \end{aligned}$$

The inequality is (A.4.15.1) and by (A.4.15.2) equality holds if and only if $\mathrm{H}_g(\mathbf{R}\mathrm{Hom}_R(X,R)) \otimes_R \mathrm{H}_i(U) \neq 0$.

(b): Again we use tensor evaluation and find:

$$\begin{aligned}
\sup\,(U \otimes^{\mathbf{L}}_R X) &= \sup\,(\mathbf{R}\mathrm{Hom}_R(\mathbf{R}\mathrm{Hom}_R(X,R),R) \otimes^{\mathbf{L}}_R U)\\
&= \sup\,(\mathbf{R}\mathrm{Hom}_R(\mathbf{R}\mathrm{Hom}_R(X,R),R \otimes^{\mathbf{L}}_R U))\\
&= \sup\,(\mathbf{R}\mathrm{Hom}_R(\mathbf{R}\mathrm{Hom}_R(X,R),U))\\
&\leq \sup U - \inf\,(\mathbf{R}\mathrm{Hom}_R(X,R))\\
&= \sup U + \mathrm{G\text{–}dim}_R\, X.
\end{aligned}$$

The inequality is (A.4.6.1) and by (A.4.6.2) equality holds if and only if $\mathrm{Hom}_R(\mathrm{H}_g(\mathbf{R}\mathrm{Hom}_R(X,R)), \mathrm{H}_s(U)) \neq 0$. □

(2.4.2) **Corollary.** *If $M$ is a finite $R$–module of finite G–dimension, i.e., $M \in \mathcal{R}_0(R)$ and $T \in \mathcal{F}_0(R)$, then*

(a) $\mathrm{Ext}^m_R(M,T) = 0$ *for* $m > \mathrm{G\text{–}dim}_R\, M$; *and*
(b) $\mathrm{Tor}^R_m(T,M) = 0$ *for* $m > \mathrm{G\text{–}dim}_R\, M$. □

(2.4.3) **Proposition.** *The following hold for $X \in \mathcal{R}(R)$ and $U \in \mathcal{I}(R)$:*

(a) $$-\inf\,(\mathbf{R}\mathrm{Hom}_R(X,U)) \leq \mathrm{G\text{–}dim}_R\, X - \inf U; \quad \textit{and}$$

(b) $$\sup\,(U \otimes^{\mathbf{L}}_R X) \leq \sup U + \mathrm{G\text{–}dim}_R\, X.$$

*Furthermore, suppose $X$ and $U$ are homologically non-trivial and set $s = \sup U$, $i = \inf U$, and $g = \inf\,(\mathbf{R}\mathrm{Hom}_R(X,R))$. Then equality holds in* (a) *if and only if $\mathrm{H}_g(\mathbf{R}\mathrm{Hom}_R(X,R)) \otimes_R \mathrm{H}_i(U) \neq 0$, and equality holds in* (b) *if and only if $\mathrm{Hom}_R(\mathrm{H}_g(\mathbf{R}\mathrm{Hom}_R(X,R)), \mathrm{H}_s(U)) \neq 0$.*

*Proof.* First note that equality holds in both (a) and (b) if $X \simeq 0$ or $U \simeq 0$. In the following we assume that $X$ and $U$ are homologically non-trivial.

(a): Since $U \in \mathcal{I}(R)$ the second equality in the next computation follows by Hom evaluation (A.4.24).

$$\begin{aligned}
-\inf\,(\mathbf{R}\mathrm{Hom}_R(X,U)) &= -\inf\,(\mathbf{R}\mathrm{Hom}_R(\mathbf{R}\mathrm{Hom}_R(\mathbf{R}\mathrm{Hom}_R(X,R),R),U))\\
&= -\inf\,(\mathbf{R}\mathrm{Hom}_R(X,R) \otimes^{\mathbf{L}}_R \mathbf{R}\mathrm{Hom}_R(R,U))\\
&= -\inf\,(\mathbf{R}\mathrm{Hom}_R(X,R) \otimes^{\mathbf{L}}_R U)\\
&\leq -\inf\,(\mathbf{R}\mathrm{Hom}_R(X,R)) - \inf U\\
&= \mathrm{G\text{–}dim}_R\, X - \inf U.
\end{aligned}$$

The inequality is (A.4.15.1) and by (A.4.15.2) equality holds if and only if $\mathrm{H}_g(\mathbf{R}\mathrm{Hom}_R(X,R)) \otimes_R \mathrm{H}_i(U) \neq 0$.

(b): The second equality below is by Hom evaluation (A.4.24).

$$\begin{aligned}\sup(U \otimes^{\mathbf{L}}_R X) &= \sup(X \otimes^{\mathbf{L}}_R \mathbf{R}\mathrm{Hom}_R(R,U))\\ &= \sup(\mathbf{R}\mathrm{Hom}_R(\mathbf{R}\mathrm{Hom}_R(X,R),U))\\ &\leq \sup U - \inf(\mathbf{R}\mathrm{Hom}_R(X,R))\\ &= \sup U + \mathrm{G\text{–}dim}_R X.\end{aligned}$$

The inequality is (A.4.6.1) and by (A.4.6.2) equality holds if and only if $\mathrm{Hom}_R(\mathrm{H}_g(\mathbf{R}\mathrm{Hom}_R(X,R)),\mathrm{H}_s(U)) \neq 0$. □

(2.4.4) **Corollary.** *If $M$ is a finite $R$–module of finite G–dimension, i.e., $M \in \mathcal{R}_0(R)$ and $T \in \mathcal{I}_0(R)$, then*

(a) $\mathrm{Ext}^m_R(M,T) = 0$ *for* $m > \mathrm{G\text{–}dim}_R M$; *and*

(b) $\mathrm{Tor}^R_m(T,M) = 0$ *for* $m > \mathrm{G\text{–}dim}_R M$. □

(2.4.5) **Theorem (Test Modules).** *Let $(R,\mathfrak{m},k)$ be local, and let $X$ be an $R$–complex of finite G–dimension, i.e., $X \in \mathcal{R}(R)$. The following hold:*

(a) *If $S$ is a finite module of finite projective dimension, and* $\mathrm{depth}_R S = 0$ *(e.g., $S = R/(x_1,\ldots,x_d)$ where $x_1,\ldots,x_d$ is a maximal $R$–sequence), then*

$$\mathrm{G\text{–}dim}_R X = \sup(S \otimes^{\mathbf{L}}_R X).$$

(b) *If $E$ is a faithfully injective $R$–module (e.g., $E = \mathrm{E}_R(k)$ the injective hull of the residue field), then*

$$\mathrm{G\text{–}dim}_R X = \sup(E \otimes^{\mathbf{L}}_R X).$$

(c) *Let $S$ and $E$ be as in* (a) *and* (b), *and set $T = \mathrm{Hom}_R(S,E)$. Then $T$ is a module of finite injective dimension, and*

$$\mathrm{G\text{–}dim}_R X = -\inf(\mathbf{R}\mathrm{Hom}_R(X,T)).$$

*Proof.* Let $X \in \mathcal{R}(R)$. All three equalities hold if $X \simeq 0$, so we assume that $X$ is homologically non-trivial, and we set $g = \inf(\mathbf{R}\mathrm{Hom}_R(X,R)) \in \mathbb{Z}$ and $H = \mathrm{H}_g(\mathbf{R}\mathrm{Hom}_R(X,R))$.

(a): Suppose $S \in \mathcal{P}^{\mathrm{f}}_0(R) = \mathcal{F}^{\mathrm{f}}_0(R)$ with $\mathrm{depth}_R S = 0$, by Proposition (2.4.1)(b) it is sufficient to prove that $\mathrm{Hom}_R(H,S) \neq 0$, and this is immediate by the Hom vanishing corollary as $\mathrm{z}_R S = \mathfrak{m}$.

(b): When $E$ is faithfully injective, we have $\mathrm{Hom}_R(H,E) \neq 0$, and the desired equality follows by Proposition (2.4.3)(b).

(c): It follows by (A.5.8.2) that $\mathrm{id}_R T \leq \mathrm{fd}_R S = \mathrm{pd}_R S$ (actually, equality holds as $E$ is faithfully injective, cf. [42, Theorem 1.4]), so $T \in \mathcal{I}_0(R)$. By Proposition (2.4.3)(a) it is sufficient to prove that $H \otimes_R T \neq 0$. By Hom evaluation for modules we have

$$H \otimes_R T = H \otimes_R \mathrm{Hom}_R(S,E) \cong \mathrm{Hom}_R(\mathrm{Hom}_R(H,S),E),$$

and the module on the right hand side is non-zero because $E$ is faithfully injective and $\mathrm{Hom}_R(H, S) \neq 0$, cf. the proof of (a). □

(2.4.6) **Corollary.** *Let $(R, \mathfrak{m}, k)$ be local, and let $M$ be a finite $R$–module of finite G–dimension, i.e., $M \in \mathcal{R}_0(R)$. The following hold:*

(a) *If $S$ is a finite module of finite projective dimension and* $\mathrm{depth}_R S = 0$, *then*

$$\text{G–dim}_R M = \sup\{m \in \mathbb{N}_0 \mid \mathrm{Tor}_m^R(S, M) \neq 0\}.$$

(b) *If $E$ is a faithfully injective $R$–module, then*

$$\text{G–dim}_R M = \sup\{m \in \mathbb{N}_0 \mid \mathrm{Tor}_m^R(E, M) \neq 0\}.$$

(c) *Let $S$ and $E$ be as in* (a) *and* (b), *and set* $T = \mathrm{Hom}_R(S, E)$. *Then $T$ is a module of finite injective dimension, and*

$$\text{G–dim}_R M = \sup\{m \in \mathbb{N}_0 \mid \mathrm{Ext}_R^m(M, T) \neq 0\}. \quad \square$$

(2.4.7) **Theorem.** *Let $X \in \mathcal{C}_{(\sqsupset)}^{(\mathrm{f})}(R)$ and consider the following three conditions:*

(*i*) $X \in \mathcal{R}(R)$.

(*ii*) $\text{G–dim}_R X < \infty$.

(*iii*) $X \in \mathcal{C}_{(\square)}^{(\mathrm{f})}(R)$ *and* $\text{G–dim}_R X \leq \sup X + \dim R$.

*Conditions* (*i*) *and* (*ii*) *are equivalent and imply* (*iii*); *and if* $\dim R < \infty$, *then all three conditions are equivalent.*

*Furthermore, if $X \in \mathcal{R}(R)$, then the next nine numbers are equal.*

| | |
|---|---|
| (D) | $\text{G–dim}_R X$, |
| (R) | $-\inf(\mathbf{R}\mathrm{Hom}_R(X, R))$, |
| (EI$_0$) | $\sup\{-\inf(\mathbf{R}\mathrm{Hom}_R(X, T)) \mid T \in \mathcal{I}_0(R)\}$, |
| (TF$_0$) | $\sup\{\sup(T \otimes_R^{\mathbf{L}} X) \mid T \in \mathcal{F}_0(R)\}$, |
| (TE) | $\sup\{\sup(\mathrm{E}_R(R/\mathfrak{p}) \otimes_R^{\mathbf{L}} X) \mid \mathfrak{p} \in \mathrm{Spec}\, R\}$, |
| (EF) | $\sup\{\inf U - \inf(\mathbf{R}\mathrm{Hom}_R(X, U)) \mid U \in \mathcal{F}(R) \wedge U \not\simeq 0\}$, |
| (EI) | $\sup\{\inf U - \inf(\mathbf{R}\mathrm{Hom}_R(X, U)) \mid U \in \mathcal{I}(R) \wedge U \not\simeq 0\}$, |
| (TF) | $\sup\{\sup(U \otimes_R^{\mathbf{L}} X) - \sup U \mid U \in \mathcal{F}(R) \wedge U \not\simeq 0\}$, *and* |
| (TI) | $\sup\{\sup(U \otimes_R^{\mathbf{L}} X) - \sup U \mid U \in \mathcal{I}(R) \wedge U \not\simeq 0\}$. |

*Proof.* Conditions (*i*) and (*ii*) are equivalent by GD Corollary (2.3.8), and if $\dim R < \infty$ then, clearly, (*iii*) implies (*ii*). Now, assume that $X$ is reflexive and set $g = \inf(\mathbf{R}\mathrm{Hom}_R(X, R)) = -\text{G–dim}_R X$. If $g = \infty$, then the inequality in (*iii*) is trivial. If $g \in \mathbb{Z}$ we can choose a prime ideal $\mathfrak{p} \in \mathrm{Supp}_R(\mathrm{H}_g(\mathbf{R}\mathrm{Hom}_R(X, R)))$, and then we have

$$\text{G–dim}_R X \geq \text{G–dim}_{R_\mathfrak{p}} X_\mathfrak{p} = -\inf(\mathbf{R}\mathrm{Hom}_{R_\mathfrak{p}}(X_\mathfrak{p}, R_\mathfrak{p})) \geq -g,$$

cf. Proposition (2.3.11), so $\text{G-dim}_R X = \text{G-dim}_{R_\mathfrak{p}} X_\mathfrak{p}$. By the AB formula (2.3.13) and (A.6.1.1) it now follows that

$$\begin{aligned}\text{G-dim}_R X &= \operatorname{depth} R_\mathfrak{p} - \operatorname{depth}_{R_\mathfrak{p}} X_\mathfrak{p}\\ &\leq \dim_R R_\mathfrak{p} + \sup X\\ &\leq \sup X + \dim R\end{aligned}$$

as wanted.

Let $X \in \mathcal{R}(R)$. By Proposition (2.4.1)(a) and GD Corollary (2.3.8) we have

$$(\text{D}) \geq (\text{EF}) \geq (\text{R}) = (\text{D}),$$

so the numbers $(\text{D})$, $(\text{EF})$, and $(\text{R})$ are equal. By Propositions (2.4.1) and (2.4.3) we also have the following inequalities:

$$(\text{D}) \geq \begin{cases}(\text{EI}) & \geq & (\text{EI}_0)\\ (\text{TF}) & \geq & (\text{TF}_0)\\ (\text{TI}) & \geq & (\text{TE})\end{cases}.$$

This leaves us three inequalities to prove. As above we can choose a prime ideal $\mathfrak{p} \in \operatorname{Spec} R$, such that $\text{G-dim}_R X = \text{G-dim}_{R_\mathfrak{p}} X_\mathfrak{p}$, and by Theorem (2.4.5)(b) we then have

$$\begin{aligned}(\text{TE}) &\geq \sup(\mathrm{E}_R(R/\mathfrak{p}) \otimes^{\mathbf{L}}_R X)\\ &\geq \sup(\mathrm{E}_R(R/\mathfrak{p}) \otimes^{\mathbf{L}}_R X)_\mathfrak{p}\\ &= \sup(\mathrm{E}_{R_\mathfrak{p}}(k(\mathfrak{p})) \otimes^{\mathbf{L}}_{R_\mathfrak{p}} X_\mathfrak{p})\\ &= \text{G-dim}_{R_\mathfrak{p}} X_\mathfrak{p}\\ &= (\text{D}).\end{aligned}$$

Let $x_1, \dots, x_d \in \mathfrak{p}_\mathfrak{p}$ be a maximal $R_\mathfrak{p}$–sequence and set $S = R_\mathfrak{p}/(\boldsymbol{x})$, then $S$ is a finite $R_\mathfrak{p}$–module of finite flat dimension, and $\operatorname{depth}_{R_\mathfrak{p}} S = 0$. Since $R_\mathfrak{p}$ is a flat $R$–algebra, $S$ is also of finite flat dimension over $R$, and by Theorem (2.4.5)(a) it follows that

$$(\text{TF}_0) \geq \sup(S \otimes^{\mathbf{L}}_R X) \geq \sup(S \otimes^{\mathbf{L}}_{R_\mathfrak{p}} X_\mathfrak{p}) = \text{G-dim}_{R_\mathfrak{p}} X_\mathfrak{p} = (\text{D}).$$

Finally, set $T = \operatorname{Hom}_{R_\mathfrak{p}}(S, \mathrm{E}_{R_\mathfrak{p}}(k(\mathfrak{p})))$. By Theorem (2.4.5)(c) the module $T$ has finite injective dimension over $R_\mathfrak{p}$, and since $R_\mathfrak{p}$ is a flat $R$–algebra, also $\operatorname{id}_R T < \infty$. It now follows that

$$\begin{aligned}(\text{EI}_0) &\geq -\inf(\mathbf{R}\mathrm{Hom}_R(X,T))\\ &\geq -\inf(\mathbf{R}\mathrm{Hom}_R(X,T)_\mathfrak{p})\\ &= -\inf(\mathbf{R}\mathrm{Hom}_{R_\mathfrak{p}}(X_\mathfrak{p},T))\\ &= \text{G-dim}_{R_\mathfrak{p}} X_\mathfrak{p}\\ &= (\text{D}),\end{aligned}$$

again by Theorem (2.4.5)(c). □

(2.4.8) **Corollary.** *Let $M$ be a finite $R$–module and consider the following three conditions:*

(*i*) $M \in \mathcal{R}_0(R)$.

(*ii*) $\text{G–dim}_R M < \infty$.

(*iii*) $\text{G–dim}_R M \leq \dim R$.

*Conditions (i) and (ii) are equivalent and imply (iii); and if $\dim R < \infty$, then all three conditions are equivalent.*

*Furthermore, if $M \in \mathcal{R}_0(R)$ then then next seven numbers are equal.*

| | |
|---|---|
| (D) | $\text{G–dim}_R M$, |
| (R) | $\sup\{m \in \mathbb{N}_0 \mid \text{Ext}_R^m(M,R) \neq 0\}$, |
| (TE) | $\sup\{m \in \mathbb{N}_0 \mid \exists\, \mathfrak{p} \in \text{Spec}\, R : \text{Tor}_m^R(\text{E}_R(R/\mathfrak{p}), M) \neq 0\}$, |
| (EF$_0$) | $\sup\{m \in \mathbb{N}_0 \mid \exists\, T \in \mathcal{F}_0(R) : \text{Ext}_R^m(M,T) \neq 0\}$, |
| (EI$_0$) | $\sup\{m \in \mathbb{N}_0 \mid \exists\, T \in \mathcal{I}_0(R) : \text{Ext}_R^m(M,T) \neq 0\}$, |
| (TF$_0$) | $\sup\{m \in \mathbb{N}_0 \mid \exists\, T \in \mathcal{F}_0(R) : \text{Tor}_m^R(T,M) \neq 0\}$, *and* |
| (TI$_0$) | $\sup\{m \in \mathbb{N}_0 \mid \exists\, T \in \mathcal{I}_0(R) : \text{Tor}_m^R(T,M) \neq 0\}$. □ |

(2.4.9) **Observation.** It is quite possible that $\text{G–dim}_R M < \dim R$ for all $M \in \mathcal{R}_0(R)$, even if $R$ is local and the Krull dimension of $R$, therefore, finite. If $\text{G–dim}_R M < \infty$, then

$$\begin{aligned}\text{G–dim}_R M &= \sup\{\text{G–dim}_{R_\mathfrak{p}} M_\mathfrak{p} \mid \mathfrak{p} \in \text{Spec}\, R\} \\ &\leq \sup\{\text{depth}\, R_\mathfrak{p} \mid \mathfrak{p} \in \text{Spec}\, R\}\end{aligned}$$

by Proposition (1.3.2) and the Auslander–Bridger formula (1.4.8). And if $R$ is local, then

$$\sup\{\text{depth}\, R_\mathfrak{p} \mid \mathfrak{p} \in \text{Spec}\, R\} = \begin{cases} \dim R & \text{if } R \text{ is Cohen–Macaulay, and} \\ \dim R - 1 & \text{otherwise.} \end{cases}$$

On the other hand, if $R$ is local and $\boldsymbol{x} = x_1, \ldots, x_d$ is a maximal $R$–sequence, then $\text{G–dim}_R R/(\boldsymbol{x}) = \text{pd}_R R/(\boldsymbol{x}) = d$, so

$$(2.4.9.1) \qquad \begin{aligned} &R \text{ is Cohen–Macaulay} \iff \\ &\sup\{\text{G–dim}_R M \mid M \in \mathcal{R}_0(R)\} = \dim R, \end{aligned}$$

still in view of the Auslander–Bridger formula (1.4.8).

## Notes

Propositions (2.4.1) and (2.4.3), and their proofs, are taken from [62], but the essence of Corollaries (2.4.2), (2.4.4), and (2.4.6) goes back to Auslander and Bridger, cf. [2, Theorem (4.13)].

Finitistic G–dimension was studied by Takeuchi in [57], and (2.4.9.1) is [57, Theorem 1].

# Chapter 3

# Auslander Categories

For a local ring $R$ with a dualizing complex (see (3.0.1) below) we will introduce two full subcategories of $R$–complexes: the Auslander class and the Bass class. They are — together with the full subcategory of reflexive complexes — known as Auslander categories, and they are linked together by Foxby equivalence. The categories are introduced and studied in the first two sections, and general Foxby equivalence is treated in section 3.3. All the main results of the first three sections have particularly nice formulations for modules over Cohen–Macaulay rings, these are summed up in the last section.

Most results in this chapter were published in [8], but the ideas had, by then, been around for some time and were used, already, in [32] and [39].

Warning! This chapter would have been considerably shorter if it had been written in the language of derived categories. Users of derived category methods are advised to study the proofs in [8, Section 3] and [15, Section 4] instead.

(3.0.1) ***Dualizing Complexes.*** Recall (from the appendix for example) that a complex $D \in \mathcal{C}_{(\square)}^{(\mathrm{f})}(R)$ is *dualizing* for a local ring $R$ if and only if it has finite injective dimension and the homothety morphism $\chi_P^R\colon R \to \operatorname{Hom}_R(P,P)$ is a quasi-isomorphism for some, equivalently every, projective resolution $P$ of $D$. In particular, we have

(3.0.1.1) $$R = \mathbf{R}\mathrm{Hom}_R(D,D)$$

(i.e., $R$ represents $\mathbf{R}\mathrm{Hom}_R(D,D)$) when $D$ is a dualizing complex for $R$.

## 3.1 The Auslander Class

In terms of Foxby equivalence it is (or rather will be) natural to view the Auslander class as an extension of the full subcategory of complexes of finite flat dimension. But it also extends the full subcategory of reflexive complexes, and both views are covered by this section.

(3.1.1) **Setup.** In this section $R$ is a **local ring with dualizing complex** $D$.

(3.1.2) **Definitions.** For $R$–complexes $X$ and $Z$ a canonical morphism $\gamma_X^Z$ is defined by requiring commutativity of the diagram

(3.1.2.1)
$$\begin{array}{ccc} X & \xrightarrow{\gamma_X^Z} & \mathrm{Hom}_R(Z, Z \otimes_R X) \\ \downarrow\cong & & \uparrow\omega_{ZZX} \\ R \otimes_R X & \xrightarrow{\chi_Z^R \otimes_R X} & \mathrm{Hom}_R(Z,Z) \otimes_R X \end{array}$$

The morphism $\gamma_X^Z$ is natural in $Z$ and $X$ and given by

$$x \longmapsto [z \mapsto (-1)^{|x||z|} z \otimes x].$$

Let $X$ be an $R$–complex, and let $P \in \mathcal{C}_{\sqsupset}^{\mathrm{P}}(R)$ be a projective resolution of the dualizing complex. Then $D \otimes_R^{\mathbf{L}} X$ is represented by $P \otimes_R X$ and $\mathbf{R}\mathrm{Hom}_R(D, D \otimes_R^{\mathbf{L}} X)$ by $\mathrm{Hom}_R(P, P \otimes_R X)$. We say that $X$ *represents* $\mathbf{R}\mathrm{Hom}_R(D, D \otimes_R^{\mathbf{L}} X)$ *canonically* if and only if the morphism

$$\gamma_X^P : X \longrightarrow \mathrm{Hom}_R(P, P \otimes_R X)$$

is a quasi-isomorphism.

(3.1.3) **Remarks.** To see that this definition of canonical representation makes sense, take two projective resolutions $P, P' \in \mathcal{C}_{\sqsupset}^{\mathrm{P}}(R)$ of $D$; there is then by (A.3.6) a quasi-isomorphism $\pi \colon P' \xrightarrow{\simeq} P$. Using the quasi-isomorphism preserving properties of the functors, we establish the following commutative diagram

$$\begin{array}{ccc} X & \xrightarrow{\gamma_X^P} & \mathrm{Hom}_R(P, P \otimes_R X) \\ \downarrow\gamma_X^{P'} & & \simeq\downarrow \mathrm{Hom}_R(\pi, P \otimes_R X) \\ \mathrm{Hom}_R(P', P' \otimes_R X) & \xrightarrow[\simeq]{\mathrm{Hom}_R(P', \pi \otimes_R X)} & \mathrm{Hom}_R(P', P \otimes_R X) \end{array}$$

and we see that $\gamma_X^P$ is a quasi-isomorphism if and only if $\gamma_X^{P'}$ is so.

Also note that if $D'$ is another dualizing complex for $R$, then $D'$ is equivalent to $D$ up to a shift, cf. (A.8.3.3), so if $P$ is a projective resolution of $D$, then, for a suitable integer $m$, $\Sigma^m P$ is a projective resolution of $D'$. By (A.2.1.1) and (A.2.1.3) there is a natural isomorphism

$$\mathrm{Hom}_R(\Sigma^m P, \Sigma^m P \otimes_R X) \cong \mathrm{Hom}_R(P, P \otimes_R X),$$

so it is easy to see that the complex $X$ represents $\mathbf{R}\mathrm{Hom}_R(D, D \otimes_R^{\mathbf{L}} X)$ canonically if and only if it represents $\mathbf{R}\mathrm{Hom}_R(D', D' \otimes_R^{\mathbf{L}} X)$ canonically.

By the defining diagram (3.1.2.1) the canonical morphism $\gamma$ is closely related to tensor evaluation. The next lemma expresses canonical representation of $\mathbf{R}\mathrm{Hom}_R(D, D \otimes_R^{\mathbf{L}} X)$ in terms of tensor evaluation.

**(3.1.4) Lemma.** *Let $P \in \mathcal{C}_{\sqsupset}^{\mathrm{P}}(R)$ be a projective resolution of $D$, and let $X \in \mathcal{C}_{(\sqsupset)}(R)$. If $F \in \mathcal{C}_{\sqsupset}^{\mathrm{F}}(R)$ is equivalent to $X$, then the following are equivalent:*

*(i) $X$ represents $\mathbf{R}\mathrm{Hom}_R(D, D \otimes_R^{\mathbf{L}} X)$ canonically.*

*(ii) The canonical morphism $\gamma_F^P$ is a quasi-isomorphism.*

*(iii) The tensor evaluation morphism $\omega_{PDF}$ is a quasi-isomorphism.*

*Proof.* Let $X \in \mathcal{C}_{(\sqsupset)}(R)$ and a projective resolution $\pi\colon P \xrightarrow{\simeq} D$ be given. Take a projective resolution $\varphi\colon Q \xrightarrow{\simeq} X$; since $F \simeq X$ there is also a quasi-isomorphism $\varphi'\colon Q \xrightarrow{\simeq} F$, cf. (A.3.6). The homothety morphism $\chi_P^R\colon R \to \mathrm{Hom}_R(P,P)$ is a quasi-isomorphism; and by the quasi-isomorphism preserving properties of the various functors we have the following diagram

$$
\begin{array}{ccccc}
X & \xrightarrow{\gamma_X^P} & \mathrm{Hom}_R(P, P \otimes_R X) & & \\
{\scriptstyle \simeq}\uparrow{\scriptstyle \varphi} & & {\scriptstyle \simeq}\uparrow{\scriptstyle \mathrm{Hom}_R(P, P\otimes_R \varphi)} & & \\
Q & \xrightarrow{\gamma_Q^P} & \mathrm{Hom}_R(P, P \otimes_R Q) & & \\
{\scriptstyle \simeq}\downarrow{\scriptstyle \varphi'} & & {\scriptstyle \simeq}\downarrow{\scriptstyle \mathrm{Hom}_R(P, P\otimes_R \varphi')} & & \\
F & \xrightarrow{\gamma_F^P} & \mathrm{Hom}_R(P, P \otimes_R F) & \xrightarrow[\simeq]{\mathrm{Hom}_R(P, \pi \otimes_R F)} & \mathrm{Hom}_R(P, D \otimes_R F) \\
{\scriptstyle \cong}\downarrow & & \uparrow{\scriptstyle \omega_{PPF}} & & \uparrow{\scriptstyle \omega_{PDF}} \\
R \otimes_R F & \xrightarrow[\simeq]{\chi_P^R \otimes_R F} & \mathrm{Hom}_R(P, P) \otimes_R F & \xrightarrow[\simeq]{\mathrm{Hom}_R(P, \pi) \otimes_R F} & \mathrm{Hom}_R(P, D) \otimes_R F
\end{array}
$$

It is straightforward to check that the diagram is commutative, and the equivalence of the three conditions follows. $\square$

**(3.1.5) Definition.** The *Auslander class* $\mathcal{A}(R)$ is the full subcategory of $\mathcal{C}(R)$, actually of $\mathcal{C}_{(\square)}(R)$, defined by specifying its objects as follows: An $R$–complex $X$ belongs to $\mathcal{A}(R)$ if and only if

(1) $X \in \mathcal{C}_{(\square)}(R)$;

(2) $D \otimes_R^{\mathbf{L}} X \in \mathcal{C}_{(\square)}(R)$; and

(3) $X$ represents $\mathbf{R}\mathrm{Hom}_R(D, D \otimes_R^{\mathbf{L}} X)$ canonically.

We also use the notation $\mathcal{A}(R)$ with subscript 0 and superscript f/(f). The definitions are as usual:

$$
\mathcal{A}_0(R) = \mathcal{A}(R) \cap \mathcal{C}_0(R);
$$
$$
\mathcal{A}_0^{\mathrm{f}}(R) = \mathcal{A}(R) \cap \mathcal{C}_0^{\mathrm{f}}(R); \quad \text{and}
$$
$$
\mathcal{A}^{(\mathrm{f})}(R) = \mathcal{A}(R) \cap \mathcal{C}^{(\mathrm{f})}(R).
$$

If $X$ belongs to $\mathcal{A}(R)$, then so does every equivalent complex $X' \simeq X$; this follows by the Lemma, as a flat resolution $F$ of $X$ is also equivalent to $X'$. For an equivalence class $\boldsymbol{X}$ of $R$–complexes the notation $\boldsymbol{X} \in \mathcal{A}(R)$ means that some, equivalently every, representative $X$ of $\boldsymbol{X}$ belongs to the Auslander class.

(3.1.6) **Remark.** The Auslander class is defined in terms of a dualizing complex $D$ for $R$, but the symbol $\mathcal{A}(R)$ makes no mention of $D$. This is justified by the last remark in (3.1.3) which shows that $\mathcal{A}(R)$ is independent of the choice of dualizing complex.

(3.1.7) **Observation.** Let $\mathfrak{p} \in \operatorname{Spec} R$, then $D_\mathfrak{p}$ is a dualizing complex for $R_\mathfrak{p}$ by (A.8.3.4), and, as in Observation (2.1.11), it is straightforward to check that

$$X \in \mathcal{A}(R) \implies X_\mathfrak{p} \in \mathcal{A}(R_\mathfrak{p}).$$

(3.1.8) **Proposition.** *Every $R$–complex of finite flat dimension belongs to the Auslander class. That is, there is a full embedding:*

$$\mathcal{F}(R) \subseteq \mathcal{A}(R).$$

*Proof.* Suppose $X \in \mathcal{F}(R)$, then both $X$ and $D \otimes^{\mathbf{L}}_R X$ belong to $\mathcal{C}_{(\square)}(R)$, cf. (A.5.6). By (3.1.6) we are free to assume that $D \in \mathcal{C}_{\square}(R)$. Choose a complex $F \in \mathcal{C}^{\mathrm{F}}_{\square}(R)$ equivalent to $X$, and take a projective resolution $P \in \mathcal{C}^{\mathrm{fP}}_{\sqsupset}(R)$ of the dualizing complex. The tensor evaluation morphism $\omega_{PDF}$ is then an isomorphism, cf. (A.2.10), in particular a quasi-isomorphism, so by Lemma (3.1.4) $X$ represents $\mathbf{R}\mathrm{Hom}_R(D, D \otimes^{\mathbf{L}}_R X)$ canonically. □

We have now established the Auslander class as an extension of the full subcategory of complexes of finite flat dimension; the next task is to prove the connection to reflexive complexes and, thereby, to G–dimension.

(3.1.9) **Lemma.** *If $X \in \mathcal{C}^{(\mathrm{f})}_{(\sqsupset)}(R)$, then $X$ represents $\mathbf{R}\mathrm{Hom}_R(\mathbf{R}\mathrm{Hom}_R(X,R),R)$ canonically if and only if $X$ represents $\mathbf{R}\mathrm{Hom}_R(D, D \otimes^{\mathbf{L}}_R X)$ canonically.*

*Proof.* Take a resolution by finite free modules $X \xleftarrow{\simeq} L \in \mathcal{C}^{\mathrm{L}}_{\sqsupset}(R)$, a projective resolution $P \in \mathcal{C}^{\mathrm{P}}_{\sqsupset}(R)$ of the dualizing complex, and an injective resolution $R \xrightarrow{\simeq} I \in \mathcal{C}^{\mathrm{I}}_{\sqsubset}(R)$. The dualizing complex is equivalent to a bounded complex of injective modules, so by (3.1.6) we are free to assume that $D \in \mathcal{C}^{\mathrm{I}}_{\square}(R)$. The complex $\mathrm{Hom}_R(P,D)$ represents $\mathbf{R}\mathrm{Hom}_R(D,D)$ and is therefore equivalent to $R$, so by (A.3.5) there is a quasi-isomorphism $\iota \colon \mathrm{Hom}_R(P,D) \xrightarrow{\simeq} I$. Note that $\mathrm{Hom}_R(P,D)$ is a complex of injective modules and bounded to the left, i.e., $\mathrm{Hom}_R(P,D) \in \mathcal{C}^{\mathrm{I}}_{\sqsubset}(R)$. The Hom evaluation morphism

$$\theta_{LRD} : L \otimes_R \mathrm{Hom}_R(R,D) \longrightarrow \mathrm{Hom}_R(L^*,D)$$

is an isomorphism by (A.2.11), and using thequasi-isomorphism preserving properties of the various functors, we set up a commutative diagram:

$$\begin{array}{ccc}
\mathrm{Hom}_R(P,D)\otimes_R L & \xrightarrow[\cong]{\tau_{\mathrm{Hom}_R(P,D)L}} & L\otimes_R \mathrm{Hom}_R(P,D) \\
\Big\downarrow \omega_{PDL} & & \simeq \Big\downarrow L\otimes_R \iota \\
\mathrm{Hom}_R(P,D\otimes_R L) & & L\otimes_R I \\
\cong \Big\downarrow \mathrm{Hom}_R(P,\tau_{DL}) & & \Big\downarrow \cong \\
\mathrm{Hom}_R(P,L\otimes_R D) & & L\otimes_R \mathrm{Hom}_R(R,I) \\
\Big\downarrow \cong & & \Big\downarrow \theta_{LRI} \\
\mathrm{Hom}_R(P,L\otimes_R \mathrm{Hom}_R(R,D)) & & \mathrm{Hom}_R(L^*,I) \\
\cong \Big\downarrow \mathrm{Hom}_R(P,\theta_{LRD}) & & \simeq \Big\uparrow \mathrm{Hom}_R(L^*,\iota) \\
\mathrm{Hom}_R(P,\mathrm{Hom}_R(L^*,D)) & \xrightarrow[\cong]{\varsigma_{PL^*D}} & \mathrm{Hom}_R(L^*,\mathrm{Hom}_R(P,D))
\end{array}$$

The diagram shows that $\omega_{PDL}$ is a quasi-isomorphism if and only if $\theta_{LRI}$ is so; that is, $X$ represents $\mathbf{R}\mathrm{Hom}_R(D, D\otimes^{\mathbf{L}}_R X)$ canonically if and only if it represents $\mathbf{R}\mathrm{Hom}_R(\mathbf{R}\mathrm{Hom}_R(X,R),R)$ canonically, cf. Lemmas (2.1.5) and (3.1.4). ☐

**(3.1.10) Theorem.** *A complex $X \in \mathcal{C}^{(\mathrm{f})}_{(\square)}(R)$ belongs to the Auslander class if and only if it is reflexive. That is, there is an equality of full subcategories:*

$$\mathcal{A}^{(\mathrm{f})}(R) = \mathcal{R}(R).$$

*Proof.* "$\subseteq$": If $X \in \mathcal{A}^{(\mathrm{f})}(R)$, then $X$ and $D\otimes^{\mathbf{L}}_R X$ belong to $\mathcal{C}^{(\mathrm{f})}_{(\square)}(R)$, and $X$ represents $\mathbf{R}\mathrm{Hom}_R(D, D\otimes^{\mathbf{L}}_R X)$ canonically. According to the Lemma $X$ also represents $\mathbf{R}\mathrm{Hom}_R(\mathbf{R}\mathrm{Hom}_R(X,R),R)$ canonically, so all we have to prove is that $\mathbf{R}\mathrm{Hom}_R(X,R)$ belongs to $\mathcal{C}^{(\mathrm{f})}_{(\square)}(R)$. We have

$$\begin{aligned}\mathbf{R}\mathrm{Hom}_R(X,R) &= \mathbf{R}\mathrm{Hom}_R(X,\mathbf{R}\mathrm{Hom}_R(D,D)) \\ &= \mathbf{R}\mathrm{Hom}_R(X\otimes^{\mathbf{L}}_R D, D)\end{aligned}$$

by (3.0.1.1) and adjointness (A.4.21), and since $D\in\mathcal{I}^{(\mathrm{f})}(R)$ it follows by (A.5.2) that $\mathbf{R}\mathrm{Hom}_R(X,R)\in\mathcal{C}^{(\mathrm{f})}_{(\square)}(R)$ as desired.

"$\supseteq$": Let $X \in \mathcal{R}(R)$, that is, $X$ and $\mathbf{R}\mathrm{Hom}_R(X,R)$ belong to $\mathcal{C}^{(\mathrm{f})}_{(\square)}(R)$, and $X$ represents $\mathbf{R}\mathrm{Hom}_R(\mathbf{R}\mathrm{Hom}_R(X,R),R)$ canonically. By the Lemma $X$ also represents $\mathbf{R}\mathrm{Hom}_R(D, D\otimes^{\mathbf{L}}_R X)$ canonically, so we only have to prove that $D\otimes^{\mathbf{L}}_R X\in\mathcal{C}_{(\square)}(R)$. Since $D$ is of finite injective dimension, we have

$$\begin{aligned}D\otimes^{\mathbf{L}}_R X &= X\otimes^{\mathbf{L}}_R \mathbf{R}\mathrm{Hom}_R(R,D) \\ &= \mathbf{R}\mathrm{Hom}_R(\mathbf{R}\mathrm{Hom}_R(X,R),D)\end{aligned}$$

by Hom evaluation (A.4.24), so $D\otimes^{\mathbf{L}}_R X$ is homologically bounded as desired, again by (A.5.2). ☐

(3.1.11) **Corollary.** *A complex $X \in \mathcal{C}^{(\mathrm{f})}_{(\sqsupset)}(R)$ has finite G–dimension if and only if it belongs to the Auslander class; that is,*

$$\text{G–dim}_R X < \infty \iff X \in \mathcal{A}^{(\mathrm{f})}(R).$$

*Furthermore, if $X \in \mathcal{A}^{(\mathrm{f})}(R)$, then*

$$\text{G–dim}_R X = \inf D - \inf(\mathbf{R}\mathrm{Hom}_R(X, D)).$$

*Proof.* The first assertion is immediate by the Theorem and GD Corollary (2.3.8). The equality follows by (A.8.5.1) and the AB formula (2.3.13):

$$\begin{aligned}\inf D - \inf(\mathbf{R}\mathrm{Hom}_R(X, D)) &= (\operatorname{depth} R - \operatorname{depth}_R D)\\ &\quad - (\operatorname{depth}_R X - \operatorname{depth}_R D)\\ &= \operatorname{depth} R - \operatorname{depth}_R X\\ &= \text{G–dim}_R X. \quad \square\end{aligned}$$

The Auslander class of a Gorenstein ring is "as large as possible" and, in fact, this characterizes these rings.

(3.1.12) **Gorenstein Theorem, $\mathcal{A}$ Version.** *Let $R$ be a local ring with residue field $k$. If $R$ admits a dualizing complex, then the following are equivalent:*

(*i*) *$R$ is Gorenstein.*
(*ii*) *$k \in \mathcal{A}_0(R)$.*
(*iii*) *$\mathcal{A}^{\mathrm{f}}_0(R) = \mathcal{C}^{\mathrm{f}}_0(R)$.*
(*iv*) *$\mathcal{A}_0(R) = \mathcal{C}_0(R)$.*
(*v*) *$\mathcal{A}(R) = \mathcal{C}_{(\square)}(R)$.*

*Proof.* In view of Theorem (3.1.10) conditions (*i*), (*ii*), and (*iii*) are just the first three equivalent conditions of the $\mathcal{R}$ version (2.3.14). Since (*v*) is stronger than (*iv*), and (*iv*) is stronger than (*iii*), it is sufficient to prove that (*i*) implies (*v*). If $R$ is Gorenstein, then $R$ is a dualizing complex for $R$, cf. (A.8.3.1), so by (3.1.6) we can assume that $D = R$. For $X \in \mathcal{C}_{(\square)}(R)$ homological boundedness of $D \otimes^{\mathbf{L}}_R X$ is then automatic, and for any $F \in \mathcal{C}^{\mathrm{F}}_{\sqsupset}(R)$ the tensor evaluation morphism $\omega_{RRF}$ is an isomorphism, cf. (A.2.10). By (A.3.2) and Lemma (3.1.4) it now follows that every complex $X \in \mathcal{C}_{(\square)}(R)$ belongs to $\mathcal{A}(R)$. $\square$

The next two results are auxiliaries needed in chapters 4 and 5, but they belong in this section.

(3.1.13) **Lemma.** *Let $0 \to X' \to X \to X'' \to 0$ be a short exact sequence in $\mathcal{C}_{(\sqsupset)}(R)$. If two of the complexes belong to the Auslander class $\mathcal{A}(R)$, then so does the third.*

*Proof.* By (A.3.4) we can choose a short exact sequence

$$(\dagger) \qquad 0 \to Q' \to Q \to Q'' \to 0$$

in $\mathcal{C}^{\mathrm{P}}_{\sqsupset}(R)$, such that $Q'$, $Q$, and $Q''$ are projective resolutions of, respectively, $X'$, $X$, and $X''$. As in the proof of Lemma (2.1.12) it follows, by inspection of the long exact sequence of homology modules associated to ($\dagger$), that homological boundedness of two complexes in the original short exact sequence implies that also the third belongs to $\mathcal{C}_{(\square)}(R)$.

Let $P \in \mathcal{C}^{\mathrm{P}}_{\sqsupset}(R)$ be a projective resolution of $D$; applying $P \otimes_R -$ to ($\dagger$) we get another short exact sequence:

$$(\ddagger) \qquad 0 \to P \otimes_R Q' \to P \otimes_R Q \to P \otimes_R Q'' \to 0.$$

The complexes in ($\ddagger$) represent $D \otimes^{\mathbf{L}}_R X'$, $D \otimes^{\mathbf{L}}_R X$, and $D \otimes^{\mathbf{L}}_R X''$, and, as above, if two of these complexes belong to $\mathcal{C}_{(\square)}(R)$, then so does the third.

Finally, applying $\mathrm{Hom}_R(P, -)$ to ($\ddagger$) we get the bottom row in the diagram.

$$\begin{array}{ccccccc} 0 \to & Q' & \to & Q & \to & Q'' & \to 0 \\ & \downarrow \gamma^P_{Q'} & & \downarrow \gamma^P_{Q} & & \downarrow \gamma^P_{Q''} & \\ 0 \to & \mathrm{Hom}_R(P, P \otimes_R Q') & \to & \mathrm{Hom}_R(P, P \otimes_R Q) & \to & \mathrm{Hom}_R(P, P \otimes_R Q'') & \to 0 \end{array}$$

The rows are exact and the diagram is commutative. As in the proof of Lemma (2.1.12) we pass to homology to see that if two of the morphisms $\gamma^P_{Q'}$, $\gamma^P_{Q}$, and $\gamma^P_{Q''}$ are quasi-isomorphisms, then so is the third. □

(3.1.14) **Proposition.** *A bounded complex of modules from $\mathcal{A}_0(R)$ belongs to the Auslander class.*

*Proof.* The proof of Proposition (2.1.13) applies verbatim, only use the Lemma above instead of Lemma (2.1.12). □

## 3.2 The Bass Class

The Bass class is the dual of the Auslander class: at least in the sense that it is an extension of the full subcategory of complexes of finite injective dimension, and duality with respect to an injective module takes complexes from one class into the other.

(3.2.1) **Setup.** In this section $R$ is a **local ring with dualizing complex** $D$.

(3.2.2) **Definitions.** For $R$–complexes $Y$ and $Z$ a canonical morphism $\xi_Y^Z$ is defined by requiring commutativity of the diagram

$$\begin{array}{ccc} Z \otimes_R \operatorname{Hom}_R(Z,Y) & \xrightarrow{\xi_Y^Z} & Y \\ \Big\downarrow \theta_{ZZY} & & \cong \Big\uparrow \\ \operatorname{Hom}_R(\operatorname{Hom}_R(Z,Z),Y) & \xrightarrow{\operatorname{Hom}_R(\chi_Z^R,Y)} & \operatorname{Hom}_R(R,Y) \end{array} \tag{3.2.2.1}$$

The morphism $\xi_Y^Z$ is natural in $Z$ and $Y$ and given by

$$z \otimes \psi \longmapsto (-1)^{|z||\psi|}\psi(z).$$

Let $Y$ be an $R$–complex, and let $P \in \mathcal{C}_{\sqsupset}^{\mathrm{P}}(R)$ be a projective resolution of the dualizing complex. Then $\mathbf{R}\mathrm{Hom}_R(D,Y)$ is represented by $\operatorname{Hom}_R(P,Y)$ and $D \otimes_R^{\mathbf{L}} \mathbf{R}\mathrm{Hom}_R(D,Y)$ by $P \otimes_R \operatorname{Hom}_R(P,Y)$. We say that *$Y$ represents $D \otimes_R^{\mathbf{L}} \mathbf{R}\mathrm{Hom}_R(D,Y)$ canonically* if and only if the morphism

$$\xi_Y^P : P \otimes_R \operatorname{Hom}_R(P,Y) \longrightarrow Y$$

is a quasi-isomorphism.

(3.2.3) **Remark.** As in (3.1.3) it is straightforward to check that this definition of canonical representation makes sense. That is, if $P$ and $P'$ are projective resolutions of $D$, and $Y$ is an $R$–complex, then $\xi_Y^P$ is a quasi-isomorphism if and only if $\xi_Y^{P'}$ is so. And if $D'$ is another dualizing complex for $R$, then $Y$ represents $D \otimes_R^{\mathbf{L}} \mathbf{R}\mathrm{Hom}_R(D,Y)$ canonically if and only if it represents $D' \otimes_R^{\mathbf{L}} \mathbf{R}\mathrm{Hom}_R(D',Y)$ canonically.

By the defining diagram (3.2.2.1) the canonical morphism $\xi$ is closely linked to Hom evaluation, and the next lemma expresses canonical representation of $D \otimes_R^{\mathbf{L}} \mathbf{R}\mathrm{Hom}_R(D,Y)$ in terms of Hom evaluation.

(3.2.4) **Lemma.** *Let $P \in \mathcal{C}_{\sqsupset}^{\mathrm{P}}(R)$ be a projective resolution of $D$, and let $Y \in \mathcal{C}_{(\sqsubset)}(R)$. If $I \in \mathcal{C}_{\sqsubset}^{\mathrm{I}}(R)$ is an injective resolution of $Y$, then the following are equivalent:*

(*i*) *$Y$ represents $D \otimes_R^{\mathbf{L}} \mathbf{R}\mathrm{Hom}_R(D,Y)$ canonically.*
(*ii*) *The canonical morphism $\xi_I^P$ is a quasi-isomorphism.*
(*iii*) *The Hom evaluation morphism $\theta_{PDI}$ is a quasi-isomorphism.*

*Proof.* Let a projective resolution $\pi\colon P \xrightarrow{\simeq} D$ be given, and take an injective resolution $\iota\colon Y \xrightarrow{\simeq} I$. The homothety morphism $\chi_P^R\colon R \to \operatorname{Hom}_R(P,P)$ is a quasi-isomorphism, and using the quasi-isomorphism preserving properties of the various functors, we set up the following diagram

$$
\begin{array}{ccccc}
 & & P \otimes_R \mathrm{Hom}_R(P,Y) & \xrightarrow{\ \xi^P_Y\ } & Y \\
 & & \simeq \Big\downarrow {\scriptstyle P\otimes_R \mathrm{Hom}_R(P,\iota)} & & \simeq \Big\downarrow {\scriptstyle \iota} \\
P \otimes_R \mathrm{Hom}_R(D,I) & \xrightarrow[\simeq]{P\otimes_R \mathrm{Hom}_R(\pi,I)} & P \otimes_R \mathrm{Hom}_R(P,I) & \xrightarrow{\ \xi^P_I\ } & I \\
\Big\downarrow {\scriptstyle \theta_{PDI}} & & \Big\downarrow {\scriptstyle \theta_{PPI}} & & \cong \Big\downarrow \\
[[P,D]I] & \xrightarrow[\simeq]{[[P,\pi]I]} & [[P,P]I] & \xrightarrow[\simeq]{\mathrm{Hom}_R(\chi^R_P,I)} & \mathrm{Hom}_R(R,I)
\end{array}
$$

where we use the abbreviated notation $[[P,-]I] = \mathrm{Hom}_R(\mathrm{Hom}_R(P,-),I)$. It is straightforward to check that the diagram is commutative, and the equivalence of the three conditions follows. □

(3.2.5) **Definition.** The *Bass class* $\mathcal{B}(R)$ is the full subcategory of $\mathcal{C}(R)$, actually of $\mathcal{C}_{(\square)}(R)$, defined by specifying its objects as follows: An $R$–complex $Y$ belongs to $\mathcal{B}(R)$ if and only if

(1) $Y \in \mathcal{C}_{(\square)}(R)$;
(2) $\mathbf{R}\mathrm{Hom}_R(D,Y) \in \mathcal{C}_{(\square)}(R)$; and
(3) $Y$ represents $D \otimes^{\mathbf{L}}_R \mathbf{R}\mathrm{Hom}_R(D,Y)$ canonically.

We also use the notation $\mathcal{B}(R)$ with subscript 0 and superscript f/(f) defined as in (3.1.5).

If $Y$ belongs to $\mathcal{B}(R)$, then so does every equivalent complex $Y' \simeq Y$; this follows by the Lemma, as an injective resolution $I$ of $Y$ is also a resolution of $Y'$, cf. (A.3.5). For an equivalence class $\mathbf{Y}$ of $R$–complexes the notation $\mathbf{Y} \in \mathcal{B}(R)$ means that some, equivalently every, representative $Y$ of $\mathbf{Y}$ belongs to the Bass class.

(3.2.6) **Remark.** The Bass class is defined in terms of a dualizing complex $D$ for $R$, but the symbol $\mathcal{B}(R)$ makes no mention of $D$. As in the case of the Auslander class, this is justified by the fact that $\mathcal{B}(R)$ is independent of the choice of dualizing complex, cf. (3.2.3).

(3.2.7) **Observation.** Let $\mathfrak{p} \in \operatorname{Spec} R$, then $D_\mathfrak{p}$ is a dualizing complex for $R_\mathfrak{p}$, cf. (A.8.3.4), and, as in Observation (2.1.11), it is straightforward to check that

$$Y \in \mathcal{B}(R) \implies Y_\mathfrak{p} \in \mathcal{B}(R_\mathfrak{p}).$$

(3.2.8) **Proposition.** *Every $R$–complex of finite injective dimension belongs to the Bass class. That is, there is a full embedding:*

$$\mathcal{I}(R) \subseteq \mathcal{B}(R).$$

*Proof.* Suppose $Y \in \mathcal{I}(R)$, then both $Y$ and $\mathbf{R}\mathrm{Hom}_R(D,Y)$ belong to $\mathcal{C}_{(\square)}(R)$, cf. (A.5.2). By (3.2.6) we are free to assume that $D \in \mathcal{C}_{\square}(R)$. Choose a bounded resolution $Y \xrightarrow{\simeq} I \in \mathcal{C}^{\mathrm{I}}_{\square}(R)$, and take a projective resolution $P \in \mathcal{C}^{\mathrm{fP}}_{\sqsupset}(R)$ of the dualizing complex. The Hom evaluation morphism $\theta_{PDI}$ is then an isomorphism, cf. (A.2.11), and by Lemma (3.2.4) $Y$, therefore, represents $D \otimes^{\mathbf{L}}_R \mathbf{R}\mathrm{Hom}_R(D,Y)$ canonically. □

We have now established the Bass class as an extension of the full subcategory of complexes of finite injective dimension. The next lemma should — for reasons to be revealed in chapter 6 — be perceived as an extension of Ishikawa's formulas (see [42] or page 7).

(3.2.9) **Lemma.** *Consider complexes $X \in \mathcal{C}_{(\sqsupset)}(R)$ and $Y \in \mathcal{C}_{(\sqsubset)}(R)$, and let $E$ be an injective $R$–module; then the following hold:*

(a) *If $X \in \mathcal{A}(R)$ then $\mathrm{Hom}_R(X,E) \in \mathcal{B}(R)$, and the converse holds if $E$ is faithfully injective.*

(b) *If $Y \in \mathcal{B}(R)$ then $\mathrm{Hom}_R(Y,E) \in \mathcal{A}(R)$, and the converse holds if $E$ is faithfully injective.*

*Proof.* We have

$$(\dagger)\qquad -\inf(\mathrm{Hom}_R(X,E)) \le \sup X,$$

and by adjointness (A.4.21) and (A.5.2.1) we have

$$\begin{aligned} -\inf(\mathbf{R}\mathrm{Hom}_R(D,\mathrm{Hom}_R(X,E))) &= -\inf(\mathbf{R}\mathrm{Hom}_R(D,\mathbf{R}\mathrm{Hom}_R(X,E))) \\ (\ddagger)\qquad &= -\inf(\mathbf{R}\mathrm{Hom}_R(D\otimes^{\mathbf{L}}_R X,E)) \\ &\le \sup(D\otimes^{\mathbf{L}}_R X) \end{aligned}$$

as $E$ is injective. Thus, if $X \in \mathcal{A}(R)$ then, in particular, $\mathrm{Hom}_R(X,E)$ and $\mathbf{R}\mathrm{Hom}_R(D,\mathrm{Hom}_R(X,E))$ are homologically bounded. On the other hand, if $E$ is faithfully injective, then equality holds in (†) and (‡), cf. (A.4.10), so $X$ and $D \otimes^{\mathbf{L}}_R X$ are homologically bounded if $\mathrm{Hom}_R(X,E) \in \mathcal{B}(R)$. Take a flat resolution $\varphi\colon F \xrightarrow{\simeq} X$, then

$$\mathrm{Hom}_R(\varphi,E): \mathrm{Hom}_R(X,E) \xrightarrow{\simeq} \mathrm{Hom}_R(F,E)$$

is an injective resolution of $\mathrm{Hom}_R(X,E)$. Let $L$ be a resolution of $D$ by finite free modules, then $L \otimes_R F$ represents the homologically bounded $D \otimes^{\mathbf{L}}_R X$, so by (A.1.14) there is a quasi-isomorphism $\omega\colon L\otimes_R F \xrightarrow{\simeq} V$, where $V \in \mathcal{C}_{\square}(R)$ is a suitable (soft left) truncation of $L \otimes_R F$. The commutative diagram

$$\begin{array}{ccc} L\otimes_R \mathrm{Hom}_R(V,E) & \xrightarrow[\cong]{\theta_{LVE}} & \mathrm{Hom}_R(\mathrm{Hom}_R(L,V),E) \\ \simeq\downarrow{\scriptstyle L\otimes_R\mathrm{Hom}_R(\omega,E)} & & \simeq\downarrow{\scriptstyle \mathrm{Hom}_R(\mathrm{Hom}_R(L,\omega),E)} \\ L\otimes_R \mathrm{Hom}_R(L\otimes_R F,E) & \xrightarrow{\theta_{LL\otimes_R FE}} & \mathrm{Hom}_R(\mathrm{Hom}_R(L,L\otimes_R F),E) \end{array}$$

where the Hom evaluation morphism $\theta_{LVE}$ is an isomorphism by (A.2.11), shows that $\theta_{LL\otimes_R FE}$ is a quasi-isomorphism. Also the next diagram is commutative.

$$\begin{array}{ccc} L \otimes_R \operatorname{Hom}_R(L \otimes_R F, E) & \xrightarrow[\simeq]{\theta_{LL\otimes_R FE}} & \operatorname{Hom}_R(\operatorname{Hom}_R(L, L \otimes_R F), E) \\ \cong \downarrow L\otimes_R \rho_{LFE} & & \downarrow \operatorname{Hom}_R(\gamma_F^L, E) \\ L \otimes_R \operatorname{Hom}_R(L, \operatorname{Hom}_R(F, E)) & \xrightarrow{\xi^L_{\operatorname{Hom}_R(F,E)}} & \operatorname{Hom}_R(F, E) \end{array}$$

If $X$ belongs to the Auslander class, then the canonical morphism $\gamma_F^L$ is a quasi-isomorphism, cf. Lemma (3.1.4), and, hence, so is $\operatorname{Hom}_R(\gamma_F^L, E)$. The diagram then shows that also $\xi^L_{\operatorname{Hom}_R(F,E)}$ is a quasi-isomorphism, so $\operatorname{Hom}_R(X, E) \in \mathcal{B}(R)$ by Lemma (3.2.4). Conversely, if $\operatorname{Hom}_R(X, E) \in \mathcal{B}(R)$, then the canonical morphism $\xi^L_{\operatorname{Hom}_R(F,E)}$ and, thereby, $\operatorname{Hom}_R(\gamma_F^L, E)$ is a quasi-isomorphism; and if $E$ is faithfully injective, then this is tantamount to $\gamma_F^L$ being a quasi-isomorphism, cf. (A.2.1.4) and (A.1.19). This proves part (a), and the proof of (b) is similar. □

The next theorem is parallel to the $\mathcal{A}$ version, Theorem (3.1.12), it characterizes Gorenstein ring as being those with the "largest possible" Bass class.

(3.2.10) **Gorenstein Theorem, $\mathcal{B}$ Version.** *Let $R$ be a local ring with residue field $k$. If $R$ admits a dualizing complex, then the following are equivalent:*

(*i*) *$R$ is Gorenstein.*
(*ii*) *$k \in \mathcal{B}_0(R)$.*
(*iii*) *$\mathcal{B}_0^{\mathrm{f}}(R) = \mathcal{C}_0^{\mathrm{f}}(R)$.*
(*iv*) *$\mathcal{B}_0(R) = \mathcal{C}_0(R)$.*
(*v*) *$\mathcal{B}(R) = \mathcal{C}_{(\square)}(R)$.*

*Proof.* Obviously, (*v*) is stronger than (*iv*), and (*iv*) is stronger than (*iii*), which in turn is stronger than (*ii*). Hence, it is sufficient to prove that (*i*) implies (*v*) and (*ii*) implies (*i*).

(*i*) ⇒ (*v*): If $R$ is Gorenstein, then $R$ is a dualizing complex for $R$, so by (3.2.6) we can assume that $D = R$. For $Y \in \mathcal{C}_{(\square)}(R)$ homological boundedness of $\mathbf{R}\operatorname{Hom}_R(D, Y)$ is then automatic, and for any complex $I \in \mathcal{C}_{\sqsubset}^{\mathrm{I}}(R)$ the Hom evaluation morphism $\theta_{RRI}$ is an isomorphism, cf. (A.2.11). By (A.3.2) and Lemma (3.2.4) every complex $Y \in \mathcal{C}_{(\square)}(R)$ then belongs to $\mathcal{B}(R)$.

(*ii*) ⇒ (*i*): The Matlis dual of $k$ is $k$, i.e., $\operatorname{Hom}_R(k, \mathrm{E}_R(k)) \cong k$, so it follows by the Lemma that the residue field belongs to the Auslander class if and only if it belongs to the Bass class. In particular, if $k \in \mathcal{B}_0(R)$ then $k \in \mathcal{A}_0(R)$, and $R$ is then Gorenstein by the the $\mathcal{A}$ version (3.1.12). □

(3.2.11) **Remark.** The proof above of the implication "(*ii*) ⇒ (*i*)" in the $\mathcal{B}$ version uses the $\mathcal{A}$ version but, of course, a direct proof also exists: if $k$ is in the

Bass class, then $\mathbf{R}\mathrm{Hom}_R(D,k)$ is homologically bounded, so $\mathrm{pd}_R\, D < \infty$ by (A.5.7.3) and, hence,

$$\mathrm{id}_R\, R = \mathrm{id}_R(\mathbf{R}\mathrm{Hom}_R(D,D)) \leq \mathrm{fd}_R\, D + \mathrm{id}_R\, D = \mathrm{pd}_R\, D + \mathrm{id}_R\, D < \infty$$

by (3.0.1.1), (A.5.8.2), and (A.5.7.2), so $R$ is Gorenstein.

The last two results of this section are auxiliaries needed in chapter 6.

(3.2.12) **Lemma.** *Let $0 \to Y' \to Y \to Y'' \to 0$ be a short exact sequence in $\mathcal{C}_{(\sqsubset)}(R)$. If two of the complexes belong to the Bass class $\mathcal{B}(R)$, then so does the third.*

*Proof.* Similar to the proof of Lemma (3.1.13), only use (A.3.3) instead of (A.3.4). □

(3.2.13) **Proposition.** *A bounded complex of modules from $\mathcal{B}_0(R)$ belongs to the Bass class.*

*Proof.* The proof of Proposition (2.1.13) applies verbatim, only use the Lemma above instead of Lemma (2.1.12). □

## 3.3 Foxby Equivalence

Over a Cohen–Macaulay local ring with a dualizing (canonical) module the full subcategories of finite modules of, respectively, finite projective dimension and finite injective dimension are equivalent. This was established by Sharp in [55]. In [34] and [8, Section 3] Foxby has extended and generalized Sharp's construction in several directions; and in [32] and [61] Enochs, Jenda, and Xu have used the name 'Foxby duality' for the version to be described in this section. The involved functors are, however, not contravariant, so we find the name *Foxby equivalence* more appropriate.

(3.3.1) **Setup.** In this section $R$ is a **local ring with dualizing complex** $D$.

(3.3.2) **Theorem (Foxby Equivalence).** *If $R$ is a local ring and $D$ is a dualizing complex for $R$, then the following hold for complexes $X$ and $Y$ in $\mathcal{C}_{(\square)}(R)$:*

(a) $X \in \mathcal{A}(R) \iff D \otimes^{\mathbf{L}}_R X \in \mathcal{B}(R)$;
(b) $Y \in \mathcal{B}(R) \iff \mathbf{R}\mathrm{Hom}_R(D,Y) \in \mathcal{A}(R)$;
(c) $X \in \mathcal{F}(R) \iff D \otimes^{\mathbf{L}}_R X \in \mathcal{I}(R)$; and
(d) $Y \in \mathcal{I}(R) \iff \mathbf{R}\mathrm{Hom}_R(D,Y) \in \mathcal{F}(R)$.

*Furthermore, $X \in \mathcal{A}(R)$ has finite homology modules, respectively, finite depth if and only if $D \otimes^{\mathbf{L}}_R X$ has the same; and $Y \in \mathcal{B}(R)$ has finite homology* *modules, respectively, finite depth if and only if* $\mathbf{R}\mathrm{Hom}_R(D,Y)$ *has the same. That is, the following hold for* $X \in \mathcal{A}(R)$ *and* $Y \in \mathcal{B}(R)$:

(e) $\quad X \in \mathcal{C}^{(\mathrm{f})}_{(\square)}(R) \iff D \otimes^{\mathbf{L}}_R X \in \mathcal{C}^{(\mathrm{f})}_{(\square)}(R);$

(f) $\quad Y \in \mathcal{C}^{(\mathrm{f})}_{(\square)}(R) \iff \mathbf{R}\mathrm{Hom}_R(D,Y) \in \mathcal{C}^{(\mathrm{f})}_{(\square)}(R);$

(g) $\quad \operatorname{depth}_R X < \infty \iff \operatorname{depth}_R(D \otimes^{\mathbf{L}}_R X) < \infty;$ *and*

(h) $\quad \operatorname{depth}_R Y < \infty \iff \operatorname{depth}_R(\mathbf{R}\mathrm{Hom}_R(D,Y)) < \infty.$

*Proof.* (a): Let $X \in \mathcal{C}_{(\square)}(R)$, take a complex $F \in \mathcal{C}^{\mathrm{F}}_{\sqsupset}(R)$ equivalent to $X$ and a projective resolution $P$ of the dualizing complex, then $D \otimes^{\mathbf{L}}_R X$ is represented by the complex $W = P \otimes_R F$. We will prove that $X \in \mathcal{A}(R)$ if and only if $W \in \mathcal{B}(R)$. If $X$ belongs to $\mathcal{A}(R)$ it follows that $W \in \mathcal{C}_{(\square)}(R)$, and $\mathbf{R}\mathrm{Hom}_R(D,W) = \mathbf{R}\mathrm{Hom}_R(D, D \otimes^{\mathbf{L}}_R X)$ is represented by $X$, so also $\mathbf{R}\mathrm{Hom}_R(D,W) \in \mathcal{C}_{(\square)}(R)$. On the other hand, if $W \in \mathcal{B}(R)$ then, in particular, $D \otimes^{\mathbf{L}}_R X \in \mathcal{C}_{(\square)}(R)$. Take an injective resolution $\iota\colon W \xrightarrow{\simeq} I$; by Lemmas (3.1.4) and (3.2.4) it is now sufficient to prove that the canonical morphism $\gamma^P_F$ is a quasi-isomorphism if and only if $\xi^P_I$ is so. The commutative diagram

$$\begin{array}{ccc}
P \otimes_R F & \xrightarrow{P \otimes_R \gamma^P_F} & P \otimes_R \mathrm{Hom}_R(P, P \otimes_R F) \\
\simeq \downarrow \iota & & \simeq \downarrow P \otimes_R \mathrm{Hom}_R(P,\iota) \\
I & \xleftarrow{\xi^P_I} & P \otimes_R \mathrm{Hom}_R(P,I)
\end{array}$$

shows that $\xi^P_I$ is a quasi-isomorphism exactly when $P \otimes_R \gamma^P_F$ is so, and since $\operatorname{Supp}_R D = \operatorname{Spec} R$ by (A.8.6.1) it follows by (A.8.12) that $P \otimes_R \gamma^P_F$ is a quasi-isomorphism if and only if $\gamma^P_F$ is so.

(b): Let $Y \in \mathcal{C}_{(\square)}(R)$, take an injective resolution $Y \xrightarrow{\simeq} I \in \mathcal{C}^{\mathrm{I}}_{\sqsubset}(R)$ and a projective resolution $P$ of the dualizing complex, then $\mathbf{R}\mathrm{Hom}_R(D,Y)$ is represented by the complex $V = \mathrm{Hom}_R(P,I)$. We will prove that $Y \in \mathcal{B}(R)$ if and only if $V \in \mathcal{A}(R)$. If $V$ belongs to $\mathcal{A}(R)$ then, certainly, $\mathbf{R}\mathrm{Hom}_R(D,Y)$ is homologically bounded. On the other hand, if $Y$ belongs to $\mathcal{B}(R)$, then $V \in \mathcal{C}_{(\square)}(R)$ and $D \otimes^{\mathbf{L}}_R V = D \otimes^{\mathbf{L}}_R \mathbf{R}\mathrm{Hom}_R(D,Y)$ is represented by $Y$, so also $D \otimes^{\mathbf{L}}_R V$ is homologically bounded. Now take a flat resolution $\varphi\colon F \xrightarrow{\simeq} V$; in view of (A.8.13) it follows from the commutative diagram

$$\begin{array}{ccc}
F & \xrightarrow[\simeq]{\varphi} & \mathrm{Hom}_R(P,I) \\
\downarrow \gamma^P_F & & \uparrow \mathrm{Hom}_R(P,\xi^P_I) \\
\mathrm{Hom}_R(P, P \otimes_R F) & \xrightarrow[\simeq]{\mathrm{Hom}_R(P, P \otimes_R \varphi)} & \mathrm{Hom}_R(P, P \otimes_R \mathrm{Hom}_R(P,I))
\end{array}$$

that $\gamma^P_F$ is a quasi-isomorphism if and only if $\xi^P_I$ is so. By Lemmas (3.1.4) and (3.2.4) the proof of (b) is then complete.

(c): If $X \in \mathcal{F}(R)$, then $D \otimes^{\mathbf{L}}_R X \in \mathcal{I}(R)$ as $D \in \mathcal{I}(R)$, cf. (A.5.8.3). On the other hand, if $D \otimes^{\mathbf{L}}_R X \in \mathcal{I}(R)$, then $X \in \mathcal{A}(R)$ by (a) and Proposition (3.2.8), so $X$ represents $\mathbf{R}\mathrm{Hom}_R(D, D \otimes^{\mathbf{L}}_R X)$, and $\mathbf{R}\mathrm{Hom}_R(D, D \otimes^{\mathbf{L}}_R X) \in \mathcal{F}(R)$ by (A.5.8.4).

(d): If $Y \in \mathcal{I}(R)$, then $\mathbf{R}\mathrm{Hom}_R(D,Y) \in \mathcal{F}(R)$ by (A.5.8.4). If $\mathbf{R}\mathrm{Hom}_R(D,Y) \in \mathcal{F}(R)$ then $Y \in \mathcal{B}(R)$ by Proposition (3.1.8) and (b), so $Y$ represents $D \otimes^{\mathbf{L}}_R \mathbf{R}\mathrm{Hom}_R(D,Y) \in \mathcal{I}(R)$, cf. (A.5.8.3).

Both (e) and (f) are immediate, cf. (A.4.4) and (A.4.13); and to prove (g) and (h) it is sufficient to prove "$\Rightarrow$" in both statements.

(g): By (A.6.4) and (A.6.3.2) we have

$$\begin{aligned}\infty > \operatorname{depth}_R X &= \operatorname{depth}_R(\mathbf{R}\mathrm{Hom}_R(D, D \otimes^{\mathbf{L}}_R X)) \\ &= \inf D + \operatorname{depth}_R(D \otimes^{\mathbf{L}}_R X),\end{aligned}$$

so $\operatorname{depth}_R(D \otimes^{\mathbf{L}}_R X) < \infty$ as wanted.

(h): Again we have

$$\operatorname{depth}_R(\mathbf{R}\mathrm{Hom}_R(D,Y)) = \inf D + \operatorname{depth}_R Y < \infty$$

by (A.6.4) and (A.6.3.2). □

(3.3.3) **Lemma.** *The following hold for* $U \in \mathcal{C}_{(\square)}(R)$, $X \in \mathcal{A}(R)$, *and* $Y \in \mathcal{B}(R)$:

(a) $\mathbf{R}\mathrm{Hom}_R(U,X) = \mathbf{R}\mathrm{Hom}_R(D \otimes^{\mathbf{L}}_R U, D \otimes^{\mathbf{L}}_R X)$;

(b) $\mathbf{R}\mathrm{Hom}_R(Y,U) = \mathbf{R}\mathrm{Hom}_R(\mathbf{R}\mathrm{Hom}_R(D,Y), \mathbf{R}\mathrm{Hom}_R(D,U))$; *and*

(c) $Y \otimes^{\mathbf{L}}_R U = (D \otimes^{\mathbf{L}}_R U) \otimes^{\mathbf{L}}_R \mathbf{R}\mathrm{Hom}_R(D,Y)$.

*Proof.* The proof of (a) is straightforward, it uses adjointness (A.4.21) and commutativity (A.4.19):

$$\begin{aligned}\mathbf{R}\mathrm{Hom}_R(U,X) &= \mathbf{R}\mathrm{Hom}_R(U, \mathbf{R}\mathrm{Hom}_R(D, D \otimes^{\mathbf{L}}_R X)) \\ &= \mathbf{R}\mathrm{Hom}_R(U \otimes^{\mathbf{L}}_R D, D \otimes^{\mathbf{L}}_R X) \\ &= \mathbf{R}\mathrm{Hom}_R(D \otimes^{\mathbf{L}}_R U, D \otimes^{\mathbf{L}}_R X).\end{aligned}$$

The proofs of (b) and (c) are similar. □

The next two theorems characterize Gorenstein rings in terms of special properties of the (almost[1]) derived functors $D \otimes^{\mathbf{L}}_R -$ and $\mathbf{R}\mathrm{Hom}_R(D,-)$ and existence of special complexes in the Auslander categories. The first part of (3.3.4) should be compared to the PD/ID version on page 6.

[1] See section A.4.

**(3.3.4) Gorenstein Theorem, Foxby Equivalence Version.** *Let $R$ be a local ring. The following are equivalent:*

(*i*) *$R$ is Gorenstein.*

(*ii*) *An $R$–complex $X \in \mathcal{C}_{(\square)}(R)$ has finite flat dimension if and only if it has finite injective dimension; that is, $\mathrm{fd}_R X < \infty \Leftrightarrow \mathrm{id}_R X < \infty$.*

*Furthermore, if $D$ is a dualizing complex for $R$, then the next three conditions are equivalent, and equivalent to those above.*

(*iii*) *There is a complex $Y \in \mathcal{C}_{(\square)}(R)$ with $\mathrm{depth}_R Y < \infty$ such that $\mathbf{R}\mathrm{Hom}_R(D,Y)$ belongs to $\mathcal{A}(R)$ and $D \otimes^{\mathbf{L}}_R Y \in \mathcal{I}(R)$.*

(*iv*) *There is a complex $X \in \mathcal{C}_{(\square)}(R)$ with $\mathrm{depth}_R X < \infty$ such that $D \otimes^{\mathbf{L}}_R X$ belongs to $\mathcal{B}(R)$ and $\mathbf{R}\mathrm{Hom}_R(D,X) \in \mathcal{F}(R)$.*

(*v*) *$D \in \mathcal{P}^{(\mathrm{f})}(R)$.*

*Proof.* If every complex of finite flat dimension has finite injective dimension, then, in particular, $\mathrm{id}_R R < \infty$ and $R$ is Gorenstein. On the other hand, if $R$ is Gorenstein, then $R$ is a dualizing complex for $R$, cf. (A.8.3.1), so it follows by Foxby equivalence (3.3.2) that the full subcategories $\mathcal{F}(R)$ and $\mathcal{I}(R)$ are equal. This proves equivalence of the first two conditions.

Now assume that $D$ is a dualizing complex for $R$.

(*ii*) $\Rightarrow$ (*iii*): Set $Y = R$, then $\mathbf{R}\mathrm{Hom}_R(D,Y) \in \mathcal{F}(R)$ and $D \otimes^{\mathbf{L}}_R Y \in \mathcal{I}(R)$ by (A.5.8.4) and (A.5.8.3).

(*ii*) $\Rightarrow$ (*iv*): Set $X = R$ and use (A.5.8) as above.

(*iii*) $\Rightarrow$ (*v*): It follows by (b), (c), and (h) in Theorem (3.3.2) that $Y$ belongs to $\mathcal{B}(R) \cap \mathcal{F}(R)$ and $\mathrm{depth}_R(\mathbf{R}\mathrm{Hom}_R(D,Y)) < \infty$. By (A.6.6) also $\mathrm{width}_R(\mathbf{R}\mathrm{Hom}_R(D,Y)) < \infty$, in particular, $-\sup(\mathbf{R}\mathrm{Hom}_R(D,Y) \otimes^{\mathbf{L}}_R k) < \infty$. By (c) in the Lemma we have $Y \otimes^{\mathbf{L}}_R k = (D \otimes^{\mathbf{L}}_R k) \otimes^{\mathbf{L}}_R \mathbf{R}\mathrm{Hom}_R(D,Y)$, so

$$\sup(Y \otimes^{\mathbf{L}}_R k) = \sup(D \otimes^{\mathbf{L}}_R k) + \sup(\mathbf{R}\mathrm{Hom}_R(D,Y) \otimes^{\mathbf{L}}_R k)$$

by (A.7.9.1). By (A.5.7.2) and (A.5.6.1) we now have

$$\begin{aligned}\mathrm{pd}_R D = \sup(D \otimes^{\mathbf{L}}_R k) &= \sup(Y \otimes^{\mathbf{L}}_R k) - \sup(\mathbf{R}\mathrm{Hom}_R(D,Y) \otimes^{\mathbf{L}}_R k)\\ &\leq \mathrm{fd}_R Y - \sup(\mathbf{R}\mathrm{Hom}_R(D,Y) \otimes^{\mathbf{L}}_R k)\\ &< \infty.\end{aligned}$$

(*iv*) $\Rightarrow$ (*v*): It follows by (a), (d), and (g) in Theorem (3.3.2) that $X$ belongs to $\mathcal{A}(R) \cap \mathcal{I}(R)$ and $\mathrm{depth}_R(D \otimes^{\mathbf{L}}_R X) < \infty$. This means, in particular, that $\inf(\mathbf{R}\mathrm{Hom}_R(k, D \otimes^{\mathbf{L}}_R X)) < \infty$. By (a) in the Lemma we now have $\mathbf{R}\mathrm{Hom}_R(k,X) = \mathbf{R}\mathrm{Hom}_R(D \otimes^{\mathbf{L}}_R k, D \otimes^{\mathbf{L}}_R X)$, so

$$\inf(\mathbf{R}\mathrm{Hom}_R(k,X)) = \inf(\mathbf{R}\mathrm{Hom}_R(k, D \otimes^{\mathbf{L}}_R X)) - \sup(D \otimes^{\mathbf{L}}_R k)$$

by (A.7.9.4). It now follows from (A.5.7.2) and (A.5.2.1) that

$$\begin{aligned}\mathrm{pd}_R D = \sup(D \otimes^{\mathbf{L}}_R k) &= \inf(\mathbf{R}\mathrm{Hom}_R(k, D \otimes^{\mathbf{L}}_R X)) - \inf(\mathbf{R}\mathrm{Hom}_R(k,X))\\ &\leq \inf(\mathbf{R}\mathrm{Hom}_R(k, D \otimes^{\mathbf{L}}_R X)) + \mathrm{id}_R X\\ &< \infty.\end{aligned}$$

$(v) \Rightarrow (i)$: Suppose $D$ belongs to $\mathcal{P}^{(\mathrm{f})}(R)$, then

$$\mathrm{id}_R\, R = \mathrm{id}_R(\mathbf{R}\mathrm{Hom}_R(D,D)) \leq \mathrm{pd}_R\, D + \mathrm{id}_R\, D < \infty$$

by (3.0.1.1), (A.5.7.2), and (A.5.8.2). □

**(3.3.5) Gorenstein Theorem, Special Complexes Version.** *Let $R$ be a local ring. If $D$ is a dualizing complex for $R$, then the following are equivalent:*

*(i)* *$R$ is Gorenstein.*
*(ii)* *$R \in \mathcal{B}(R)$.*
*(ii')* *$D \in \mathcal{A}(R)$.*
*(iii)* *$\mathrm{depth}_R\, Y < \infty$ for some $Y \in \mathcal{B}(R) \cap \mathcal{F}(R)$.*
*(iii')* *$\mathrm{depth}_R\, X < \infty$ for some $X \in \mathcal{A}(R) \cap \mathcal{I}(R)$.*
*(iv)* *$\mathcal{A}(R) = \mathcal{B}(R)$.*

*Proof.* The following implications are immediate by the Foxby equivalence version (3.3.4), Propositions (3.1.8) and (3.2.8), the $\mathcal{A}$ version (3.1.12), and the $\mathcal{B}$ version (3.2.10):

$$\begin{array}{ccccc} (ii') & \Leftarrow & (i) & \Rightarrow & (ii) \\ \Downarrow & & \Downarrow & & \Downarrow \\ (iii') & \Leftarrow & (iv) & \Rightarrow & (iii) \end{array}$$

It is now sufficient to prove that *(iii)* and *(iii')* imply *(i)*.

*(iii)* ⇒ *(i)*: If $Y \in \mathcal{B}(R) \cap \mathcal{F}(R)$, then $\mathbf{R}\mathrm{Hom}_R(D,Y) \in \mathcal{A}(R)$ and $D \otimes^{\mathbf{L}}_R Y \in \mathcal{I}(R)$, so it follows by the Foxby equivalence version (3.3.4) that $R$ is Gorenstein.

*(iii')* ⇒ *(i)*: If $X \in \mathcal{A}(R) \cap \mathcal{I}(R)$, then $D \otimes^{\mathbf{L}}_R X \in \mathcal{B}(R)$ and $\mathbf{R}\mathrm{Hom}_R(D,X) \in \mathcal{F}(R)$, so it follows, again by (3.3.4), that $R$ is Gorenstein. □

The next two results answer the question of 'how much the homological size of a complex can change under Foxby equivalence'; and it paves the way for a description of Foxby equivalence over Cohen–Macaulay rings in terms of classical homological algebra.

**(3.3.6) Lemma.** *For $X \in \mathcal{C}_{(\sqsupset)}(R)$ and $Y \in \mathcal{C}_{(\sqsubset)}(R)$ the next inequalities hold.*

(a) $\sup Y - \sup D \leq \sup(\mathbf{R}\mathrm{Hom}_R(D,Y)) \leq \sup Y - \inf D$; *and*

(b) $\inf X + \inf D \leq \inf(D \otimes^{\mathbf{L}}_R X) \leq \inf X + \sup D$.

*Proof.* The second inequality in (a) and the first one in (b) are the standard inequalities (A.4.6.1) and (A.4.15.1). Since $\mathrm{Supp}_R\, D = \mathrm{Spec}\, R$ by (A.8.6.1) the first inequality in (a) and the second in (b) follow by, respectively, (A.8.7) and (A.8.8). □

**(3.3.7) Proposition (Amplitude Inequalities).** *For complexes $X \in \mathcal{A}(R)$ and $Y \in \mathcal{B}(R)$ there are inequalities:*

(a) $\sup X + \inf D \leq \sup (D \otimes^{\mathbf{L}}_R X) \leq \sup X + \sup D$;

(b) $\operatorname{amp} X - \operatorname{amp} D \leq \operatorname{amp}(D \otimes^{\mathbf{L}}_R X) \leq \operatorname{amp} X + \operatorname{amp} D$;

(c) $\inf Y - \sup D \leq \inf (\mathbf{R}\mathrm{Hom}_R(D,Y)) \leq \inf Y - \inf D$; *and*

(d) $\operatorname{amp} Y - \operatorname{amp} D \leq \operatorname{amp}(\mathbf{R}\mathrm{Hom}_R(D,Y)) \leq \operatorname{amp} Y + \operatorname{amp} D$.

*Proof.* (a): Since $X = \mathbf{R}\mathrm{Hom}_R(D, D \otimes^{\mathbf{L}}_R X)$ it follows by Lemma (3.3.6)(a) that

$$\sup (D \otimes^{\mathbf{L}}_R X) - \sup D \leq \sup X \leq \sup (D \otimes^{\mathbf{L}}_R X) - \inf D;$$

and, therefore,

$$-\sup X - \sup D \leq -\sup (D \otimes^{\mathbf{L}}_R X) \leq -\sup X - \inf D.$$

(b): Using the inequalities in (a) and Lemma (3.3.6)(b) we find:

$$\begin{aligned}\operatorname{amp}(D \otimes^{\mathbf{L}}_R X) &= \sup (D \otimes^{\mathbf{L}}_R X) - \inf (D \otimes^{\mathbf{L}}_R X)\\ &\leq \sup X + \sup D - \inf (D \otimes^{\mathbf{L}}_R X)\\ &\leq \sup X + \sup D - (\inf X + \inf D)\\ &= \operatorname{amp} X + \operatorname{amp} D; \quad \text{and}\\ \operatorname{amp}(D \otimes^{\mathbf{L}}_R X) &= \sup (D \otimes^{\mathbf{L}}_R X) - \inf (D \otimes^{\mathbf{L}}_R X)\\ &\geq \sup X + \inf D - \inf (D \otimes^{\mathbf{L}}_R X)\\ &\geq \sup X + \inf D - (\inf X + \sup D)\\ &= \operatorname{amp} X - \operatorname{amp} D.\end{aligned}$$

The proof of (c) is similar to that of (a), only it uses Lemma (3.3.6)(b). The proof of (d) uses (c) and Lemma (3.3.6)(a), otherwise it is analogous to the proof of (b). □

**(3.3.8) Theorem.** *Let $R$ be a local ring. If $D$ is a dualizing complex for $R$, then the following are equivalent:*

(*i*) *$R$ is Cohen–Macaulay.*

(*ii*) $\operatorname{amp}(D \otimes^{\mathbf{L}}_R X) = \operatorname{amp} X$ *for all $X \in \mathcal{A}(R)$.*

(*ii'*) $\operatorname{amp}(D \otimes^{\mathbf{L}}_R M) = 0$ *for all $M \in \mathcal{A}_0(R)$.*

(*iii*) $\operatorname{amp}(\mathbf{R}\mathrm{Hom}_R(D,Y)) = \operatorname{amp} Y$ *for all $Y \in \mathcal{B}(R)$.*

(*iii'*) $\operatorname{amp}(\mathbf{R}\mathrm{Hom}_R(D,N)) = 0$ *for all $N \in \mathcal{B}_0(R)$.*

(*iv*) $\operatorname{amp} D = 0$.

*Proof.* The equivalence of (*i*) and (*iv*) is well-known, cf. (A.8.5.3), and the implications (*iv*)$\Rightarrow$(*ii*) and (*iv*)$\Rightarrow$(*iii*) follow by the amplitude inequalities (b) and (d) in (3.3.7). It is also clear that (*ii*) $\Rightarrow$ (*ii'*) and (*iii*) $\Rightarrow$ (*iii'*); the remaining implications are proved as follows:

(*ii'*) $\Rightarrow$ (*iv*): The ring $R$ belongs to $\mathcal{A}_0(R)$, so

$$\operatorname{amp} D = \operatorname{amp}(D \otimes_R^{\mathbf{L}} R) = 0.$$

(*iii'*) $\Rightarrow$ (*iv*): The injective hull of the residue field, $\mathrm{E}_R(k)$, is a faithfully injective $R$–module, so it belongs to $\mathcal{B}_0(R)$ by Proposition (3.2.8), and

$$\operatorname{amp} D = \operatorname{amp}(\mathbf{R}\mathrm{Hom}_R(D, E)) = 0$$

by (A.4.10). □

## Notes

Theorem (3.3.5) is due to Foxby; a module version (see (3.4.12)) was announced in [39, Theorem (5.1)], and generalized versions are found in [15, Section 8].

The last two results of this section are extended versions of [8, Lemmas (1.2.3)(a) and (3.3)]; they are generalized in [15, Section 4], and so is Theorem (3.3.2).

The fact that we do not work in the derived category has forced a quite unsatisfactory formulation of the Foxby equivalence Theorem (3.3.2): indeed, the word 'equivalence' only appears in the label and not in the statement! This will be partially remedied in the next section, but to justify the name of the current section we also state the "correct" version [8, Theorem (3.2)]:

**Theorem.** *Let $R$ be a local ring. If $D$ is a dualizing complex for $R$, then there is a commutative diagram*

$$\begin{array}{ccc}
\mathcal{D}(R) & \underset{\mathbf{R}\mathrm{Hom}_R(D,-)}{\overset{D\otimes_R^{\mathbf{L}}-}{\rightleftarrows}} & \mathcal{D}(R) \\
\cup\! | & & \cup\! | \\
\mathcal{A}(R) & \rightleftarrows & \mathcal{B}(R) \\
\cup\! | & & \cup\! | \\
\mathcal{F}(R) & \rightleftarrows & \mathcal{I}(R)
\end{array}$$

*in which the vertical inclusions are full embeddings, and the unlabeled horizontal arrows are quasi-inverse equivalences of categories.*

*Furthermore, for bounded complexes $X$ and $Y$ the following hold:*

(a) $D \otimes_R^{\mathbf{L}} X \in \mathcal{B}(R) \Longrightarrow X \in \mathcal{A}(R)$;

(b) $\mathbf{R}\mathrm{Hom}_R(D,Y) \in \mathcal{A}(R) \Longrightarrow Y \in \mathcal{B}(R)$;

(c) $D \otimes_R^{\mathbf{L}} X \in \mathcal{I}(R) \Longrightarrow X \in \mathcal{F}(R)$; *and*

(d) $\mathbf{R}\mathrm{Hom}_R(D,Y) \in \mathcal{F}(R) \Longrightarrow Y \in \mathcal{I}(R)$.

Here $\mathcal{D}(R)$ is the derived category, and $\mathbf{R}\mathrm{Hom}_R(D,-)$ and $D \otimes_R^{\mathbf{L}} -$ are derived functors, cf. section A.4.

## 3.4 Cohen–Macaulay Rings

We now return to Sharp's [55] original setting: a Cohen–Macaulay local ring with a dualizing module. We start by describing — in classical homological terms — the modules in the Auslander and Bass classes of such a ring, and then we can state and prove a more satisfactory version of the Foxby equivalence Theorem (3.3.2).

(3.4.1) ***Dualizing Modules.*** Let $R$ be a local ring with a dualizing complex $D$. If $R$ is Cohen–Macaulay, then $\operatorname{amp} D = 0$ by Theorem (3.3.8), so we can assume (after a shift) that the complex $D$ has homology concentrated in degree zero and identify it with the module $\mathrm{H}_0(D)$, cf. (A.1.15). For obvious reasons $D$ is then called a *dualizing module.*

In the literature, in [12] for example, a dualizing module is also called a canonical module. Recalling from the appendix, or [12, Section 3.3], a dualizing module $D$ is a finite $R$–module of finite injective dimension, $\operatorname{Ext}^m_R(D,D) = 0$ for $m > 0$, and the canonical map

$$\chi^R_D : R \longrightarrow \operatorname{Hom}_R(D,D),$$

given by $\chi^R_D(r)(d) = rd$, is an isomorphism, i.e., $R \cong \operatorname{Hom}_R(D,D)$.

Note that if $R$ is a Cohen–Macaulay local ring with a dualizing module $D$, then

(3.4.1.1) $$\operatorname{id}_R D = \operatorname{depth} R = \dim R.$$

by the Bass formula.

(3.4.2) **Setup.** In this section $R$ is a **Cohen–Macaulay local ring with a dualizing module** $D$.

(3.4.3) **Lemma.** *The following hold for* $U \in \mathcal{C}_{(\square)}(R)$, $X \in \mathcal{A}(R)$, *and* $Y \in \mathcal{B}(R)$:

(a) $\inf (D \otimes^{\mathbf{L}}_R U) = \inf U$;

(b) $\sup (\mathbf{R}\mathrm{Hom}_R(D,U)) = \sup U$;

(c) $\sup (D \otimes^{\mathbf{L}}_R X) = \sup X$; *and*

(d) $\inf (\mathbf{R}\mathrm{Hom}_R(D,Y)) = \inf Y$.

*Proof.* Since $\sup D = \inf D = 0$ the equalities in (a) and (b) follow by the inequalities in Lemma (3.3.6); and the equalities in (c) and (d) follow by (a) and (c) in Proposition (3.3.7). □

(3.4.4) ***Canonical maps.*** Let $M$ and $N$ be $R$–modules, then the canonical maps from (3.1.2.1) and (3.2.2.1),

$$\gamma^D_M : M \to \operatorname{Hom}_R(D, D \otimes_R M) \quad \text{and} \quad \xi^D_N : D \otimes_R \operatorname{Hom}_R(D,N) \to N,$$

are natural homomorphisms of $R$-modules defined by:

$$\gamma_M^D(m)(d) = d \otimes m \quad \text{and} \quad \xi_N^D(d \otimes \psi) = \psi(d)$$

for $m \in M$, $d \in D$, and $\psi \in \mathrm{Hom}_R(D,N)$.

(3.4.5) **Proposition.** *Let $M$ be an $R$-module. The following hold:*

(a) *$D \otimes_R^{\mathbf{L}} M$ has homology concentrated in degree zero if and only if $\mathrm{Tor}_m^R(D,M) = 0$ for $m > 0$.*

(b) *If $D \otimes_R^{\mathbf{L}} M \in \mathcal{C}_{(0)}(R)$, then $D \otimes_R^{\mathbf{L}} M$ is represented by the module $D \otimes_R M$, and $\mathbf{R}\mathrm{Hom}_R(D, D \otimes_R^{\mathbf{L}} M)$ belongs to $\mathcal{C}_{(0)}(R)$ if and only if $\mathrm{Ext}_R^m(D, D \otimes_R M)] = 0$ for $m > 0$.*

(c) *If $D \otimes_R^{\mathbf{L}} M$ and $\mathbf{R}\mathrm{Hom}_R(D, D \otimes_R^{\mathbf{L}} M)$ have homology concentrated in degree zero, then the canonical map $\gamma_M^D \colon M \to \mathrm{Hom}_R(D, D \otimes_R M)$ is an isomorphism if and only if $M$ represents $\mathbf{R}\mathrm{Hom}_R(D, D \otimes_R^{\mathbf{L}} M)$ canonically.*

*Proof.* (a) is immediate by (A.4.12) and (A.4.15.1).

Take resolutions $\varphi \colon F \xrightarrow{\simeq} M$ and $\pi \colon P \xrightarrow{\simeq} D$, where $F \in \mathcal{C}_{\sqsupset}^{\mathrm{F}}(R)$ and $P \in \mathcal{C}_{\sqsupset}^{\mathrm{P}}(R)$ have $F_\ell = 0$ and $P_\ell = 0$ for $\ell < 0$.

(b): The complex $D \otimes_R F$ represents $D \otimes_R^{\mathbf{L}} M$, and the induced homomorphism

$$\mathrm{H}_0(D \otimes_R \varphi) : \mathrm{H}_0(D \otimes_R F) \longrightarrow D \otimes_R M$$

is invertible, cf. (A.4.15). Therefore, if $D \otimes_R^{\mathbf{L}} M \in \mathcal{C}_{(0)}(R)$ then $D \otimes_R \varphi$ is a quasi-isomorphism, in particular, $D \otimes_R M$ represents $D \otimes_R^{\mathbf{L}} M$. By Lemma (3.4.3)(b) it now follows that $\mathbf{R}\mathrm{Hom}_R(D, D \otimes_R^{\mathbf{L}} M)$ has homology concentrated in non-positive degrees, and for $m \in \mathbb{N}_0$ we have

$$\begin{aligned} \mathrm{H}_{-m}(\mathbf{R}\mathrm{Hom}_R(D, D \otimes_R^{\mathbf{L}} M)) &= \mathrm{H}_{-m}(\mathbf{R}\mathrm{Hom}_R(D, D \otimes_R M)) \\ &= \mathrm{Ext}_R^m(D, D \otimes_R M). \end{aligned}$$

So, indeed, $\mathbf{R}\mathrm{Hom}_R(D, D \otimes_R^{\mathbf{L}} M)$ has homology concentrated in degree zero if and only if the modules $\mathrm{Ext}_R^m(D, D \otimes_R M)$ vanish for $m > 0$.

(c): From what we have already proved it follows that $\mathbf{R}\mathrm{Hom}_R(D, D \otimes_R^{\mathbf{L}} M)$ is represented by $\mathrm{Hom}_R(P, D \otimes_R M)$, and since $\mathbf{R}\mathrm{Hom}_R(D, D \otimes_R^{\mathbf{L}} M)$ has homology concentrated in degree zero, the induced morphism

$$\mathrm{Hom}_R(\pi, D \otimes_R M) : \mathrm{Hom}_R(D, D \otimes_R M) \longrightarrow \mathrm{Hom}_R(P, D \otimes_R M)$$

is a quasi-isomorphism. We have now established a commutative diagram

$$
\begin{array}{ccc}
F & \xrightarrow{\ \gamma_F^P\ } & \mathrm{Hom}_R(P, P \otimes_R F) \\
\simeq \big\downarrow \varphi & & \simeq \big\downarrow \mathrm{Hom}_R(P, \pi \otimes_R F) \\
M & & \mathrm{Hom}_R(P, D \otimes_R F) \\
\big\downarrow \gamma_M^D & & \simeq \big\downarrow \mathrm{Hom}_R(P, D \otimes_R \varphi) \\
\mathrm{Hom}_R(D, D \otimes_R M) & \xrightarrow[\simeq]{\ \mathrm{Hom}_R(\pi, D \otimes_R M)\ } & \mathrm{Hom}_R(P, D \otimes_R M)
\end{array}
$$

from which it is evident that $\gamma_M^D$ is a quasi-isomorphism if and only if $\gamma_F^P$ is so. That is, $\gamma_M^D$ is an isomorphism of modules if and only if $M$ represents $\mathbf{R}\mathrm{Hom}_R(D, D \otimes^{\mathbf{L}}_R M)$ canonically, cf. Lemma (3.1.4). □

(3.4.6) **Theorem.** *An $R$–module $M$ belongs to $\mathcal{A}_0(R)$ if and only if it satisfies the following three conditions:*

(1) $\mathrm{Tor}^R_m(D, M) = 0$ *for* $m > 0$*;*
(2) $\mathrm{Ext}^m_R(D, D \otimes_R M) = 0$ *for* $m > 0$*; and*
(3) *the canonical map* $\gamma_M^D : M \to \mathrm{Hom}_R(D, D \otimes_R M)$ *is an isomorphism.*

*In particular: if $M \in \mathcal{A}_0(R)$ then the module $D \otimes_R M$ represents $D \otimes^{\mathbf{L}}_R M$.*

*Proof.* "If": Using the Proposition, we see that it follows from (1) that $D \otimes^{\mathbf{L}}_R M$ has homology concentrated in degree zero, then from (2) that also $\mathbf{R}\mathrm{Hom}_R(D, D \otimes^{\mathbf{L}}_R M) \in \mathcal{C}_{(0)}(R)$, and finally from (3) that $M$ represents $\mathbf{R}\mathrm{Hom}_R(D, D \otimes^{\mathbf{L}}_R M)$ canonically, so $M \in \mathcal{A}_0(R)$.

"Only if": Let $M \in \mathcal{A}_0(R)$, then $D \otimes^{\mathbf{L}}_R M$ has homology concentrated in degree zero by Lemma (3.4.3), so $\mathrm{Tor}^R_m(D, M) = 0$ for $m > 0$, cf. (a) in the Proposition. Furthermore, $M$ represents $\mathbf{R}\mathrm{Hom}_R(D, D \otimes^{\mathbf{L}}_R M)$ canonically, in particular, $\mathbf{R}\mathrm{Hom}_R(D, D \otimes^{\mathbf{L}}_R M) \in \mathcal{C}_{(0)}(R)$, so it follows by (b) and (c) in the Proposition that $\mathrm{Ext}^m_R(D, D \otimes_R M) = 0$ for $m > 0$ and that $\gamma_M^D$ is an isomorphism.

The last assertion is now immediate by (b) in the Proposition. □

(3.4.7) **Corollary.** *Let $0 \to M' \to M \to M'' \to 0$ be a short exact sequence of $R$–modules. The following hold:*

(a) *If two of the modules belong to $\mathcal{A}_0(R)$, then so does the third.*
(b) *If the sequence splits, then $M \in \mathcal{A}_0(R)$ if and only if both $M'$ and $M''$ belong to $\mathcal{A}_0(R)$.*

*Proof.* Part (a) is a special case of Lemma (3.1.13).

If the sequence $0 \to M' \to M \to M'' \to 0$ splits, then so do the sequences

$$0 \to D \otimes_R M' \to D \otimes_R M \to D \otimes_R M'' \to 0$$

and

$$0 \to \mathrm{Hom}_R(D, D \otimes_R M') \to \mathrm{Hom}_R(D, D \otimes_R M) \to \mathrm{Hom}_R(D, D \otimes_R M'') \to 0.$$

Furthermore, there are isomorphisms

$$\operatorname{Tor}_m^R(D,M) \cong \operatorname{Tor}_m^R(D,M') \oplus \operatorname{Tor}_m^R(D,M'')$$

and

$$\operatorname{Ext}_R^m(D,D\otimes_R M) \cong \operatorname{Ext}_R^m(D,D\otimes_R M') \oplus \operatorname{Ext}_R^m(D,D\otimes_R M'')$$

for $m > 0$. It is immediate from these isomorphisms that $M$ satisfies conditions (1) and (2) in the Theorem if and only if both $M'$ and $M''$ do so. Consider the diagram

$$\begin{array}{ccccccccc} 0 \to & M' & \to & M & \to & M'' & \to 0 \\ & \downarrow \gamma^D_{M'} & & \downarrow \gamma^D_M & & \downarrow \gamma^D_{M''} & \\ 0 \to & \operatorname{Hom}_R(D,D\otimes_R M') & \to & \operatorname{Hom}_R(D,D\otimes_R M) & \to & \operatorname{Hom}_R(D,D\otimes_R M'') & \to 0 \end{array}$$

The canonical maps $\gamma$ are natural, so the diagram is commutative. Furthermore, the rows split, so it follows that $\gamma^D_M$ is an isomorphism if and only if both $\gamma^D_{M'}$ and $\gamma^D_{M''}$ are so. Part (b) now follows by the Theorem. □

The next three results are parallel to (3.4.5), (3.4.6), and (3.4.7).

(3.4.8) **Proposition.** *Let $N$ be an $R$–module. The following hold:*

(a) *$\mathbf{R}\operatorname{Hom}_R(D,N)$ has homology concentrated in degree zero if and only if $\operatorname{Ext}_R^m(D,N) = 0$ for $m > 0$.*

(b) *If $\mathbf{R}\operatorname{Hom}_R(D,N) \in \mathcal{C}_{(0)}(R)$, then $\mathbf{R}\operatorname{Hom}_R(D,N)$ is represented by the module $\operatorname{Hom}_R(D,N)$, and $D\otimes^{\mathbf{L}}_R \mathbf{R}\operatorname{Hom}_R(D,N)$ belongs to $\mathcal{C}_{(0)}(R)$ if and only if $\operatorname{Tor}_m^R(D,\operatorname{Hom}_R(D,N)) = 0$ for $m > 0$.*

(c) *If $\mathbf{R}\operatorname{Hom}_R(D,N)$ and $D\otimes^{\mathbf{L}}_R \mathbf{R}\operatorname{Hom}_R(D,N)$ have homology concentrated in degree zero, then the canonical map $\xi^D_N\colon D\otimes_R \operatorname{Hom}_R(D,N) \to N$ is an isomorphism if and only if $N$ represents $D\otimes^{\mathbf{L}}_R \mathbf{R}\operatorname{Hom}_R(D,N)$ canonically.*

*Proof.* Similar to the proof of Proposition (3.4.5). □

(3.4.9) **Theorem.** *An $R$–module $N$ belongs to $\mathcal{B}_0(R)$ if and only if it satisfies the following three conditions:*

(1) *$\operatorname{Ext}_R^m(D,N) = 0$ for $m > 0$;*

(2) *$\operatorname{Tor}_m^R(D,\operatorname{Hom}_R(D,N)) = 0$ for $m > 0$;  and*

(3) *the canonical map $\xi^D_N\colon D\otimes_R \operatorname{Hom}_R(D,N) \to N$ is an isomorphism.*

*In particular: if $N \in \mathcal{B}_0(R)$ then $\operatorname{Hom}_R(D,N)$ represents $\mathbf{R}\operatorname{Hom}_R(D,N)$.*

*Proof.* Similar to the proof of Theorem (3.4.6). □

(3.4.10) **Corollary.** *Let $0 \to N' \to N \to N'' \to 0$ be a short exact sequence of $R$–modules. The following hold:*

(a) *If two of the modules belong to $\mathcal{B}_0(R)$, then so does the third.*
(b) *If the sequence splits, then $N \in \mathcal{B}_0(R)$ if and only if both $N'$ and $N''$ belong to $\mathcal{B}_0(R)$.*

*Proof.* Similar to the proof of Corollary (3.4.7). □

We can now express Foxby equivalence in terms of usual module functors.

(3.4.11) **Theorem (Foxby Equivalence for Modules over CM Rings).** *Let $R$ be a Cohen–Macaulay local ring. If $D$ is a dualizing module for $R$, then there is a commutative diagram of categories of $R$–modules:*

$$\begin{array}{ccc} \mathcal{C}_0(R) & \underset{\mathrm{Hom}_R(D,-)}{\overset{D\otimes_R -}{\rightleftarrows}} & \mathcal{C}_0(R) \\ \cup\!| & & \cup\!| \\ \mathcal{A}_0(R) & \rightleftarrows & \mathcal{B}_0(R) \\ \cup\!| & & \cup\!| \\ \mathcal{F}_0(R) & \rightleftarrows & \mathcal{I}_0(R) \end{array}$$

*where the vertical inclusions are full embeddings, and the unlabeled horizontal arrows are quasi-inverse equivalences of categories.*

*Furthermore, the following hold for $R$–modules $M$ and $N$:*

(a) $D \otimes_R M \in \mathcal{B}_0(R) \implies M \in \mathcal{A}_0(R)$;
(b) $\mathrm{Hom}_R(D, N) \in \mathcal{A}_0(R) \implies N \in \mathcal{B}_0(R)$;
(c) $D \otimes_R M \in \mathcal{I}_0(R) \implies M \in \mathcal{F}_0(R)$; *and*
(d) $\mathrm{Hom}_R(D, N) \in \mathcal{F}_0(R) \implies N \in \mathcal{I}_0(R)$.

*Also the restrictions of the functors $D \otimes_R -$ and $\mathrm{Hom}_R(D, -)$ to the full subcategory of finite $R$–modules give quasi-inverse equivalences. That is, there is a commutative diagram of categories of $R$–modules:*

$$\begin{array}{ccc} \mathcal{C}_0^{\mathrm{f}}(R) & \underset{\mathrm{Hom}_R(D,-)}{\overset{D\otimes_R -}{\rightleftarrows}} & \mathcal{C}_0^{\mathrm{f}}(R) \\ \cup\!| & & \cup\!| \\ \mathcal{A}_0^{\mathrm{f}}(R) & \rightleftarrows & \mathcal{B}_0^{\mathrm{f}}(R) \\ \cup\!| & & \cup\!| \\ \mathcal{F}_0^{\mathrm{f}}(R) & \rightleftarrows & \mathcal{I}_0^{\mathrm{f}}(R) \end{array}$$

*where, as above, the vertical inclusions are full embeddings, and the unlabeled horizontal arrows are quasi-inverse equivalences of categories.*

*Proof.* The full embeddings were established in Propositions (3.1.8) and (3.2.8). In view of the characterization of $\mathcal{A}_0(R)$ and $\mathcal{B}_0(R)$ given in Theorems (3.4.6) and (3.4.9) all the remaining assertions are immediate from the Foxby equivalence Theorem (3.3.2). For example: if $M \in \mathcal{A}_0(R)$ then $D \otimes_R M \in \mathcal{B}_0(R)$ because $D \otimes_R M$ represents $D \otimes^{\mathbf{L}}_R M$ by (3.4.6), and $D \otimes^{\mathbf{L}}_R M \in \mathcal{B}(R)$ by (3.3.2). Now the module $\mathrm{Hom}_R(D, D \otimes_R M)$ represents $\mathbf{R}\mathrm{Hom}_R(D, D \otimes_R M)$ by (3.4.9), so it belongs to $\mathcal{A}_0(R)$ by (3.3.2) and is canonically isomorphic to $M$ by (3.4.6). □

(3.4.12) **Gorenstein Theorem, Special Modules Version.** *Let $R$ be a Cohen–Macaulay local ring. If $D$ is a dualizing module for $R$, then the following are equivalent:*

(*i*) *$R$ is Gorenstein.*
(*ii*) *$R \in \mathcal{B}_0(R)$.*
(*ii'*) *$D \in \mathcal{A}_0(R)$.*
(*iii*) *$\mathrm{depth}_R N < \infty$ for some $N \in \mathcal{B}_0(R) \cap \mathcal{F}_0(R)$.*
(*iii'*) *$\mathrm{depth}_R M < \infty$ for some $M \in \mathcal{A}_0(R) \cap \mathcal{I}_0(R)$.*
(*iv*) *$\mathcal{A}_0(R) = \mathcal{B}_0(R)$.*

*Proof.* Immediate from the special complexes version (3.3.5). □

The last results in this section will be needed in the chapters to come; the nature of their proofs suggests that they should be placed here.

(3.4.13) **Lemma.** *Let $X \in \mathcal{A}(R)$ and $Y \in \mathcal{B}(R)$. The following hold:*

(a) *If $M \in \mathcal{F}_0(R)$, then*
$$-\inf(\mathbf{R}\mathrm{Hom}_R(X, M)) \le \sup X + \dim R.$$
(b) *If $N \in \mathcal{I}_0(R)$, then*
$$\sup(N \otimes^{\mathbf{L}}_R X) \le \sup X + \dim R.$$
(c) *If $N \in \mathcal{I}_0(R)$, then*
$$-\inf(\mathbf{R}\mathrm{Hom}_R(N, Y)) \le -\inf Y + \dim R.$$

*Proof.* (a): If $M \in \mathcal{F}_0(R)$, then $M$ belongs to the Auslander class, so by Lemma (3.3.3)(a) and (A.5.2.1) we have
$$\begin{aligned}-\inf(\mathbf{R}\mathrm{Hom}_R(X, M)) &= -\inf(\mathbf{R}\mathrm{Hom}_R(D \otimes^{\mathbf{L}}_R X, D \otimes^{\mathbf{L}}_R M)) \\ &\le \sup(D \otimes^{\mathbf{L}}_R X) + \mathrm{id}_R(D \otimes^{\mathbf{L}}_R M).\end{aligned}$$
By Theorem (3.4.6) $D \otimes^{\mathbf{L}}_R M$ is represented by $D \otimes_R M$, and $D \otimes_R M \in \mathcal{I}_0(R)$ by Foxby equivalence (3.4.11), so $\mathrm{id}_R(D \otimes^{\mathbf{L}}_R M) \le \dim R$. Furthermore, we have $\sup(D \otimes^{\mathbf{L}}_R X) = \sup X$ by Lemma (3.4.3)(c), so
$$-\inf(\mathbf{R}\mathrm{Hom}_R(X, M)) \le \sup X + \dim R$$

as wanted.

(b): As above it follows by (3.3.3)(c), (A.5.6.1), (3.4.3)(c), (3.4.9), and (3.4.11) that

$$\begin{aligned}\sup(N \otimes^{\mathbf{L}}_R X) &= \sup((D \otimes^{\mathbf{L}}_R X) \otimes^{\mathbf{L}}_R \mathbf{R}\mathrm{Hom}_R(D,N))\\ &\leq \sup(D \otimes^{\mathbf{L}}_R X) + \mathrm{fd}_R(\mathbf{R}\mathrm{Hom}_R(D,N))\\ &= \sup X + \mathrm{fd}_R(\mathrm{Hom}_R(D,N))\\ &\leq \sup X + \dim R.\end{aligned}$$

(c): As above it follows by (3.3.3)(b), (A.5.4.1), (3.4.3)(d), and (3.4.9) that

$$\begin{aligned}-\inf(\mathbf{R}\mathrm{Hom}_R(N,Y)) &= -\inf(\mathbf{R}\mathrm{Hom}_R(\mathbf{R}\mathrm{Hom}_R(D,N),\mathbf{R}\mathrm{Hom}_R(D,Y)))\\ &\leq -\inf(\mathbf{R}\mathrm{Hom}_R(D,Y)) + \mathrm{pd}_R(\mathbf{R}\mathrm{Hom}_R(D,N))\\ &= -\inf Y + \mathrm{pd}_R(\mathrm{Hom}_R(D,N)).\end{aligned}$$

It follows by Foxby equivalence Theorem (3.4.11) that $\mathrm{Hom}_R(D,N) \in \mathcal{F}_0(R)$, so by Theorem (3.4.14) we have $\mathrm{pd}_R(\mathrm{Hom}_R(D,N)) \leq \dim R$ and, hence, the desired inequality holds. (The proof of Theorem (3.4.14) actually uses this lemma, but only part (a).) □

(3.4.14) **Theorem.** *If $X \in \mathcal{F}(R)$, then*

$$\mathrm{pd}_R X \leq \sup X + \dim R.$$

*In particular, there is an equality of full subcategories:*

$$\mathcal{P}(R) = \mathcal{F}(R).$$

*Proof.* Let $X \in \mathcal{F}(R)$, set $n = \sup X + \dim R$, and take a projective resolution $X \xleftarrow{\simeq} P \in \mathcal{C}^{\mathrm{P}}_{\sqsupset}(R)$. We want to prove that the cokernel $\mathrm{C}^P_n$ is projective, then $\mathrm{pd}_R X \leq n$ by (A.3.9). Since $n+1 > \sup X$, we have a short exact sequence

$$(\dagger) \qquad 0 \to \mathrm{C}^P_{n+1} \xrightarrow{\partial} P_n \to \mathrm{C}^P_n \to 0,$$

cf. (A.1.7.2), and it is sufficient to prove that (†) splits. By assumption $\mathrm{fd}_R X < \infty$, and for $\ell \geq \mathrm{fd}_R X$ the cokernel $\mathrm{C}^P_\ell$ is flat, cf. (A.5.5). Thus, for a suitable $\ell > n+1$ we have an exact sequence,

$$0 \to \mathrm{C}^P_\ell \to P_{\ell-1} \to \cdots \to P_{n+1} \to \mathrm{C}^P_{n+1} \to 0,$$

showing that $\mathrm{C}^P_{n+1}$ is a module of finite flat dimension. By Lemma (3.4.13)(a) we now have $-\inf(\mathbf{R}\mathrm{Hom}_R(X,\mathrm{C}^P_{n+1})) \leq n$, so it follows by (A.5.9) that

$$\mathrm{Ext}^1_R(\mathrm{C}^P_n,\mathrm{C}^P_{n+1}) = \mathrm{H}_{-(n+1)}(\mathbf{R}\mathrm{Hom}_R(X,\mathrm{C}^P_{n+1})) = 0;$$

and from (†) we, therefore, get an exact sequence

$$\mathrm{Hom}_R(P_n,\mathrm{C}^P_{n+1}) \xrightarrow{\mathrm{Hom}_R(\partial,\mathrm{C}^P_{n+1})} \mathrm{Hom}_R(\mathrm{C}^P_{n+1},\mathrm{C}^P_{n+1}) \to 0.$$

Thus, there is a homomorphism $\sigma\colon P_n \to \mathrm{C}^P_{n+1}$ such that $\sigma\partial = 1_{C^P_n}$, i.e., (†) splits.

The equality of full subcategories is now immediate as finite projective dimension certainly implies finite flat dimension, cf. (A.3.10). □

## Notes

Theorem (3.4.11) appeared as [8, Corollary (3.6)]; the equivalence of the full subcategories $\mathcal{F}^{\mathrm{f}}_0(R)$ and $\mathcal{I}^{\mathrm{f}}_0(R)$ was first established by Sharp [55, Theorem 2.9].

Theorem (3.4.12) is due to Foxby; it first appeared in [39], and so did Corollaries (3.4.7) and (3.4.10).

The elegant proof of Theorem (3.4.14) is due to Foxby and, actually, it works for any local ring with a dualizing complex; it just takes a few extra computations, see [33, Chapter 21]. The origin of the Theorem is a result by Jensen [45, Proposition 6]: any module of finite flat dimension over a Noetherian ring of finite Krull dimension is also of finite projective dimension.

# Chapter 4

# G–projectivity

The central notion in this chapter is that of 'Gorenstein projective modules'; it was introduced by Enochs and Jenda in [25][1]. We first present a different view (from that taken in chapter 1) on the G–class, then we move on to define Gorenstein projective modules and prove that the finite ones among them are exactly the modules in the G–class. In the last two sections we focus on Cohen–Macaulay local rings with dualizing modules. Over such rings the Gorenstein projective modules can be identified as special modules in the Auslander class; this view — also due to Enochs et al. — proves to be very fruitful, and a neat theory for Gorenstein projective dimension becomes available.

## 4.1 The G–class Revisited

From Auslander's original definition — see (1.1.2) — it is not obvious how to define "non-finite modules in the G–class", let alone how to dualize the notion. In this section we show how to characterize modules in the G–class in terms of complete resolutions by finite free modules, and this will be the starting point for our future generalizations and dualizations.

Let $M$ be some $R$-module, and let

$$0 \to K_n \to P_{n-1} \to P_{n-2} \to \cdots \to P_1 \to P_0 \to C \to 0$$

be an exact sequence where the $P_\ell$-s are projective modules. It is easy to prove that $\mathrm{Ext}_R^m(K_n, M) = \mathrm{Ext}_R^{m+n}(C, M)$ for $m > 0$; it is done by breaking the long exact sequence into short ones, and using the fact that $\mathrm{Ext}_R^m(P_\ell, M) = 0$ for all $m > 0$ because the $P_\ell$-s are projective.

[1] Strictly speaking, only finite Gorenstein projective modules were defined in this paper. But in [32], and other later papers, the same authors have tacitly understood the definition to encompass also non-finite modules; of course, we do the same.

For the application of this classical technique — sometimes called "dimension shift" — it is, of course, not vital that the $P_\ell$-s are projective, only that the modules $\mathrm{Ext}_R^m(P_\ell, M)$ vanish for the module $M$ in question. We already used the technique in Observation (1.2.5), and we will resort to it frequently in this and the next two chapters. To avoid redundancy we will apply the technique, once and for all, in a very general setting. This gives us a lemma — (4.1.1) and two parallels: (4.1.6) and (4.1.7) — to which we can then refer. To justify this approach, let us point out that it is tempting in Lemma (4.1.1)(a) to think of $X$ as a complex of injective modules; in most applications, however, $X$ will be a complex of projective modules!

(4.1.1) **Lemma.** *Let $X$ be an $R$–complex and let $M$ be an $R$–module. The following hold:*

(a) *If $\mathrm{Ext}_R^m(M, X_\ell) = 0$ for all $m > 0$ and $\ell \geq \sup X$, then*

$$\mathrm{Ext}_R^m(M, \mathrm{C}_\ell^X) = \mathrm{Ext}_R^{m+n}(M, \mathrm{C}_{\ell+n}^X)$$

*for all $m, n > 0$ and $\ell \geq \sup X$.*

(b) *If $X$ is homologically trivial, then*

$$\mathrm{Z}_{-\ell}^{\mathrm{Hom}_R(X,M)} \cong \mathrm{Hom}_R(\mathrm{C}_\ell^X, M)$$

*for all $\ell \in \mathbb{Z}$.*

(c) *If $X$ is homologically trivial, and $\mathrm{Ext}_R^m(X_\ell, M) = 0$ for all $m > 0$ and $\ell \in \mathbb{Z}$, then*

$$\mathrm{Ext}_R^m(\mathrm{C}_\ell^X, M) = \mathrm{Ext}_R^{m+n}(\mathrm{C}_{\ell-n}^X, M)$$

*for all $m, n > 0$ and $\ell \in \mathbb{Z}$. Furthermore, the following are equivalent:*

(*i*) *$\mathrm{Hom}_R(X, M)$ is homologically trivial.*

(*ii*) *$\mathrm{Ext}_R^1(\mathrm{C}_\ell^X, M) = 0$ for all $\ell \in \mathbb{Z}$.*

(*iii*) *$\mathrm{Ext}_R^m(\mathrm{C}_\ell^X, M) = 0$ for all $m > 0$ and $\ell \in \mathbb{Z}$.*

*Proof.* (a): For each $\ell \geq \sup X$ we have a short exact sequence

$$(\dagger) \qquad 0 \to \mathrm{C}_{\ell+1}^X \to X_\ell \to \mathrm{C}_\ell^X \to 0,$$

cf. (A.1.7.2). Since $\mathrm{Ext}_R^m(M, X_\ell) = 0$ for $m > 0$, the associated long exact sequence,

$$\cdots \to \mathrm{Ext}_R^m(M, X_\ell) \to \mathrm{Ext}_R^m(M, \mathrm{C}_\ell^X) \to \mathrm{Ext}_R^{m+1}(M, \mathrm{C}_{\ell+1}^X) \to \mathrm{Ext}_R^{m+1}(M, X_\ell) \to \cdots,$$

yields identities

$$\mathrm{Ext}_R^m(M, \mathrm{C}_\ell^X) = \mathrm{Ext}_R^{m+1}(M, \mathrm{C}_{\ell+1}^X)$$

for $m > 0$. Piecing these together we get the desired identity.

(b): Applying the left-exact functor $\mathrm{Hom}_R(-, M)$ to the right-exact sequence $X_{\ell+1} \to X_\ell \to \mathrm{C}_\ell^X \to 0$, we get a left-exact sequence:

$$0 \to \mathrm{Hom}_R(\mathrm{C}_\ell^X, M) \to \mathrm{Hom}_R(X_\ell, M) \to \mathrm{Hom}_R(X_{\ell+1}, M).$$

Evidently, the kernel $\mathrm{Z}_{-\ell}^{\mathrm{Hom}_R(X,M)}$, i.e., $\mathrm{Ker}(\mathrm{Hom}_R(X_\ell, M) \to \mathrm{Hom}_R(X_{\ell+1}, M))$ is isomorphic to $\mathrm{Hom}_R(\mathrm{C}_\ell^X, M)$ as wanted.

(c): The equivalence of (*i*) and (*ii*) is the fact that the complex $\mathrm{Hom}_R(X, M)$ is homologically trivial if and only if the functor $\mathrm{Hom}_R(-, M)$ leaves all the short exact sequences $0 \to \mathrm{Z}_\ell^X \to X_\ell \to \mathrm{C}_\ell^X \to 0$ exact. As in (a) the identity of Ext modules follows from (†); and now that we have $\mathrm{Ext}_R^m(\mathrm{C}_\ell^X, M) = \mathrm{Ext}_R^1(\mathrm{C}_{\ell+m-1}^X, M)$ for all $m > 0$ and $\ell \in \mathbb{Z}$, we see that also (*ii*) and (*iii*) are equivalent. □

(4.1.2) **Definition.** Let $L \in \mathcal{C}^{\mathrm{L}}(R)$ be homologically trivial. We say that $L$ is a *complete resolution by finite free modules* if and only if the dual complex $L^* = \mathrm{Hom}_R(L, R)$ is homologically trivial.

We can now apply the "general dimension shift lemma", (4.1.1), to show that modules in the G–class and complete resolutions by finite free modules are close kin.

(4.1.3) **Proposition.** *Let $L \in \mathcal{C}^{\mathrm{L}}(R)$ be homologically trivial. The following are equivalent:*

(*i*) *$L$ is a complete resolution by finite free modules.*

(*ii*) *All the cokernels $\mathrm{C}_\ell^L$, $\ell \in \mathbb{Z}$, belong to $\mathrm{G}(R)$.*

(*iii*) *$\mathrm{Hom}_R(L, T)$ is homologically trivial for every module $T \in \mathcal{F}_0(R)$.*

*Proof.* It is clear by the Definition that (*iii*) is stronger than (*i*).

(*i*) ⇒ (*ii*): Fix an $n \in \mathbb{Z}$ and set $C = \mathrm{C}_n^L$; we want to prove that $C \in \mathrm{G}(R)$. Since both $L$ and $L^*$ are homologically trivial, we have $\mathrm{Ext}_R^m(C, R) = 0$ for $m > 0$ by Lemma (4.1.1)(c), and by (A.1.7.3) and (b) in the same Lemma it follows that the dualized complex $L^*$ has

$$(\dagger) \qquad \mathrm{C}_{-n+1}^{L^*} \cong \mathrm{Z}_{-n}^{L^*} \cong (\mathrm{C}_n^L)^* = C^*.$$

Dualizing once more yields a complex $L^{**}$ which is isomorphic to $L$; in particular, it is homologically trivial and, as above, it follows that $\mathrm{Ext}_R^m(C^*, R) = 0$ for $m > 0$. The isomorphism between $L$ and $L^{**}$ is the canonical one, $\delta_L^R$, which in degree $\ell$ is just the biduality map $\delta_{L_\ell} : L_\ell \to L_\ell^{**}$. By (A.1.7.3), Lemma (4.1.1)(b), and (†) the complex $L^{**}$ has

$$\mathrm{C}_n^{L^{**}} \cong \mathrm{Z}_{n-1}^{L^{**}} \cong (\mathrm{C}_{-n+1}^{L^*})^* \cong C^{**},$$

so we have an exact ladder

$$\begin{array}{ccccccccc} \cdots & \longrightarrow & L_{n+1} & \longrightarrow & L_n & \longrightarrow & C & \longrightarrow & 0 \\ & & \cong\big\downarrow \delta_{L_{n+1}} & & \cong\big\downarrow \delta_{L_n} & & \big\downarrow \delta_C & & \\ \cdots & \longrightarrow & {L_{n+1}}^{**} & \longrightarrow & {L_n}^{**} & \longrightarrow & C^{**} & \longrightarrow & 0 \end{array}$$

and it follows by the five lemma that the biduality map $\delta_C$ is an isomorphism.

(*ii*) $\Rightarrow$ (*iii*): To prove that $\mathrm{Hom}_R(L,T)$ is homologically trivial, it is by Lemma (4.1.1)(c) sufficient to see that $\mathrm{Ext}^1_R(\mathrm{C}^L_\ell, T) = 0$ for all $\ell \in \mathbb{Z}$. For $T$ in $\mathcal{F}_0(R)$ this is immediate by Corollary (2.4.2) as $\text{G-dim}_R\,\mathrm{C}^L_\ell \leq 0$. □

The main result of this section describes modules in the G–class as infinite syzygies of finite free modules.

(4.1.4) **Theorem.** *A finite $R$–module $M$ belongs to* $\mathrm{G}(R)$ *if and only if there exists a complete resolution by finite free modules $L$ with* $\mathrm{C}^L_0 \cong M$.

*Proof.* The "if" part follows by the Proposition.

To prove "only if" we assume that $M \in \mathrm{G}(R)$, and set out to construct a complete resolution by finite free modules $L \in \mathcal{C}^{\mathrm{L}}(R)$ with $\mathrm{C}^L_0 \cong M$: When $M$ belongs to the G–class, then so does the dual module $M^*$, cf. Observation (1.1.7). Take a resolution $L'$ of $M^*$ by finite free modules, then we have an exact sequence

$$(\dagger) \qquad \cdots \xrightarrow{\partial^{L'}_{\ell+1}} L'_\ell \xrightarrow{\partial^{L'}_\ell} L'_{\ell-1} \xrightarrow{\partial^{L'}_{\ell-1}} \cdots \xrightarrow{\partial^{L'}_1} L'_0 \xrightarrow{\lambda'} M^* \to 0.$$

Also the dualized sequence,

$$0 \to M^{**} \xrightarrow{\lambda'^*} {L'_0}^* \xrightarrow{(\partial^{L'}_1)^*} \cdots \xrightarrow{(\partial^{L'}_{\ell-1})^*} {L'_{\ell-1}}^* \xrightarrow{(\partial^{L'}_\ell)^*} {L'_\ell}^* \xrightarrow{(\partial^{L'}_{\ell+1})^*} \cdots,$$

is exact, because its homology modules are $\mathrm{Ext}^m_R(M^*,R)$, cf. (A.4.3). Also take a resolution of $M$ by finite free modules:

$$\cdots \xrightarrow{\partial^{L''}_{\ell+1}} L''_\ell \xrightarrow{\partial^{L''}_\ell} L''_{\ell-1} \xrightarrow{\partial^{L''}_{\ell-1}} \cdots \xrightarrow{\partial^{L''}_1} L''_0 \xrightarrow{\lambda''} M \to 0.$$

Let $L$ be the complex in $\mathcal{C}^{\mathrm{L}}(R)$ obtained by pasting $L''$ and $L'^*$. That is, $L$ has modules

$$L_\ell = L''_\ell \quad \text{for } \ell \geq 0, \quad \text{and}$$
$$L_\ell = (L'^*)_{\ell+1} = (L'_{-(\ell+1)})^* \quad \text{for } \ell < 0;$$

and differentials

$$\partial^L_\ell = \partial^{L''}_\ell \quad \text{for } \ell > 0,$$
$$\partial^L_\ell = \partial^{L'^*}_{\ell+1} = (\partial^{L'}_{-\ell})^* \quad \text{for } \ell < 0, \quad \text{and}$$
$$\partial^L_0 = \lambda'^* \delta_M \lambda''.$$

In degrees 0, −1, and −2 the complex $L$ looks as follows:

$$\cdots \xrightarrow{\partial_1^{L''}} L_0'' \xrightarrow{\lambda'^*\delta_M\lambda''} L_0'^* \xrightarrow{(\partial_1^{L'})^*} L_1'^* \xrightarrow{(\partial_2^{L'})^*} \cdots.$$

To see that $L$ is homologically trivial, we note that

$$\begin{aligned} \mathrm{B}_\ell^L = \mathrm{B}_\ell^{L''} &= \mathrm{Z}_\ell^{L''} = \mathrm{Z}_\ell^L \quad \text{for } \ell > 0; \\ \mathrm{B}_0^L = \mathrm{B}_0^{L''} &= \operatorname{Ker}\lambda'' = \operatorname{Ker}\lambda'^*\delta_M\lambda'' = \mathrm{Z}_0^L; \\ \mathrm{B}_{-1}^L = \operatorname{Im}\lambda'^*\delta_M\lambda'' &= \operatorname{Im}\lambda'^* = \operatorname{Ker}(\partial_1^{L'})^* = \mathrm{Z}_{-1}^L; \quad \text{and} \\ \mathrm{B}_\ell^L = \mathrm{B}_{\ell+1}^{L'^*} &= \mathrm{Z}_{\ell+1}^{L'^*} = \mathrm{Z}_\ell^L \quad \text{for } \ell < -1. \end{aligned}$$

Now that $L \simeq 0$, we have $\mathrm{C}_0^L \cong \mathrm{B}_{-1}^L = \operatorname{Im}\lambda'^* \cong M^{**} \cong M$ as wanted, cf. (A.1.7.3).

It is equally straightforward to see that $L^*$ is homologically trivial. One can, namely, consider it as the splice of the sequences

$$\cdots \xrightarrow{(\partial_{\ell+1}^{L'})^{**}} L_\ell'^{**} \xrightarrow{(\partial_\ell^{L'})^{**}} L_{\ell-1}'^{**} \xrightarrow{(\partial_{\ell-1}^{L'})^{**}} \cdots \xrightarrow{(\partial_1^{L'})^{**}} L_0'^{**} \xrightarrow{\lambda'^{**}} (M^*)^{**} \to 0$$

and

$$0 \to M^* \xrightarrow{\lambda''^*} L_0''^* \xrightarrow{(\partial_1^{L''})^*} \cdots \xrightarrow{(\partial_{\ell-1}^{L''})^*} L_{\ell-1}''^* \xrightarrow{(\partial_\ell^{L''})^*} L_\ell''^* \xrightarrow{(\partial_{\ell+1}^{L''})^*} \cdots.$$

The first one is isomorphic to (†) and, in particular, exact. The second is exact because its homology modules are $\operatorname{Ext}_R^m(M,R)$. This concludes the proof. □

(4.1.5) **Example.** Let $R$ be a local ring, and assume that $x$ and $y$ are elements in the maximal ideal with

$$(\dagger) \qquad \operatorname{Ann}_R(x) = (y) \quad \text{and} \quad \operatorname{Ann}_R(y) = (x).$$

The complex

$$L = \cdots \xrightarrow{x} R \xrightarrow{y} R \xrightarrow{x} R \xrightarrow{y} R \xrightarrow{x} R \xrightarrow{y} \cdots$$

is then homologically trivial, and $\operatorname{Hom}_R(L,R) \cong \Sigma^1 L$, so $L$ is a complete resolution by finite free modules. The modules $R/(x)$ and $R/(y)$ are not projective, but all the cokernels in $L$ have this form, so it follows by the Theorem that $R/(x)$ and $R/(y)$ belong to $\mathrm{G}(R)$.

The immediate concrete example of such a ring is the Gorenstein ring $R = k[\![X,Y]\!]/(XY)$, where $k$ is a field. More generally we can set $R = R'[\![X,Y]\!]/(XY)$, where $R'$ is any local ring; then the residue classes $x$ and $y$ of, respectively, $X$ and $Y$ have the property (†). It follows by [49, Theorem 23.5] and [12, Proposition 3.1.19(b)] that $R$ is Gorenstein if and only if $R'$ is so, and by [12, Theorems 2.1.2 and 2.1.9] the same holds for the Cohen–Macaulay property. In particular we now have examples of non-projective modules in the G–class of non-Gorenstein rings.

The last two lemmas are parallel to Lemma (4.1.1); they will come in handy at a later point.

(4.1.6) **Lemma.** *Let $X$ be an $R$–complex and let $M$ be an $R$–module. The following hold:*

(a) *If* $\mathrm{Ext}_R^m(X_\ell, M) = 0$ *for all* $m > 0$ *and* $\ell \leq \inf X$, *then*
$$\mathrm{Ext}_R^m(\mathrm{Z}_\ell^X, M) = \mathrm{Ext}_R^{m+n}(\mathrm{Z}_{\ell-n}^X, M)$$
*for all* $m, n > 0$ *and* $\ell \leq \inf X$.

(b) *If $X$ is homologically trivial, then*
$$\mathrm{Z}_\ell^{\mathrm{Hom}_R(M,X)} \cong \mathrm{Hom}_R(M, \mathrm{Z}_\ell^X)$$
*for all* $\ell \in \mathbb{Z}$.

(c) *If $X$ is homologically trivial, and* $\mathrm{Ext}_R^m(M, X_\ell) = 0$ *for all* $m > 0$ *and* $\ell \in \mathbb{Z}$, *then*
$$\mathrm{Ext}_R^m(M, \mathrm{Z}_\ell^X) = \mathrm{Ext}_R^{m+n}(M, \mathrm{Z}_{\ell+n}^X)$$
*for all* $m, n > 0$ *and* $\ell \in \mathbb{Z}$. *Furthermore, the following are equivalent:*

(*i*) $\mathrm{Hom}_R(M, X)$ *is homologically trivial.*

(*ii*) $\mathrm{Ext}_R^1(M, \mathrm{Z}_\ell^X) = 0$ *for all* $\ell \in \mathbb{Z}$.

(*iii*) $\mathrm{Ext}_R^m(M, \mathrm{Z}_\ell^X) = 0$ *for all* $m > 0$ *and* $\ell \in \mathbb{Z}$.

*Proof.* Similar to the proof of Lemma (4.1.1). □

(4.1.7) **Lemma.** *Let $X$ be an $R$–complex and let $M$ be an $R$–module. The following hold:*

(a) *If* $\mathrm{Tor}_m^R(X_\ell, M) = 0$ *for all* $m > 0$ *and* $\ell \leq \inf X$, *then*
$$\mathrm{Tor}_m^R(\mathrm{Z}_\ell^X, M) = \mathrm{Tor}_{m+n}^R(\mathrm{Z}_{\ell-n}^X, M)$$
*for all* $m, n > 0$ *and* $\ell \leq \inf X$.

(b) *If $X$ is homologically trivial, then*
$$\mathrm{C}_\ell^{M \otimes_R X} \cong M \otimes_R \mathrm{C}_\ell^X$$
*for all* $\ell \in \mathbb{Z}$.

(c) *If $X$ is homologically trivial, and* $\mathrm{Tor}_m^R(M, X_\ell) = 0$ *for all* $m > 0$ *and* $\ell \in \mathbb{Z}$, *then*
$$\mathrm{Tor}_m^R(M, \mathrm{C}_\ell^X) = \mathrm{Tor}_{m+n}^R(M, \mathrm{C}_{\ell-n}^X)$$
*for all* $m, n > 0$ *and* $\ell \in \mathbb{Z}$. *Furthermore, the following are equivalent:*

(*i*) $M \otimes_R X$ *is homologically trivial.*

(*ii*) $\mathrm{Tor}_1^R(M, \mathrm{C}_\ell^X) = 0$ *for all* $\ell \in \mathbb{Z}$.

(*iii*) $\mathrm{Tor}_m^R(M, \mathrm{C}_\ell^X) = 0$ *for all* $m > 0$ *and* $\ell \in \mathbb{Z}$.

*Proof.* Similar to the proof of Lemma (4.1.1). □

**Notes**

The view we have taken on the G–class in this section is quite different from that taken in chapter 1, but it is still part of Auslander's original work: the hard part of Theorem (4.1.4) is covered by [1, Proposition 8, p. 67].

## 4.2 Gorenstein Projective Modules

We introduce Gorenstein projective modules — a notion that includes the usual projective modules — and we prove that a finite module belongs to the G–class if and only if it is Gorenstein projective.

(4.2.1) **Definitions.** Let $P \in \mathcal{C}^{\mathrm{P}}(R)$ be homologically trivial. We say that $P$ is a *complete projective resolution* if and only if the complex $\mathrm{Hom}_R(P,Q)$ is homologically trivial for every projective $R$–module $Q$.

A module $M$ is said to be *Gorenstein projective* if and only if there exists a complete projective resolution $P$ with $\mathrm{C}_0^P \cong M$.

(4.2.2) **Observation.** Let $P'$ be a projective module, then the complex $P = 0 \to P' \xrightarrow{=} P' \to 0$, concentrated in degrees 0 and $-1$, is a complete projective resolution with $\mathrm{C}_0^P \cong P'$. Thus, every projective module is Gorenstein projective.

By Theorem (4.1.4) and Proposition (4.1.3) it follows that all modules in the G–class are Gorenstein projective. The converse also holds, that is, finite Gorenstein projective modules belong to the G–class. This is the contents of Theorem (4.2.6).

(4.2.3) **Remark.** If $M$ is a Gorenstein projective $R$–module and $\mathfrak{p}$ is a prime ideal in $R$, then it is not obvious from the definition that $M_\mathfrak{p}$ is a Gorenstein projective $R_\mathfrak{p}$–module. It is, however, so (at least) if $R$ is a Cohen–Macaulay local ring with a dualizing module; we prove this in Proposition (4.4.14).

(4.2.4) **Lemma.** *Let $M$ be an $R$–module and assume that $\mathrm{Ext}_R^m(M,Q) = 0$ for all $m > 0$ and all projective modules $Q$. If $T$ is a module of finite flat dimension, then $\mathrm{Ext}_R^m(M,T) = 0$ for $m > 0$.*

*Proof.* If $T \in \mathcal{F}_0(R)$, then $T$ has finite projective dimension; this follows by Jensen's [45, Proposition 6], see also Theorem (3.4.14). Let

$$Q = 0 \to Q_u \to \cdots \to Q_1 \to Q_0 \to 0$$

be a projective resolution of $T$, then $\sup Q = 0$, $\mathrm{C}_0^Q \cong T$, and $\mathrm{C}_u^Q = Q_u$. For $m > 0$ we then have

$$\mathrm{Ext}_R^m(M,T) = \mathrm{Ext}_R^{m+u}(M,Q_u)$$

by Lemma (4.1.1)(a), so $\mathrm{Ext}_R^m(M,T) = 0$ for $m > 0$ as wanted. □

(4.2.5) **Proposition.** *If $P \in \mathcal{C}^{\mathrm{P}}(R)$ is homologically trivial, then the following are equivalent:*

(*i*) *$P$ is a complete projective resolution.*

(*ii*) *All the cokernels $\mathrm{C}^P_\ell$, $\ell \in \mathbb{Z}$, are Gorenstein projective modules.*

(*iii*) $\mathrm{Hom}_R(P,T)$ *is homologically trivial for every module $T \in \mathcal{F}_0(R)$.*

*In particular: if $M$ is Gorenstein projective and $T \in \mathcal{F}_0(R)$, then $\mathrm{Ext}^m_R(M,T) = 0$ for $m > 0$.*

*Proof.* It is clear from the definitions in (4.2.1) that (*i*) $\Rightarrow$ (*ii*), and that (*iii*) is stronger than (*i*). If all the cokernels in $P$ are Gorenstein projective, then, by (4.2.1) and Lemma (4.1.1)(c), we have $\mathrm{Ext}^m_R(\mathrm{C}^P_\ell, Q) = 0$ for all $m > 0$, all $\ell \in \mathbb{Z}$, and all projective modules $Q$. For every $\ell \in \mathbb{Z}$ and $T \in \mathcal{F}_0(R)$ it then follows by Lemma (4.2.4) that $\mathrm{Ext}^m_R(\mathrm{C}^P_\ell, T) = 0$ for $m > 0$. This proves the last assertion, and it follows by Lemma (4.1.1)(c) that $\mathrm{Hom}_R(P,T)$ is homologically trivial, so (*ii*) implies (*iii*). □

The last assertion in (4.2.5) can be interpreted as saying that, as far as modules of finite flat dimension are concerned, Gorenstein projective modules behave as projectives.

(4.2.6) **Theorem.** *A finite $R$–module is Gorenstein projective if and only if it belongs to the G–class. That is,*

$$M \text{ is finite and Gorenstein projective} \iff M \in \mathrm{G}(R).$$

*Proof.* The "if" part is, as observed in (4.2.2), immediate by Theorem (4.1.4) and Proposition (4.1.3); the converse, however, requires a little more work. Let $M$ be a finite Gorenstein projective $R$–module, we want to construct a complete resolution $L$ by finite free $R$–modules such that $\mathrm{C}^L_0 \cong M$. We get the left half of a complex $L \in \mathcal{C}^{\mathrm{L}}(R)$ by taking a resolution of $M$ by finite free modules:

$$\cdots \to L_\ell \to \cdots \to L_1 \to L_0 \to M \to 0.$$

It is now sufficient to prove that $M$ fits in a short exact sequence

$$(\dagger) \qquad 0 \to M \to L_{-1} \to C_{-1} \to 0,$$

where $L_{-1}$ is a finite free module and $C_{-1}$ is a finite Gorenstein projective module. The right half of $L$ can then be constructed recursively: the $n$-th step supplies a finite free module $L_{-n}$ (and an obvious differential) and a finite Gorenstein projective module $C_{-n}$. A complex $L$ constructed this way is homologically trivial and has $\mathrm{C}^L_0 \cong M$. For $\ell < 0$ we have $\mathrm{Ext}^1_R(\mathrm{C}^L_\ell, R) = 0$ by Proposition (4.2.5), because the cokernel $\mathrm{C}^L_\ell = C_\ell$ is Gorenstein projective; and for $\ell \geq 0$ it follows by Lemma (4.1.1)(c) that

$$\mathrm{Ext}^1_R(\mathrm{C}^L_\ell, R) = \mathrm{Ext}^{1+\ell}_R(M,R) = 0,$$

so $L$ is a complete resolution by finite free modules. That is, the proof is complete when the short exact sequence (†) is established.

Since $M$ is Gorenstein projective there exists a complete projective resolution $P$ with $\mathrm{Z}_{-1}^{P} \cong \mathrm{C}_{0}^{P} \cong M$, cf. (A.1.7.3). That is, there is a short exact sequence

$$(\ddagger) \qquad 0 \to M \to P_{-1} \to \mathrm{C}_{-1}^{P} \to 0,$$

where $P_{-1}$ is projective and $\mathrm{C}_{-1}^{P}$ is Gorenstein projective, cf. Proposition (4.2.5). For a suitable projective module $Q$ the sum $P_{-1} \oplus Q$ is free, and adding to $P$ the homologically trivial complex $0 \to Q \xrightarrow{=} Q \to 0$ (concentrated in degrees $-1$ and $-2$), we get a new complete projective resolution $P'$ with $\mathrm{C}_{0}^{P'} \cong M$ and a free module in degree $-1$. Thus, we can assume that $P_{-1}$ is free. Since $M$ is finite, the image of $M$ in $P_{-1}$ is contained in a finite free submodule $L_{-1}$ of $P_{-1}$. We now have a short exact ladder

$$(\star) \qquad \begin{array}{ccccccccc} 0 & \longrightarrow & M & \longrightarrow & L_{-1} & \longrightarrow & C_{-1} & \longrightarrow & 0 \\ & & \big\downarrow{=} & & \big\downarrow & & \big\downarrow & & \\ 0 & \longrightarrow & M & \longrightarrow & P_{-1} & \longrightarrow & \mathrm{C}_{-1}^{P} & \longrightarrow & 0 \end{array}$$

To see that $C_{-1}$ is Gorenstein projective, it is sufficient to prove that $\mathrm{Ext}_R^1(C_{-1}, Q) = 0$ for every projective $R$–module $Q$. This follows by a result[2] similar to [25, Theorem 2.13] (see the remarks on p. 626 ibid.). But this is easy: $\mathrm{Ext}_R^1(\mathrm{C}_{-1}^{P}, Q) = 0$ and $\mathrm{Ext}_R^1(L_{-1}, Q) = 0$ for every projective module $Q$, so we have a commutative diagram

$$\begin{array}{ccccccccc} 0 \to & \mathrm{Hom}_R(\mathrm{C}_{-1}^{P}, Q) & \to & \mathrm{Hom}_R(P_{-1}, Q) & \to & \mathrm{Hom}_R(M, Q) & \to & 0 & \\ & \big\downarrow & & \big\downarrow & & \big\downarrow{=} & & & \\ 0 \to & \mathrm{Hom}_R(C_{-1}, Q) & \to & \mathrm{Hom}_R(L_{-1}, Q) & \to & \mathrm{Hom}_R(M, Q) & \to & \mathrm{Ext}_R^1(C_{-1}, Q) & \to 0 \end{array}$$

and we can immediately see that the map $\mathrm{Hom}_R(L_{-1}, Q) \to \mathrm{Hom}_R(M, Q)$ is surjective and, therefore, $\mathrm{Ext}_R^1(C_{-1}, Q) = 0$ as desired. □

### Notes

The proof of Theorem (4.2.6) is due to Avramov et al.; it will appear in [6]. In (5.1.11) we will use the same technique to prove that finite Gorenstein flat modules belong to the G–class.

## 4.3 G–projectives over Cohen–Macaulay Rings

The purpose of this section is to characterize Gorenstein projective modules over Cohen–Macaulay rings as distinguished modules in the Auslander class. This view is due to Enochs, Jenda, and Xu [32].

[2] It is spelled out in Corollary (4.3.5) and proved for modules over a Cohen–Macaulay local ring with a dualizing module.

(4.3.1) **Setup.** In this section $R$ is a **Cohen–Macaulay local ring with a dualizing module** $D$.

Enochs' notion of flat preenvelopes plays a key role in the proof of the main theorem, so we start by recalling the definition. The extra assumptions on $R$ are irrelevant for (4.3.2) and (4.3.3) but, needless to say, crucial for (4.3.4).

(4.3.2) ***Flat Preenvelopes.*** Let $M$ be an $R$–module. A homomorphism $\phi\colon M \to F$, where $F$ is a flat $R$–module, is said to be a *flat preenvelope* of $M$ if and only if the sequence

$$\mathrm{Hom}_R(F,F') \xrightarrow{\mathrm{Hom}_R(\phi,F')} \mathrm{Hom}_R(M,F') \longrightarrow 0$$

is exact for every flat $R$–module $F'$. That is, if $F'$ is flat and $\nu\colon M \to F'$ is a homomorphism, then there exists a $\nu' \in \mathrm{Hom}_R(F,F')$ such that $\nu = \nu'\phi$.

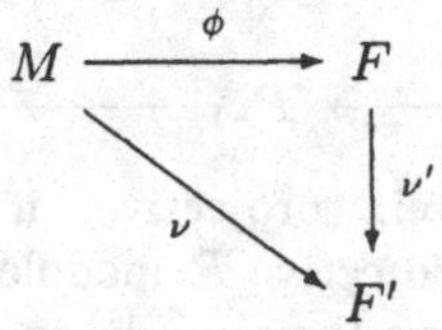


Every module over a Noetherian ring has a flat preenvelope, cf. [21, Proposition 5.1].

(4.3.3) **Lemma.** *Let $M$ be an $R$–module. If there exists an injective homomorphism from $M$ into a flat $R$–module, then every flat preenvelope of $M$ is injective.*

*Proof.* Let $\nu\colon M \to F'$ be an injective homomorphism from $M$ into a flat module $F'$, and let $\phi\colon M \to F$ be a flat preenvelope of $M$. There exists then a homomorphism $\nu'\colon F \to F'$ such that $\nu = \nu'\phi$, and since $\nu$ is injective so is $\phi$. □

(4.3.4) **Theorem.** *For an $R$–module $M$ the next three conditions are equivalent.*

*(i)* *$M$ is Gorenstein projective.*

*(ii)* *$M \in \mathcal{A}_0(R)$ and $\mathrm{Ext}_R^m(M,Q) = 0$ for all $m > 0$ and all projective modules $Q$.*

*(iii)* *$M \in \mathcal{A}_0(R)$ and $\mathrm{Ext}_R^m(M,T) = 0$ for all $m > 0$ and all $T \in \mathcal{F}_0(R)$.*

*Proof.* The third condition is stronger than the second; this leaves us two implications to prove.

*(i)* ⇒ *(iii)*: It was proved in Proposition (4.2.5) that $\mathrm{Ext}_R^m(M,T) = 0$ for all $m > 0$ and $T \in \mathcal{F}_0(R)$; now we prove that $M$ meets conditions (1)–(3) of Theorem (3.4.6). Let $E$ be a faithfully injective $R$–module, then $T = \mathrm{Hom}_R(D,E)$

belongs to $\mathcal{F}_0(R)$. Let $P$ be a complete projective resolution with $C_0^P \cong M$; by commutativity and adjointness (A.2.8) we have

$$\mathrm{Hom}_R(D \otimes_R P, E) \cong \mathrm{Hom}_R(P, \mathrm{Hom}_R(D, E)) = \mathrm{Hom}_R(P, T),$$

and the latter complex is homologically trivial by Proposition (4.2.5). By faithfulness of $\mathrm{Hom}_R(-, E)$ it then follows that $D \otimes_R P$ is homologically trivial; in particular, $\mathrm{Tor}_m^R(D, M) = \mathrm{Tor}_m^R(D, C_0^P) = 0$ for $m > 0$, cf. Lemma (4.1.7)(c), so $M$ meets the first condition in Theorem (3.4.6). Furthermore, we have

$$(\ddagger) \qquad Z_{-1}^{D \otimes_R P} \cong C_0^{D \otimes_R P} \cong D \otimes_R C_0^P \cong D \otimes_R M$$

by (A.1.7.3) and Lemma (4.1.7)(b). Also the complex $\mathrm{Hom}_R(D, D \otimes_R P)$ is homologically trivial; this follows because it is isomorphic to the complete projective resolution $P$: the isomorphism is the natural one, $\gamma_P^D$, where the $\ell$-th component $(\gamma_P^D)_\ell = \gamma_{P_\ell}^D$ is invertible as $P_\ell \in \mathcal{A}_0(R)$. For the same reason, for each $\ell \in \mathbb{Z}$ we have $\mathrm{Ext}_R^m(D, D \otimes_R P_\ell) = 0$ for $m > 0$, and $D \otimes_R P_\ell = (D \otimes_R P)_\ell$, so by Lemma (4.1.6)(c) it follows that $\mathrm{Ext}_R^m(D, Z_\ell^{D \otimes_R P}) = 0$ for all $\ell \in \mathbb{Z}$ and $m > 0$. In particular, $\mathrm{Ext}_R^m(D, D \otimes_R M) = 0$ for $m > 0$, cf. ($\ddagger$), so $M$ satisfies also the second condition in (3.4.6). In view of ($\ddagger$) it follows by Lemma (4.1.6)(b) that

$$Z_{-1}^{\mathrm{Hom}_R(D, D \otimes_R P)} \cong \mathrm{Hom}_R(D, D \otimes_R M);$$

and $Z_{-1}^P \cong M$, cf. (A.1.7.3), so we have an exact ladder

$$\begin{array}{ccccccc}
0 \to & M & \to & P_{-1} & \to & P_{-2} & \to \cdots \\
& \downarrow \gamma_M^D & & \cong \downarrow \gamma_{P_{-1}}^D & & \cong \downarrow \gamma_{P_{-2}}^D & \\
0 \to & \mathrm{Hom}_R(D, D \otimes_R M) & \to & \mathrm{Hom}_R(D, D \otimes_R P_{-1}) & \to & \mathrm{Hom}_R(D, D \otimes_R P_{-2}) & \to \cdots
\end{array}$$

and the five lemma applies to show that the canonical map $\gamma_M^D$ is an isomorphism. Hereby, also the third condition in Theorem (3.4.6) is met, and it follows that $M \in \mathcal{A}_0(R)$.

(*ii*) $\Rightarrow$ (*i*): We assume that $M$ belongs to the Auslander class and has $\mathrm{Ext}_R^m(M, Q) = 0$ for all integers $m > 0$ and all projective modules $Q$. Our target is construction of a complete projective resolution $P$ with $C_0^P \cong M$. First, note that we get the left half of a complex $P \in \mathcal{C}^{\mathrm{P}}(R)$ for free by taking a projective resolution of $M$:

$$\cdots \to P_\ell \to \cdots \to P_1 \to P_0 \to M \to 0.$$

Next, note that to establish the right half of $P$ it is sufficient to prove the existence of a short exact sequence

$$(\star) \qquad 0 \to M \to P_{-1} \to C_{-1} \to 0,$$

where $P_{-1}$ is projective and $C_{-1}$ is a module with the same properties as $M$. Then the right half can be constructed recursively: the $n$-th step supplies a projective module $P_{-n}$ (and an obvious differential) and a module $C_{-n} \in \mathcal{A}_0(R)$ with $\operatorname{Ext}_R^m(C_{-n}, Q) = 0$ for $m > 0$ and $Q$ projective. A complex $P$ established this way is homologically trivial with $\mathrm{C}_0^P \cong M$. Let $Q$ be a projective $R$-module; for $\ell \geq 0$ we have $\operatorname{Ext}_R^1(\mathrm{C}_\ell^P, Q) = \operatorname{Ext}_R^{\ell+1}(M, Q) = 0$ by Lemma (4.1.1)(c) and the assumptions on $M$, and for $\ell < 0$ we have $\operatorname{Ext}_R^1(\mathrm{C}_\ell^P, Q) = 0$ because $\mathrm{C}_\ell^P = C_\ell$ is a module with the same properties as $M$. Thus, $P$ will be a complete projective resolution, and the Theorem is, therefore, proved when we have established the short exact sequence $(\star)$.

First, choose an injective module $I$ such that $D \otimes_R M$ can be embedded in $I$, and apply $\operatorname{Hom}_R(D, -)$ to the sequence $0 \to D \otimes_R M \to I$. This yields an exact sequence

$$(*) \qquad 0 \to M \xrightarrow{\mu} T,$$

where we have used that $\operatorname{Hom}_R(D, D \otimes_R M) \cong M$ as $M \in \mathcal{A}_0(R)$, and we have set $T = \operatorname{Hom}_R(D, I)$. Next, choose a flat module $F'$ such that $T$ is a homomorphic image of $F'$, and consider the short exact sequence

$$(\dagger\dagger) \qquad 0 \to K \to F' \xrightarrow{\varphi} T \to 0.$$

Applying $\operatorname{Hom}_R(M, -)$ to $(\dagger\dagger)$ we get an exact sequence

$$\operatorname{Hom}_R(M, F') \xrightarrow{\operatorname{Hom}_R(M,\varphi)} \operatorname{Hom}_R(M, T) \to \operatorname{Ext}_R^1(M, K).$$

Since $F'$ is flat and $T \in \mathcal{F}_0(R)$, by Foxby equivalence (3.4.11), also $K \in \mathcal{F}_0(R)$ and, therefore, $\operatorname{Ext}_R^1(M, K) = 0$ by Lemma (4.2.4) and the assumptions on $M$. The composition map $\operatorname{Hom}_R(M, \varphi)$ is, consequently, surjective, so there exists a homomorphism $\nu \in \operatorname{Hom}_R(M, F')$ such that $\mu = \varphi\nu$, and since $\mu$ is injective so is $\nu$. Now take a flat preenvelope $\phi\colon M \to F$, cf. (4.3.2). Since $F'$ is flat and $\nu$ is injective, also $\phi$ is injective, cf. Lemma (4.3.3), so we have an exact sequence

$$(\ddagger\ddagger) \qquad 0 \to M \xrightarrow{\phi} F.$$

Choose a projective module $P_{-1}$ such that $F$ is a homomorphic image of $P_{-1}$, that is,

$$(\star\star) \qquad 0 \to Z \to P_{-1} \xrightarrow{\pi} F \to 0$$

is exact. Arguing on $(\ddagger\ddagger)$ and $(\star\star)$ as we did above on $(*)$ and $(\dagger\dagger)$, we prove the existence of an injective homomorphism $\partial\colon M \to P_{-1}$ such that $\phi = \pi\partial$, and setting $C_{-1} = \operatorname{Coker} \partial$, we have a short exact sequence

$$(**) \qquad 0 \to M \xrightarrow{\partial} P_{-1} \to C_{-1} \to 0.$$

What now remains to be proved is that $C_{-1}$ has the same properties as $M$. The projective module $P_{-1}$ belongs to the Auslander class, and by assumption so does $M$; by Corollary (3.4.7)(a) it then follows from (**) that also $C_{-1} \in \mathcal{A}_0(R)$. Let $Q$ be projective; for $m > 0$ we have $\mathrm{Ext}_R^m(M,Q) = 0 = \mathrm{Ext}_R^m(P_{-1},Q)$, so it follows from the long exact sequence of Ext modules associated to (**) that $\mathrm{Ext}_R^m(C_{-1},Q) = 0$ for $m > 1$. To prove that $\mathrm{Ext}_R^1(C_{-1},Q) = 0$, we consider the right-exact sequence

$$\mathrm{Hom}_R(P_{-1},Q) \xrightarrow{\mathrm{Hom}_R(\partial,Q)} \mathrm{Hom}_R(M,Q) \to \mathrm{Ext}_R^1(C_{-1},Q) \to 0.$$

Since $Q$ is flat and $\phi\colon M \to F$ is a flat preenvelope, there exists, for each $\eta \in \mathrm{Hom}_R(M,Q)$, a homomorphism $\eta'\colon F \to Q$ such that $\eta = \eta'\phi$; that is, $\eta = \eta'\pi\partial = \mathrm{Hom}_R(\partial,Q)(\eta'\pi)$.

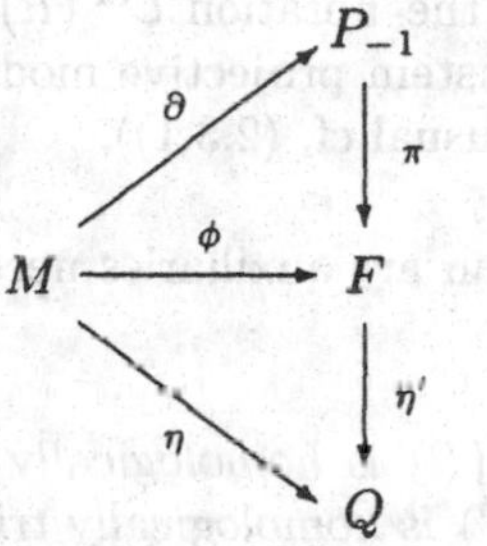


Thus, the induced map $\mathrm{Hom}_R(\partial,Q)$ is surjective and, therefore, $\mathrm{Ext}_R^1(C_{-1},Q) = 0$. This concludes the proof. □

The next result holds over Noetherian rings in general (it is the dual of [25, Theorem 2.13]), but the general version has a different proof.

(4.3.5) **Corollary.** *Let $0 \to M' \to M \to M'' \to 0$ be a short exact sequence of $R$-modules. The following hold:*

(a) *If $M''$ is Gorenstein projective, then $M$ is Gorenstein projective if and only if $M'$ is so.*

(b) *If $M'$ and $M$ are Gorenstein projective, then $M''$ is Gorenstein projective if and only if $\mathrm{Ext}_R^1(M'',Q) = 0$ for every projective module $Q$.*

(c) *If the sequence splits, then $M$ is Gorenstein projective if and only if both $M'$ and $M''$ are so.*

*Proof.* (a): Assume that $M''$ is Gorenstein projective, then, in particular, $M''$ belongs to the Auslander class, and it follows by Corollary (3.4.7)(a) that $M \in \mathcal{A}_0(R)$ if and only if $M' \in \mathcal{A}_0(R)$. Let $Q$ be a projective $R$-module; inspection of the long exact sequence

$$(\dagger)\qquad \cdots \to \mathrm{Ext}_R^m(M'',Q) \to \mathrm{Ext}_R^m(M,Q) \to \mathrm{Ext}_R^m(M',Q) \to \mathrm{Ext}_R^{m+1}(M'',Q) \to \cdots$$

shows that $\mathrm{Ext}_R^m(M,Q) = \mathrm{Ext}_R^m(M',Q)$ for $m > 0$, as $\mathrm{Ext}_R^m(M'',Q) = 0$ for $m > 0$. It now follows by the Theorem that $M$ is Gorenstein projective if and only if $M'$ is so.

(b): It follows by Corollary (3.4.7)(a) that $M''$ belongs to the Auslander class because $M'$ and $M$ do so. Let $Q$ be a projective module, then $\mathrm{Ext}_R^m(M,Q) = 0 = \mathrm{Ext}_R^m(M',Q)$ for $m > 0$, so from (†) it follows that $\mathrm{Ext}_R^m(M'',Q) = 0$ for $m > 1$. The assertion is now immediate by the Theorem.

(c): If the sequence $0 \to M' \to M \to M'' \to 0$ splits, we have isomorphisms

$$\mathrm{Ext}_R^m(M,Q) \cong \mathrm{Ext}_R^m(M',Q) \oplus \mathrm{Ext}_R^m(M'',Q)$$

for all integers $m > 0$ and all projective modules $Q$. The assertion is then evident by the Theorem and Corollary (3.4.7)(b). □

(4.3.6) **Definition.** We use the notation $\mathcal{C}^{\mathrm{GP}}(R)$ for the full subcategory (of $\mathcal{C}(R)$) of complexes of Gorenstein projective modules, and we use it with subscripts $\square$ and $\sqsupset$ (defined as usual cf. (2.3.1)).

The last results of this section are auxiliaries needed for the proof of the main theorem in section 4.4.

(4.3.7) **Lemma.** *If $A \in \mathcal{C}_{\sqsupset}^{\mathrm{GP}}(R)$ is homologically trivial and $F \in \mathcal{C}_{\square}^{\mathrm{F}}(R)$, then also the complex $\mathrm{Hom}_R(A,F)$ is homologically trivial.*

*Proof.* If $F = 0$ the assertion is trivial, so we assume that $F$ is non-zero. Furthermore, we can, without loss of generality, assume that $A_\ell = 0$ and $F_\ell = 0$ for $\ell < 0$. Set $u = \sup\{\ell \in \mathbb{Z} \mid F_\ell \neq 0\}$; we proceed by induction on $u$.

If $u = 0$ then $F$ is a flat module, and $\mathrm{Ext}_R^m(A_\ell, F) = 0$ for all $m > 0$ and $\ell \in \mathbb{Z}$, cf. Theorem (4.3.4). Note that $\mathrm{C}_\ell^A = 0$ for $\ell \leq 0$; it follows by Lemma (4.1.1)(c) that

$$\mathrm{Ext}_R^1(\mathrm{C}_\ell^A, F) = \mathrm{Ext}_R^{1+\ell}(\mathrm{C}_0^A, F) = 0$$

for $\ell \geq 0$, so $\mathrm{Hom}_R(A,F)$ is homologically trivial, again by (4.1.1)(c).

Let $u > 0$ and assume that $\mathrm{Hom}_R(A,\widetilde{F})$ is homologically trivial for all complexes $\widetilde{F} \in \mathcal{C}_{\square}^{\mathrm{F}}(R)$ concentrated in at most $u-1$ degrees. The short exact sequence of complexes $0 \to \sqsubset_{u-1} F \to F \to \Sigma^u F_u \to 0$ is degree-wise split, cf. (A.1.17), so it stays exact after application of $\mathrm{Hom}_R(A,-)$. The complexes $\mathrm{Hom}_R(A,F_u)$ and $\mathrm{Hom}_R(A, \sqsubset_{u-1} F)$ are homologically trivial by, respectively, the induction base and hypothesis, so it follows that also $\mathrm{Hom}_R(A,F)$ is homologically trivial. □

(4.3.8) **Proposition.** *If $X$ is equivalent to $A \in \mathcal{C}_{\sqsupset}^{\mathrm{GP}}(R)$ and $U \simeq F \in \mathcal{C}_{\square}^{\mathrm{F}}(R)$, then $\mathbf{R}\mathrm{Hom}_R(X,U)$ is represented by $\mathrm{Hom}_R(A,F)$.*

*Proof.* Take a projective resolution $P \in \mathcal{C}_{\sqsupset}^{\mathrm{P}}(R)$ of $X$, then $\mathbf{R}\mathrm{Hom}_R(X,U)$ is represented by the complex $\mathrm{Hom}_R(P,F)$. Since $P \simeq X \simeq A$ there is by (A.3.6) a quasi-isomorphism $\alpha\colon P \xrightarrow{\simeq} A$, and hence a morphism

$$\mathrm{Hom}_R(\alpha,F) : \mathrm{Hom}_R(A,F) \longrightarrow \mathrm{Hom}_R(P,F).$$

The mapping cone $\mathcal{M}(\alpha)$ is homologically trivial, and it follows by Corollary (4.3.5)(c) that it belongs to $\mathcal{C}_{\sqsupset}^{\mathrm{GP}}(R)$. By (A.2.1.4) we have

$$\mathcal{M}(\mathrm{Hom}_R(\alpha, F)) \cong \Sigma^1 \mathrm{Hom}_R(\mathcal{M}(\alpha), F),$$

so it follows from the Lemma that the mapping cone $\mathcal{M}(\mathrm{Hom}_R(\alpha, F))$ is homologically trivial, and $\mathrm{Hom}_R(\alpha, F)$ is, therefore, a quasi-isomorphism. In particular, the two complexes $\mathrm{Hom}_R(A, F)$ and $\mathrm{Hom}_R(P, F)$ are equivalent, so also $\mathrm{Hom}_R(A, F)$ represents $\mathbf{R}\mathrm{Hom}_R(X, U)$. □

(4.3.9) **Lemma.** *Let $F$ be a flat $R$-module. If $X \in \mathcal{C}_{(\square)}(R)$ is equivalent to $A \in \mathcal{C}_{\sqsupset}^{\mathrm{GP}}(R)$ and $n \geq \sup X$, then*

$$\mathrm{Ext}_R^m(\mathrm{C}_n^A, F) = \mathrm{H}_{-(m+n)}(\mathbf{R}\mathrm{Hom}_R(X, F))$$

*for $m > 0$. In particular, there is an inequality:*

$$\inf(\mathbf{R}\mathrm{Hom}_R(\mathrm{C}_n^A, F)) \geq \inf(\mathbf{R}\mathrm{Hom}_R(X, F)) + n.$$

*Proof.* Since $n \geq \sup X = \sup A$ we have $A_n\sqsupset \simeq \Sigma^n \mathrm{C}_n^A$, cf. (A.1.14.3), and since $F$ is flat it follows by the Proposition that $\mathbf{R}\mathrm{Hom}_R(\mathrm{C}_n^A, F)$ is represented by $\mathrm{Hom}_R(\Sigma^{-n}(A_n\sqsupset), F)$. For $m > 0$ the isomorphism class $\mathrm{Ext}_R^m(\mathrm{C}_n^A, F)$ is then represented by

$$\begin{aligned}
\mathrm{H}_{-m}(\mathrm{Hom}_R(\Sigma^{-n}(A_n\sqsupset), F)) &= \mathrm{H}_{-m}(\Sigma^n \mathrm{Hom}_R(A_n\sqsupset, F)) \\
&= \mathrm{H}_{-(m+n)}(\mathrm{Hom}_R(A_n\sqsupset, F)) \\
&= \mathrm{H}_{-(m+n)}(\sqsubset_{-n} \mathrm{Hom}_R(A, F)) \\
&= \mathrm{H}_{-(m+n)}(\mathrm{Hom}_R(A, F)),
\end{aligned}$$

cf. (A.2.1.3), (A.1.3.1), and (A.1.20.2). It also follows from the Proposition that the complex $\mathrm{Hom}_R(A, F)$ represents $\mathbf{R}\mathrm{Hom}_R(X, F)$, so $\mathrm{Ext}_R^m(\mathrm{C}_n^A, F) = \mathrm{H}_{-(m+n)}(\mathbf{R}\mathrm{Hom}_R(X, F))$ as wanted, and the inequality of infima follows. □

**Notes**

The proof of Theorem (4.3.4) is based on an idea due to Enochs and Xu; it was communicated to the author by Foxby.

The Auslander class is defined for every local ring with a dualizing complex, but for non-Cohen–Macaulay rings the relation to Gorenstein projective modules is yet to be uncovered.

## 4.4 Gorenstein Projective Dimension

Since every projective module is Gorenstein projective, cf. Observation (4.2.2), the definition of Gorenstein projective dimension — (4.4.2) below — makes sense over any Noetherian ring. However, the only successful approach (that we know of) to a nice functorial description goes via the Auslander class, and to make it work it is (so far) necessary to take the base ring Cohen–Macaulay.

(4.4.1) **Setup.** In this section $R$ is a **Cohen–Macaulay local ring with a dualizing module $D$.**

(4.4.2) **Definition.** The *Gorenstein projective dimension*, $\mathrm{Gpd}_R\,X$, of a complex $X \in \mathcal{C}_{(\sqsupset)}(R)$ is defined as

$$\mathrm{Gpd}_R\,X = \inf\,\{\sup\,\{\ell \in \mathbb{Z} \mid A_\ell \neq 0\} \mid X \simeq A \in \mathcal{C}_{\sqsupset}^{\mathrm{GP}}(R)\}.$$

Note that the set over which infimum is taken is non-empty: any complex $X \in \mathcal{C}_{(\sqsupset)}(R)$ has a projective resolution $X \xleftarrow{\simeq} P \in \mathcal{C}_{\sqsupset}^{\mathrm{P}}(R)$, and $\mathcal{C}_{\sqsupset}^{\mathrm{P}}(R) \subseteq \mathcal{C}_{\sqsupset}^{\mathrm{GP}}(R)$.

(4.4.3) **Observation.** We note the following facts about the Gorenstein projective dimension of $X \in \mathcal{C}_{(\sqsupset)}(R)$:

$$\begin{gathered}\mathrm{Gpd}_R\,X \in \{-\infty\} \cup \mathbb{Z} \cup \{\infty\};\\ \mathrm{pd}_R\,X \geq \mathrm{Gpd}_R\,X \geq \sup X; \quad \text{and}\\ \mathrm{Gpd}_R\,X = -\infty \;\Leftrightarrow\; X \simeq 0.\end{gathered}$$

While the Definition and the Observation above make perfect sense over any Noetherian ring, the proof (at least) of the next theorem relies heavily on the fact that the base ring is local Cohen–Macaulay and has a dualizing module.

(4.4.4) **GPD Theorem.** *Let $X \in \mathcal{C}_{(\sqsupset)}(R)$ and $n \in \mathbb{Z}$. The following are equivalent:*

*(i)* *$X$ is equivalent to a complex $A \in \mathcal{C}_{\square}^{\mathrm{GP}}(R)$ concentrated in degrees at most $n$; and $A$ can be chosen with $A_\ell = 0$ for $\ell < \inf X$.*

*(ii)* $\mathrm{Gpd}_R\,X \leq n$.

*(iii)* *$X \in \mathcal{A}(R)$ and $n \geq \inf U - \inf\,(\mathbf{R}\mathrm{Hom}_R(X,U))$ for all $U \not\simeq 0$ in $\mathcal{F}(R)$.*

*(iv)* *$X \in \mathcal{A}(R)$, $n \geq \sup X$, and $n \geq -\inf\,(\mathbf{R}\mathrm{Hom}_R(X,Q))$ for all projective modules $Q$.*

*(v)* *$n \geq \sup X$ and the module $\mathrm{C}_n^A$ is Gorenstein projective whenever $A \in \mathcal{C}_{\sqsupset}^{\mathrm{GP}}(R)$ is equivalent to $X$.*

*Proof.* It is immediate by Definition (4.4.2) that *(i)* implies *(ii)*.

*(ii)* ⇒ *(iii)*: Choose a complex $A$ in $\mathcal{C}_{\square}^{\mathrm{GP}}(R)$ concentrated in degrees at most $n$ and equivalent to $X$. It follows by Proposition (3.1.14) that $A$, and thereby $X$, belongs to the Auslander class. Let $U \in \mathcal{F}(R)$ be homologically non-trivial, set $i = \inf U$, and choose by (A.5.5) a complex $F \simeq U$ in $\mathcal{C}_{\square}^{\mathrm{F}}(R)$ with $F_\ell = 0$ for $\ell < i$. By Proposition (4.3.8) the complex $\mathrm{Hom}_R(A,F)$ represents $\mathbf{R}\mathrm{Hom}_R(X,U)$, in particular, $\inf\,(\mathbf{R}\mathrm{Hom}_R(X,U)) = \inf\,(\mathrm{Hom}_R(A,F))$. For $\ell < i - n$ and $p \in \mathbb{Z}$ either $p > n$ or $p + \ell \leq n + \ell < i$, so the module

$$\mathrm{Hom}_R(A,F)_\ell = \prod_{p\in\mathbb{Z}} \mathrm{Hom}_R(A_p, F_{p+\ell})$$

vanishes. In particular, the homology modules $\mathrm{H}_\ell(\mathrm{Hom}_R(A,F))$ vanish for $\ell < i-n$, so $\inf(\mathbf{R}\mathrm{Hom}_R(X,U)) \geq i-n = \inf U - n$ as desired.

*(iii)* $\Rightarrow$ *(iv)*: Let $E$ be a faithfully injective $R$-module, then $\mathrm{Hom}_R(D,E) \in \mathcal{F}_0(R)$, and by Lemma (3.4.3)(c), (A.4.10), and adjointness (A.4.21) we have

$$\begin{aligned}\sup X &= \sup(D \otimes^{\mathbf{L}}_R X)\\ &= -\inf(\mathbf{R}\mathrm{Hom}_R(X \otimes^{\mathbf{L}}_R D, E))\\ &= -\inf(\mathbf{R}\mathrm{Hom}_R(X, \mathbf{R}\mathrm{Hom}_R(D,E)))\\ &= -\inf(\mathbf{R}\mathrm{Hom}_R(X, \mathrm{Hom}_R(D,E)))\\ &\leq n.\end{aligned}$$

*(iv)* $\Rightarrow$ *(v)*: Choose a complex $A \in \mathcal{C}^{\mathrm{GP}}_{\sqsupset}(R)$ equivalent to $X$, and consider the short exact sequence of complexes $0 \to \sqsubset_{n-1} A \to \sqsubset_n A \to \Sigma^n \mathrm{C}^A_n \to 0$. By Proposition (3.1.14) the complex $\sqsubset_{n-1} A$ belongs to $\mathcal{A}(R)$, and since $n \geq \sup X$ we have $\sqsubset_n A \simeq A \simeq X \in \mathcal{A}(R)$; it, therefore, follows by Lemma (3.1.13) that $\mathrm{C}^A_n \in \mathcal{A}_0(R)$. For projective modules $Q$ we have $-\inf(\mathbf{R}\mathrm{Hom}_R(\mathrm{C}^A_n, Q)) \leq -\inf(\mathbf{R}\mathrm{Hom}_R(X,Q)) - n \leq 0$ by Lemma (4.3.9), so it follows by Theorem (4.3.4) that $\mathrm{C}^A_n$ is Gorenstein projective.

*(v)* $\Rightarrow$ *(i)*: Choose by (A.3.2) a projective resolution $A \in \mathcal{C}^{\mathrm{P}}_{\sqsupset}(R) \subseteq \mathcal{C}^{\mathrm{GP}}_{\sqsupset}(R)$ of $X$ with $A_\ell = 0$ for $\ell < \inf X$. Since $n \geq \sup X = \sup A$ it follows by (A.1.14.2) that $X \simeq \sqsubset_n A$, and $\sqsubset_n A \in \mathcal{C}^{\mathrm{GP}}_{\square}(R)$ as $\mathrm{C}^A_n$ is Gorenstein projective. □

(4.4.5) **GPD Corollary.** *For a complex $X \in \mathcal{C}_{(\sqsupset)}(R)$ the next three conditions are equivalent.*

*(i)* $X \in \mathcal{A}(R)$.

*(ii)* $\mathrm{Gpd}_R X < \infty$.

*(iii)* $X \in \mathcal{C}_{(\square)}(R)$ *and* $\mathrm{Gpd}_R X \leq \sup X + \dim R$.

*Furthermore, if $X \in \mathcal{A}(R)$, then*

$$\begin{aligned}\mathrm{Gpd}_R X &= \sup\{\inf U - \inf(\mathbf{R}\mathrm{Hom}_R(X,U)) \mid U \in \mathcal{F}(R) \wedge U \not\simeq 0\}\\ &= \sup\{-\inf(\mathbf{R}\mathrm{Hom}_R(X,Q)) \mid Q \in \mathcal{C}^{\mathrm{P}}_0(R)\}.\end{aligned}$$

*Proof.* It follows by the Theorem that *(ii)* implies *(i)*, and *(iii)* is clearly stronger than *(ii)*. For $X \in \mathcal{A}(R)$ and $Q$ projective it follows by Lemma (3.4.13)(a) that

$$-\inf(\mathbf{R}\mathrm{Hom}_R(X,Q)) \leq \sup X + \dim R,$$

so by the equivalence of *(ii)* and *(iv)* in the Theorem we have $\mathrm{Gpd}_R X \leq \sup X + \dim R$ as wanted. This proves the equivalence of the three conditions.

For $X \in \mathcal{A}(R)$ the equalities now follow by the equivalence of *(ii)*, *(iii)*, and *(iv)* in the Theorem. □

As one would expect by now, the Gorenstein projective dimension agrees with the G-dimension for complexes with finite homology.

**(4.4.6) Corollary (GD–GPD Equality).** *For every $X \in \mathcal{C}_{(\sqsupset)}^{(\mathrm{f})}(R)$ there is an equality:*

$$\text{G-dim}_R X = \mathrm{Gpd}_R X.$$

*Proof.* It follows by GD Corollary (2.3.8), Theorem (3.1.10), and GPD Corollary (4.4.5) that the two dimensions are simultaneously finite, namely when $X$ belongs to $\mathcal{R}(R) = \mathcal{A}^{(\mathrm{f})}(R)$. The equality is now immediate by (*EF*) in Theorem (2.4.7) and the equalities in (4.4.5). □

The next proposition shows that Gorenstein projective dimension is a refinement of projective dimension.

**(4.4.7) Proposition (GPD–PD Inequality).** *For every $X \in \mathcal{C}_{(\sqsupset)}(R)$ there is an inequality:*

$$\mathrm{Gpd}_R X \leq \mathrm{pd}_R X,$$

*and equality holds if $\mathrm{pd}_R X < \infty$.*

*Proof.* The inequality is, as we have already observed, immediate because projective modules are Gorenstein projective. Furthermore, equality holds if $X \simeq 0$, so we assume that $\mathrm{pd}_R X = p \in \mathbb{Z}$ and choose, by (A.5.4.1), an $R$–module $T$ such that $p = -\inf(\mathbf{R}\mathrm{Hom}_R(X,T))$. Also choose a projective module $Q$ such that $T$ is a homomorphic image of $Q$. The short exact sequence of modules $0 \to K \to Q \to T \to 0$ induces, cf. (A.4.7), a long exact sequence of homology modules:

$$\cdots \to \mathrm{H}_{-p}(\mathbf{R}\mathrm{Hom}_R(X,Q)) \to \mathrm{H}_{-p}(\mathbf{R}\mathrm{Hom}_R(X,T)) \to \\ \mathrm{H}_{-(p+1)}(\mathbf{R}\mathrm{Hom}_R(X,K)) \to \cdots.$$

Since, by (A.5.4.1), $\mathrm{H}_{-(p+1)}(\mathbf{R}\mathrm{Hom}_R(X,K)) = 0$ while $\mathrm{H}_{-p}(\mathbf{R}\mathrm{Hom}_R(X,T)) \neq 0$, we conclude that also $\mathrm{H}_{-p}(\mathbf{R}\mathrm{Hom}_R(X,Q))$ is non-zero. This proves, in view of GPD Corollary (4.4.5), that $\mathrm{Gpd}_R X \geq p$, and hence equality holds. □

By GPD Corollary (4.4.5) the next theorem is just a rewrite of the $\mathcal{A}$ version (3.1.12).

**(4.4.8) Gorenstein Theorem, GPD Version.** *Let $R$ be a Cohen–Macaulay local ring with residue field $k$. If $R$ admits a dualizing module, then the following are equivalent:*

(*i*) *$R$ is Gorenstein.*
(*ii*) *$\mathrm{Gpd}_R k < \infty$.*
(*iii*) *$\mathrm{Gpd}_R M < \infty$ for all finite $R$–modules $M$.*
(*iv*) *$\mathrm{Gpd}_R M < \infty$ for all $R$–modules $M$.*
(*v*) *$\mathrm{Gpd}_R X < \infty$ for all complexes $X \in \mathcal{C}_{(\square)}(R)$.* □

In (4.4.9)–(4.4.13) we treat Gorenstein projective dimension for modules: we rewrite (4.4.4) and (4.4.5) in classical terms of resolutions and Ext modules.

(4.4.9) **Definition.** A *Gorenstein projective resolution* of a module $M$ is defined the usual way, cf. (1.2.1). All modules have a projective resolution and, hence, a Gorenstein projective one.

(4.4.10) **Lemma.** *Let $M$ be an $R$–module. If $M$ is equivalent to $A \in \mathcal{C}^{\mathrm{GP}}_{\sqsupset}(R)$, then the truncated complex*

$$A_{0\supset} = \cdots \to A_\ell \to \cdots \to A_2 \to A_1 \to \mathrm{Z}_0^A \to 0$$

*is a Gorenstein projective resolution of $M$.*

*Proof.* Suppose $M$ is equivalent to $A \in \mathcal{C}^{\mathrm{GP}}_{\sqsupset}(R)$, then $\inf A = 0$, so $A_{0\supset} \simeq A \simeq M$ by (A.1.14.4), and we have an exact sequence of modules:

$$(\dagger) \qquad \cdots \to A_\ell \to \cdots \to A_2 \to A_1 \to \mathrm{Z}_0^A \to M \to 0$$

Set $v = \inf\{\ell \in \mathbb{Z} \mid A_\ell \neq 0\}$, then also the sequence

$$0 \to \mathrm{Z}_0^A \to A_0 \to \cdots \to A_{v+1} \to A_v \to 0$$

is exact. All the modules $A_0, \ldots, A_v$ are Gorenstein projective, so it follows by repeated applications of Corollary (4.3.5)(a) that $\mathrm{Z}_0^A$ is Gorenstein projective, and therefore $A_{0\supset}$ is a Gorenstein projective resolution of $M$, cf. ($\dagger$). □

(4.4.11) **Remark.** It follows by the Lemma and Definition (4.4.2) that an $R$–module $M$ is Gorenstein projective if and only if $\operatorname{Gpd}_R M \leq 0$. That is,

$$M \text{ is Gorenstein projective} \iff \operatorname{Gpd}_R M = 0 \vee M = 0.$$

(4.4.12) **GPD Theorem for Modules.** *Let $M$ be an $R$–module and $n \in \mathbb{N}_0$. The following are equivalent:*

(*i*) *$M$ has a Gorenstein projective resolution of length at most $n$. I.e., there is an exact sequence of modules $0 \to A_n \to \cdots \to A_1 \to A_0 \to M \to 0$, where $A_0, A_1, \ldots, A_n$ are Gorenstein projective.*

(*ii*) $\operatorname{Gpd}_R M \leq n$.

(*iii*) *$M \in \mathcal{A}_0(R)$ and $\operatorname{Ext}_R^m(M,T) = 0$ for all $m > n$ and all $T \in \mathcal{F}_0(R)$.*

(*iv*) *$M \in \mathcal{A}_0(R)$ and $\operatorname{Ext}_R^m(M,Q) = 0$ for all $m > n$ and all projective modules $Q$.*

(*v*) *In any Gorenstein projective resolution of $M$,*

$$\cdots \to A_\ell \to A_{\ell-1} \to \cdots \to A_0 \to M \to 0,$$

*the kernel*[3] *$K_n = \operatorname{Ker}(A_{n-1} \to A_{n-2})$ is a Gorenstein projective module.*

[3] Appropriately interpreted for small $n$ as $K_0 = M$ and $K_1 = \operatorname{Ker}(A_0 \to M)$.

*Proof.* If the sequence $\cdots \to A_\ell \to A_{\ell-1} \to \cdots \to A_0 \to M \to 0$ is exact, then $M$ is equivalent to $A = \cdots \to A_\ell \to A_{\ell-1} \to \cdots \to A_0 \to 0$. The complex $A$ belongs to $\mathcal{C}_{\sqsupset}^{\mathrm{GP}}(R)$, and it has $\mathrm{C}_0^A \cong M$, $\mathrm{C}_1^A \cong \mathrm{Ker}(A_0 \to M)$, and $\mathrm{C}_\ell^A \cong \mathrm{Z}_{\ell-1}^A = \mathrm{Ker}(A_{\ell-1} \to A_{\ell-2})$ for $\ell \geq 2$. In view of the Lemma the equivalence of the five conditions now follows from Theorem (4.4.4). □

(4.4.13) **GPD Corollary for Modules.** *For an R–module M the next three conditions are equivalent.*

(*i*) $M \in \mathcal{A}_0(R)$.

(*ii*) $\mathrm{Gpd}_R\, M < \infty$.

(*iii*) $\mathrm{Gpd}_R\, M \leq \dim R$.

*Furthermore, if* $M \in \mathcal{A}_0(R)$, *then*

$$\begin{aligned}\mathrm{Gpd}_R\, M &= \sup\{m \in \mathbb{N}_0 \mid \exists\, T \in \mathcal{F}_0(R) \colon \mathrm{Ext}_R^m(M,T) \neq 0\} \\ &= \sup\{m \in \mathbb{N}_0 \mid \exists\, Q \in \mathcal{C}_0^{\mathrm{P}}(R) \colon \mathrm{Ext}_R^m(M,Q) \neq 0\}.\end{aligned}$$

*Proof.* Immediate from Corollary (4.4.5). □

The next proposition shows that the Gorenstein projective dimension cannot grow under localization. In particular, it follows that $M_\mathfrak{p}$ is Gorenstein projective over $R_\mathfrak{p}$ if $M$ is Gorenstein projective over $R$ and, as we remarked in (4.2.3), this is not obvious from the definition.

(4.4.14) **Proposition.** *Let* $X \in \mathcal{C}_{(\sqsupset)}(R)$. *For every* $\mathfrak{p} \in \mathrm{Spec}\, R$ *there is an inequality:*

$$\mathrm{Gpd}_{R_\mathfrak{p}}\, X_\mathfrak{p} \leq \mathrm{Gpd}_R\, X.$$

*Proof.* Let $\mathfrak{p}$ be a prime ideal. If $X$ is equivalent to $A \in \mathcal{C}_{\sqsupset}^{\mathrm{GP}}(R)$, then $X_\mathfrak{p}$ is equivalent to $A_\mathfrak{p}$. It is therefore sufficient to prove that a localized module $M_\mathfrak{p}$ is Gorenstein projective over $R_\mathfrak{p}$ if $M$ is Gorenstein projective over $R$.

Let $M$ be a Gorenstein projective $R$–module and set $d = \dim R_\mathfrak{p}$. It follows from the definitions in (4.2.1) that there is an exact sequence

$$(\dagger)\qquad 0 \to M \to P_{-1} \to P_{-2} \to \cdots \to P_{-d} \to C \to 0,$$

where the modules $P_{-1}, \ldots, P_{-d}$ are projective. Since $M$ and the projective modules all belong to the Auslander class, it follows by repeated applications of Corollary (3.4.7)(a) that also $C \in \mathcal{A}_0(R)$. Localizing at $\mathfrak{p}$ we get an exact sequence

$$(\ddagger)\qquad 0 \to M_\mathfrak{p} \to (P_{-1})_\mathfrak{p} \to (P_{-2})_\mathfrak{p} \to \cdots \to (P_{-d})_\mathfrak{p} \to C_\mathfrak{p} \to 0,$$

where the modules $(P_\ell)_\mathfrak{p}$ are projective over $R_\mathfrak{p}$, and $M_\mathfrak{p}$ and $C_\mathfrak{p}$ belong to $\mathcal{A}(R_\mathfrak{p})$, cf. Observation (3.1.7). From GPD Corollary (4.4.13) it follows that $\mathrm{Gpd}_{R_\mathfrak{p}}\, C_\mathfrak{p} \leq d$, and since ($\ddagger$) is exact it follows by GPD Theorem (4.4.12) that $M_\mathfrak{p}$ is Gorenstein projective. □

Finally we will now use Foxby equivalence to establish a series of test expressions for the Gorenstein projective dimension.

**(4.4.15) Lemma.** *If $X \in \mathcal{A}(R)$ and $U \in \mathcal{I}(R)$, then*

$$\inf(\mathbf{R}\mathrm{Hom}_R(X,U)) = \inf(\mathbf{R}\mathrm{Hom}_R(X,\mathbf{R}\mathrm{Hom}_R(D,U))).$$

*Proof.* In the calculation below the first equality follows as $X \in \mathcal{A}(R)$, the second follows by Hom evaluation (A.4.24) as $U \in \mathcal{I}(R)$, the third by Lemma (3.4.3)(a), the fourth by commutativity (A.4.19) and the last one by adjointness (A.4.21).

$$\begin{aligned}\inf(\mathbf{R}\mathrm{Hom}_R(X,U)) &= \inf(\mathbf{R}\mathrm{Hom}_R(\mathbf{R}\mathrm{Hom}_R(D,D\otimes^{\mathbf{L}}_R X),U))\\ &= \inf(D\otimes^{\mathbf{L}}_R \mathbf{R}\mathrm{Hom}_R(D\otimes^{\mathbf{L}}_R X,U))\\ &= \inf(\mathbf{R}\mathrm{Hom}_R(D\otimes^{\mathbf{L}}_R X,U))\\ &= \inf(\mathbf{R}\mathrm{Hom}_R(X\otimes^{\mathbf{L}}_R D,U))\\ &= \inf(\mathbf{R}\mathrm{Hom}_R(X,\mathbf{R}\mathrm{Hom}_R(D,U))). \quad \square\end{aligned}$$

**(4.4.16) Theorem.** *If $X$ is a complex of finite Gorenstein projective dimension, i.e., $X \in \mathcal{A}(R)$, then the next five numbers are equal.*

(D) $\mathrm{Gpd}_R X$,

(EF) $\sup\{\inf U - \inf(\mathbf{R}\mathrm{Hom}_R(X,U)) \mid U \in \mathcal{F}(R) \wedge U \not\simeq 0\}$,

(EI) $\sup\{\inf U - \inf(\mathbf{R}\mathrm{Hom}_R(X,U)) \mid U \in \mathcal{I}(R) \wedge U \not\simeq 0\}$,

(EI₀) $\sup\{-\inf(\mathbf{R}\mathrm{Hom}_R(X,T)) \mid T \in \mathcal{I}_0(R)\}$, *and*

(EQ) $\sup\{-\inf(\mathbf{R}\mathrm{Hom}_R(X,Q)) \mid Q \in \mathcal{C}_0^{\mathrm{P}}(R)\}$.

*Proof.* The numbers (D), (EF), and (EQ) are equal by GPD Corollary (4.4.5), and it is obvious that $(\mathrm{EI}_0) \leq (\mathrm{EI})$. This leaves us two inequalities to prove.

"$(\mathrm{EQ}) \leq (\mathrm{EI}_0)$": Let $Q$ be a projective $R$-module, then, by Foxby equivalence (3.4.11), the module $T = D \otimes_R Q$ has finite injective dimension, and $Q \cong \mathrm{Hom}_R(D,T)$ represents $\mathbf{R}\mathrm{Hom}_R(D,T)$, cf. Theorems (3.4.6) and (3.4.9). It now follows by the Lemma that

$$-\inf(\mathbf{R}\mathrm{Hom}_R(X,T)) = -\inf(\mathbf{R}\mathrm{Hom}_R(X,Q)),$$

and hence the inequality follows.

"$(\mathrm{EI}) \leq (\mathrm{EF})$": If $U \in \mathcal{I}(R)$, then

$$\inf(\mathbf{R}\mathrm{Hom}_R(X,U)) = \inf(\mathbf{R}\mathrm{Hom}_R(X,\mathbf{R}\mathrm{Hom}_R(D,U)))$$

by the Lemma, and $\mathbf{R}\mathrm{Hom}_R(D,U) \in \mathcal{F}(R)$ by (d) in the Foxby equivalence Theorem (3.3.2). Furthermore, we have $\inf U = \inf(\mathbf{R}\mathrm{Hom}_R(D,U))$ by Lemma (3.4.3)(d), so

$$\begin{aligned}&\inf U - \inf(\mathbf{R}\mathrm{Hom}_R(X,U))\\ &\qquad = \inf(\mathbf{R}\mathrm{Hom}_R(D,U)) - \inf(\mathbf{R}\mathrm{Hom}_R(X,\mathbf{R}\mathrm{Hom}_R(D,U))).\end{aligned}$$

This proves the desired inequality, and with that the five numbers are equal. $\square$

(4.4.17) **Corollary.** *If $M$ is a module of finite Gorenstein projective dimension, i.e., $M \in \mathcal{A}_0(R)$, then the next four numbers are equal.*

(*D*) $\quad \operatorname{Gpd}_R M$,

(*EF₀*) $\quad \sup\{m \in \mathbb{N}_0 \mid \exists\, T \in \mathcal{F}_0(R) : \operatorname{Ext}_R^m(M,T) \neq 0\}$,

(*EI₀*) $\quad \sup\{m \in \mathbb{N}_0 \mid \exists\, T \in \mathcal{I}_0(R) : \operatorname{Ext}_R^m(M,T) \neq 0\}$, *and*

(*EQ*) $\quad \sup\{m \in \mathbb{N}_0 \mid \exists\, Q \in \mathcal{C}_0^{\mathrm{P}}(R) : \operatorname{Ext}_R^m(M,Q) \neq 0\}$. □

(4.4.18) **Observation.** The test expression (*EQ*) in Corollary (4.4.17) can be traced back to the definition of Gorenstein projective modules, and one might, therefore, expect it to hold over Noetherian rings in general. The test expression (*EI₀*), on the other hand, was established through Foxby equivalence, so it is possible that it will only hold over Cohen–Macaulay rings. Both test expressions are valid for G–dimension of finite modules over general Noetherian rings, cf. (*R*) and (*EF₀*) in Corollary (2.4.8). But the crux of the matter is that the two test expressions can only agree for non-finite modules if the ring is Cohen–Macaulay: suppose $R$ is local and not Cohen–Macaulay, by [10, Proposition 5.4] there exists an $R$–module $M$ with $\operatorname{pd}_R M = \dim R$, but $\operatorname{id}_R T \leq \dim R - 1$ for all $T \in \mathcal{I}_0(R)$ (see page 13). Thus,

$$\sup\{m \in \mathbb{N}_0 \mid \exists\, T \in \mathcal{I}_0(R) : \operatorname{Ext}_R^m(M,T) \neq 0\} \leq \dim R - 1,$$

but, proceeding as in the proof of Proposition (4.4.7), it is easy to see that

$$\sup\{m \in \mathbb{N}_0 \mid \exists\, Q \in \mathcal{C}_0^{\mathrm{P}}(R) : \operatorname{Ext}_R^m(M,Q) \neq 0\} = \dim R.$$

Here we tacitly assume that the Gorenstein projective dimension is defined as in (4.4.2), so that $\operatorname{Gpd}_R M \leq \operatorname{pd}_R M = \dim R < \infty$.

## Notes

The GPD Theorem (4.4.12) — and (4.4.4) — is modeled on Cartan and Eilenberg's characterization of projective dimension [13, Proposition VI.2.1]. The proof of (4.4.4) follows the pattern from Foxby's notes [33], and this includes the auxiliary results (4.3.7), (4.3.8), and (4.3.9).

The proof of Proposition (4.4.14) is due to Foxby; it appeared in [39], and so did the equalities in GPD Corollary (4.4.13).

# Chapter 5

# G–flatness

Gorenstein flat modules were introduced by Enochs, Jenda, and Torrecillas in [31], and that paper is, together with Foxby's [39], the principal published source for this chapter.

In the first two sections we establish the basic properties of Gorenstein flat modules (over general Noetherian rings) and Gorenstein flat dimension (over Cohen–Macaulay local rings). The third section is devoted to a series of, largely unpublished, results by Foxby; they deal with a functorial dimension called the restricted Tor–dimension. The reason for this detour is revealed in the final section 5.4, where the results from 5.3 are used to establish a series of test expressions for the Gorenstein flat dimension and last, but not least, a formula of the Auslander–Buchsbaum type.

## 5.1 Gorenstein Flat Modules

We introduce Gorenstein flat modules: a notion which includes both usual flat modules and Gorenstein projective modules. We prove that the finite Gorenstein flat modules are exactly the modules in the G–class, and over local Cohen–Macaulay rings we can characterize the Gorenstein flat modules as distinguished modules in the Auslander class.

(5.1.1) **Definitions.** Let $F \in \mathcal{C}^{\mathrm{F}}(R)$ be homologically trivial. We say that $F$ is a *complete flat resolution* if and only if the complex $J \otimes_R F$ is homologically trivial for every injective $R$–module $J$.

A module $M$ is said to be *Gorenstein flat* if and only if there exists a complete flat resolution $F$ with $\mathrm{C}_0^F \cong M$.

(5.1.2) **Observation.** Every flat module is Gorenstein flat: let $F'$ be flat, then the complex $F = 0 \to F' \xrightarrow{=} F' \to 0$, concentrated in degrees 0 and $-1$, is a complete flat resolution with $\mathrm{C}_0^F \cong F'$.

While it is not clear from Definition (4.2.1) if Gorenstein projectivity is preserved under localization, everything works out smoothly for Gorenstein flatness. The reason is, of course, that complete projective resolutions are defined in terms of the Hom functor, which does not (always) commute with localization, cf. (A.2.3); but complete flat resolutions are defined in terms of the tensor product, which does commute with localization, cf. (A.2.5).

(5.1.3) **Lemma.** *Let $\mathfrak{p}$ be a prime ideal in $R$. If $M$ is a Gorenstein flat $R$–module, then $M_\mathfrak{p}$ is Gorenstein flat over $R_\mathfrak{p}$.*

*Proof.* Let $F$ be a complete flat resolution with $\mathrm{C}_0^F \cong M$. The localized complex $F_\mathfrak{p}$ is homologically trivial with $\mathrm{C}_0^{F_\mathfrak{p}} \cong M_\mathfrak{p}$, and it consists of $R_\mathfrak{p}$–flat modules. If $J$ is an injective $R_\mathfrak{p}$–module, then, because $R_\mathfrak{p}$ is $R$–flat, $J$ is also injective over $R$ and, therefore,

$$J \otimes_{R_\mathfrak{p}} F_\mathfrak{p} \cong J \otimes_{R_\mathfrak{p}} (R_\mathfrak{p} \otimes_R F) \cong (J \otimes_{R_\mathfrak{p}} R_\mathfrak{p}) \otimes_R F \cong J \otimes_R F$$

is homologically trivial. Thus, $F_\mathfrak{p}$ is a complete flat resolution over $R_\mathfrak{p}$, and $M_\mathfrak{p}$ is Gorenstein flat. □

(5.1.4) **Proposition.** *A complete projective resolution is a complete flat resolution. In particular, a Gorenstein projective module is Gorenstein flat.*

*Proof.* Let $P$ be a complete projective resolution. Since $P$ is a complex of flat modules, it is sufficient to prove that $J \otimes_R P$ is homologically trivial for every injective module $J$. Let $E$ be a faithfully injective $R$–module, then $J \otimes_R P$ is homologically trivial if and only if $\mathrm{Hom}_R(J \otimes_R P, E)$ is so, and by commutativity and adjointness we have

$$\mathrm{Hom}_R(J \otimes_R P, E) \cong \mathrm{Hom}_R(P \otimes_R J, E) \cong \mathrm{Hom}_R(P, \mathrm{Hom}_R(J, E)).$$

When $J$ is injective the module $\mathrm{Hom}_R(J, E)$ is flat, so $\mathrm{Hom}_R(P, \mathrm{Hom}_R(J, E))$, and thereby $J \otimes_R P$, is homologically trivial by Proposition (4.2.5). Thus, $P$ is a complete flat resolution, and the last assertion is immediate by the definitions. □

(5.1.5) **Lemma.** *Let $M$ be an $R$–module and assume that $\mathrm{Tor}_m^R(J, M) = 0$ for all $m > 0$ and all injective modules $J$. If $T$ is a module of finite injective dimension, then $\mathrm{Tor}_m^R(T, M) = 0$ for $m > 0$.*

*Proof.* Let

$$J = 0 \to J_0 \to J_{-1} \to \cdots \to J_{-v} \to 0$$

be an injective resolution of $T$, then $\inf J = 0$, $\mathrm{Z}_0^J \cong T$, and $\mathrm{Z}_{-v}^J = J_{-v}$. For $m > 0$ we then have

$$\mathrm{Tor}_m^R(T, M) = \mathrm{Tor}_{m+v}^R(J_{-v}, M),$$

by Lemma (4.1.7)(a), so $\mathrm{Tor}_m^R(T, M) = 0$ for $m > 0$. □

(5.1.6) **Proposition.** *If $F \in \mathcal{C}^{\mathrm{F}}(R)$ is homologically trivial, then the following are equivalent:*

*(i) $F$ is a complete flat resolution.*

*(ii) All the cokernels $\mathrm{C}_\ell^F$, $\ell \in \mathbb{Z}$, are Gorenstein flat modules.*

*(iii) $T \otimes_R F$ is homologically trivial for every module $T \in \mathcal{I}_0(R)$.*

*In particular: if $M$ is Gorenstein flat and $T \in \mathcal{I}_0(R)$, then $\mathrm{Tor}_m^R(T, M) = 0$ for $m > 0$.*

*Proof.* It is clear from the definitions in (5.1.1) that $(i)\Rightarrow(ii)$ and $(iii)\Rightarrow(i)$. If all the cokernels in $F$ are Gorenstein flat, then, by (5.1.1) and Lemma (4.1.7)(c), we have $\mathrm{Tor}_m^R(J, \mathrm{C}_\ell^F) = 0$ for all $m > 0$, all $\ell \in \mathbb{Z}$, and all injective modules $J$. For every $\ell \in \mathbb{Z}$ and $T \in \mathcal{I}_0(R)$ it, therefore, follows by Lemma (5.1.5) that $\mathrm{Tor}_m^R(T, \mathrm{C}_\ell^F) = 0$ for $m > 0$. This proves the last assertion, and by Lemma (4.1.7)(c) it follows that $T \otimes_R F$ is homologically trivial, so $(ii)$ implies $(iii)$. □

The last assertion in (5.1.6) can be interpreted as saying that, as far as modules of finite injective dimension are concerned, Gorenstein flat modules behave as flat ones.

The proof of the next theorem is quite similar to the proof of Theorem (4.3.4); it is, in fact, a little easier. The key ingredient is still Enochs' flat preenvelopes, cf. (4.3.2).

(5.1.7) **Theorem.** *Let $R$ be a Cohen–Macaulay local ring with a dualizing module. For an $R$–module $M$ the next three conditions are then equivalent.*

*(i) $M$ is Gorenstein flat.*

*(ii) $M \in \mathcal{A}_0(R)$ and $\mathrm{Tor}_m^R(J, M) = 0$ for all $m > 0$ and all injective modules $J$.*

*(iii) $M \in \mathcal{A}_0(R)$ and $\mathrm{Tor}_m^R(T, M) = 0$ for all $m > 0$ and all $T \in \mathcal{I}_0(R)$.*

*Proof.* The third condition is stronger than the second; this leaves us two implications to prove.

$(i)\Rightarrow(iii)$: For $T \in \mathcal{I}_0(R)$ it follows by Proposition (5.1.6) that $\mathrm{Tor}_m^R(T, M) = 0$ for $m > 0$. The dualizing module $D$ has finite injective dimension so, in particular, $\mathrm{Tor}_m^R(D, M) = 0$ for $m > 0$; that is, $M$ meets the first condition in Theorem (3.4.6). Let $F$ be a complete flat resolution with $\mathrm{C}_0^F \cong M$. The complex $D \otimes_R F$ is homologically trivial, again by Proposition (5.1.6), and the modules in $F$ all belong to $\mathcal{A}_0(R)$, so the proof now continues verbatim as the proof of $(i) \Rightarrow (iii)$ in Theorem (4.3.4).

$(ii) \Rightarrow (i)$: We assume that $M$ belongs to the Auslander class and has $\mathrm{Tor}_m^R(J, M) = 0$ for all integers $m > 0$ and all injective modules $J$. Our target is construction of a complete flat resolution $F$ with $\mathrm{C}_0^F \cong M$. The left half of a complex $F \in \mathcal{C}^{\mathrm{F}}(R)$ we get for free by taking a flat resolution of $M$:

$$\cdots \to F_\ell \to \cdots \to F_1 \to F_0 \to M \to 0.$$

To establish the right half of $F$ it is sufficient to prove the existence of a short exact sequence

$$(\ddagger) \qquad 0 \to M \to F_{-1} \to C_{-1} \to 0,$$

where $F_{-1}$ is flat and $C_{-1}$ is a module with the same properties as $M$. Then the right half can be constructed recursively: the $n$-th step supplies a flat module $F_{-n}$ (and an obvious differential) and a module $C_{-n} \in \mathcal{A}_0(R)$ with $\operatorname{Tor}_m^R(J, C_{-n}) = 0$ for $m > 0$ and $J$ injective. A complex $F$ established this way is homologically trivial, and it has $\mathrm{C}_0^F \cong M$. Let $J$ be an injective $R$–module; for $\ell \geq 0$ we have $\operatorname{Tor}_1^R(J, \mathrm{C}_\ell^F) = \operatorname{Tor}_{1+\ell}^R(J, M) = 0$ by Lemma (4.1.7)(c) and the assumptions on $M$, and for $\ell < 0$ we have $\operatorname{Tor}_1^R(J, \mathrm{C}_\ell^F) = 0$ because $\mathrm{C}_\ell^F = C_\ell$ is a module with the same properties as $M$. Thus, $F$ will be a complete flat resolution, and the Theorem is, therefore, proved when we have established the short exact sequence ($\ddagger$).

First, choose an injective module $I$ such that $D \otimes_R M$ can be embedded in $I$, and apply $\operatorname{Hom}_R(D, -)$ to the sequence $0 \to D \otimes_R M \to I$. This yields an exact sequence

$$(\star) \qquad 0 \to M \xrightarrow{\mu} T,$$

where we have used that $\operatorname{Hom}_R(D, D \otimes_R M) \cong M$ as $M \in \mathcal{A}_0(R)$, and we have set $T = \operatorname{Hom}_R(D, I)$. By Foxby equivalence (3.4.11) it follows that $T \in \mathcal{F}_0(R)$. We want to prove the existence of an injective homomorphism from $M$ into a flat module. For this end, choose a flat module $F$ such that $T$ is a homomorphic image of $F$, and consider the short exact sequence

$$(*) \qquad 0 \to K \to F \xrightarrow{\varphi} T \to 0,$$

where also $K$ is of finite flat dimension. Let $E$ be a faithfully injective $R$–module and apply the exact functor $-^\vee = \operatorname{Hom}_R(-, E)$ twice to ($*$) to get another exact sequence

$$(\dagger\dagger) \qquad 0 \to K^{\vee\vee} \to F^{\vee\vee} \xrightarrow{\varphi^{\vee\vee}} T^{\vee\vee} \to 0.$$

Now, $K^\vee$ is a module of finite injective dimension, so it follows by Lemma (5.1.5), adjointness, and the assumptions on $M$ that $\operatorname{Ext}_R^1(M, K^{\vee\vee}) = \operatorname{Tor}_1^R(K^\vee, M)^\vee = 0$, so when we apply $\operatorname{Hom}_R(M, -)$ to ($\dagger\dagger$), we get an exact sequence

$$\operatorname{Hom}_R(M, F^{\vee\vee}) \xrightarrow{\operatorname{Hom}_R(M, \varphi^{\vee\vee})} \operatorname{Hom}_R(M, T^{\vee\vee}) \longrightarrow 0.$$

That is, the composition map $\operatorname{Hom}_R(M, \varphi^{\vee\vee})$ is surjective, so there exists a homomorphism $\nu \in \operatorname{Hom}_R(M, F^{\vee\vee})$ such that $\varphi^{\vee\vee}\nu = \delta_T^E \mu$. The biduality homomorphism $\delta_T^E$ is injective, because $E$ is faithfully injective, and $\mu$ is injective, so it follows that also $\nu$ is injective. Now, let $\phi \colon M \to F_{-1}$ be a flat preenvelope, cf. (4.3.2). Since $F^{\vee\vee}$ is flat and $\nu$ is injective, it follows by Lemma (4.3.3) that also $\phi$ is injective. We set $C_{-1} = \operatorname{Coker}\phi$, and then we have a short exact sequence

$$(\ddagger\ddagger) \qquad 0 \to M \xrightarrow{\phi} F_{-1} \to C_{-1} \to 0.$$

What now remains to be proved is that $C_{-1}$ has the same properties as $M$. Both $M$ and the flat module $F_{-1}$ belong to the Auslander class, so by Corollary (3.4.7)(a) it follows from ($\ddagger\ddagger$) that also $C_{-1} \in \mathcal{A}_0(R)$. Let $J$ be injective; for $m > 0$ we have $\operatorname{Tor}_m^R(J, M) = 0 = \operatorname{Tor}_m^R(J, F_{-1})$, so it follows from the long exact sequence of Tor modules associated to ($\ddagger\ddagger$) that $\operatorname{Tor}_m^R(J, C_{-1}) = 0$ for $m > 1$. There is an exact sequence

$$0 \to \operatorname{Tor}_1^R(C_{-1}, J) \to M \otimes_R J \xrightarrow{\phi \otimes_R J} F_{-1} \otimes_R J,$$

so to prove that $\operatorname{Tor}_1^R(J, C_{-1}) = \operatorname{Tor}_1^R(C_{-1}, J) = 0$ it is sufficient to show that $\phi \otimes_R J$ is injective or, equivalently, that $\operatorname{Hom}_R(\phi \otimes_R J, E)$ is surjective. Consider the commutative diagram

$$\begin{array}{ccc}
\operatorname{Hom}_R(F_{-1} \otimes_R J, E) & \xrightarrow{\operatorname{Hom}_R(\phi \otimes_R J, E)} & \operatorname{Hom}_R(M \otimes_R J, E) \\
\cong \big\downarrow \rho_{F_{-1}JE} & & \cong \big\downarrow \rho_{MJE} \\
\operatorname{Hom}_R(F_{-1}, \operatorname{Hom}_R(J, E)) & \xrightarrow{\operatorname{Hom}_R(\phi, \operatorname{Hom}_R(J,E))} & \operatorname{Hom}_R(M, \operatorname{Hom}_R(J, E))
\end{array}$$

The module $\operatorname{Hom}_R(J, E)$ is flat, and $\phi$ is a flat preenvelope of $M$, so $\operatorname{Hom}_R(\phi, \operatorname{Hom}_R(J, E))$ is surjective, cf. (4.3.2), and hence so is $\operatorname{Hom}_R(\phi \otimes_R J, E)$. This concludes the proof. □

(5.1.8) **Remark.** If $R$ is local Cohen–Macaulay with a dualizing module, and $M$ is a finite Gorenstein flat $R$–module, then it is now immediate that $M$ belongs to the G–class: by Theorem (5.1.7) $M$ belongs to $\mathcal{A}_0^{\mathrm{f}}(R)$ and has

$$\sup \{m \in \mathbb{Z} \mid \operatorname{Tor}_m^R(T, M) \neq 0\} \leq 0$$

for every module $T \in \mathcal{I}_0(R)$. By Theorem (3.1.10) and ($TI_0$) in Corollary (2.4.8) we then conclude that $\operatorname{G-dim}_R M \leq 0$, that is, $M \in \mathrm{G}(R)$.

In Theorem (5.1.11) this result is proved for general Noetherian rings.

(5.1.9) **Corollary.** *Let $0 \to M' \to M \to M'' \to 0$ be a short exact sequence of $R$–modules. The following hold:*

(a) *If $M''$ is Gorenstein flat, then $M$ is Gorenstein flat if and only if $M'$ is the same.*

(b) *If $M'$ and $M$ are Gorenstein flat, then $M''$ is Gorenstein flat if and only if $\operatorname{Tor}_1^R(J, M'') = 0$ for every injective module $J$.*

(c) *If the sequence splits, then $M$ is Gorenstein flat if and only if both $M'$ and $M''$ are so.*

*Proof.* In view of Theorem (6.4.2) all three assertions follow immediately by [25, Theorem 2.13]; see also Corollary (6.1.8). If the base ring is local Cohen–Macaulay with a dualizing module, then a direct and easy proof is available, and since this is the typical setting for our applications of the Corollary, we spell out the proof in this special case.

We now assume that $R$ is a Cohen–Macaulay local ring with a dualizing module.

(a): Assume that $M''$ is Gorenstein flat, then, in particular, $M''$ belongs to the Auslander class, and it follows by Corollary (3.4.7)(a) that $M \in \mathcal{A}_0(R)$ if and only if $M' \in \mathcal{A}_0(R)$. Let $J$ be an injective $R$-module; inspection of the long exact sequence

$$(\dagger) \qquad \cdots \to \operatorname{Tor}^R_{m+1}(J, M'') \to \operatorname{Tor}^R_m(J, M') \to \operatorname{Tor}^R_m(J, M) \to \operatorname{Tor}^R_m(J, M'') \to \cdots$$

shows that $\operatorname{Tor}^R_m(J, M') = \operatorname{Tor}^R_m(J, M)$ for $m > 0$, as $\operatorname{Tor}^R_m(J, M'') = 0$ for $m > 0$. It now follows by the Theorem that $M$ is Gorenstein flat if and only if $M'$ is so.

(b): It follows by Corollary (3.4.7)(a) that $M''$ belongs to the Auslander class. Let $J$ be an injective module, then $\operatorname{Tor}^R_m(J, M') = 0 = \operatorname{Tor}^R_m(J, M)$ for $m > 0$, so from ($\dagger$) it follows that $\operatorname{Tor}^R_m(J, M'') = 0$ for $m > 1$. The assertion is now immediate by the Theorem.

(c): If the sequence $0 \to M' \to M \to M'' \to 0$ splits, we have isomorphisms

$$\operatorname{Tor}^R_m(J, M) \cong \operatorname{Tor}^R_m(J, M') \oplus \operatorname{Tor}^R_m(J, M'')$$

for all integers $m > 0$ and all injective modules $J$. The assertion is then evident by the Theorem and Corollary (3.4.7)(b). □

(5.1.10) **Lemma.** *Let $L$ be a homologically trivial complex of finite free $R$-modules. The following are equivalent:*

(*i*) *$L$ is a complete resolution by finite free modules.*

(*ii*) *$L$ is a complete projective resolution.*

(*iii*) *$L$ is a complete flat resolution.*

*Proof.* A complete resolution by finite free modules is also a complete projective resolution, cf. Proposition (4.1.3), so (*i*) implies (*ii*). By Proposition (5.1.4) every complete projective resolution is a complete flat resolution, so (*ii*) implies (*iii*). Now, assume that $L$ is a complete flat resolution, and let $E$ be a faithfully injective $R$-module. We want to see that $\operatorname{Hom}_R(L, R)$ is homologically trivial, and this is the case if and only if $\operatorname{Hom}_R(\operatorname{Hom}_R(L, R), E)$ is homologically trivial. The isomorphism

$$\operatorname{Hom}_R(\operatorname{Hom}_R(L, R), E) \cong L \otimes_R \operatorname{Hom}_R(R, E)$$

follows by applying Hom evaluation for modules in each degree, and

$$L \otimes_R \operatorname{Hom}_R(R, E) \cong L \otimes_R E \cong E \otimes_R L.$$

By assumption $E \otimes_R L$ is homologically trivial and, hence, so is $\operatorname{Hom}_R(L, R)$. This concludes the proof. □

(5.1.11) **Theorem.** *A finite R–module is Gorenstein flat if and only if it belongs to the G–class. That is,*

$$M \text{ is finite and Gorenstein flat} \quad \Longleftrightarrow \quad M \in \mathrm{G}(R).$$

*Proof.* The "if" part is immediate by the Lemma and Theorem (4.1.4). To prove the converse we proceed as in the proof of Theorem (4.2.6). Let $M$ be a finite Gorenstein flat $R$–module, we want to construct a complete resolution $L$ by finite free $R$–modules such that $\mathrm{C}_0^L \cong M$. We get the left half of a complex $L \in \mathcal{C}^{\mathrm{L}}(R)$ by taking a resolution of $M$ by finite free modules:

$$\cdots \to L_\ell \to \cdots \to L_1 \to L_0 \to M \to 0.$$

To establish the right half of $L$ it is sufficient to prove that $M$ fits in a short exact sequence

$$(\dagger) \qquad 0 \to M \to L_{-1} \to C_{-1} \to 0,$$

where $L_{-1}$ is a finite free module and $C_{-1}$ is a finite Gorenstein flat module. Then the right half of $L$ can be constructed recursively: the $n$-th step supplies a finite free module $L_{-n}$ (and an obvious differential) and a finite Gorenstein flat module $C_{-n}$. A complex $L$ constructed this way is homologically trivial, and it has $\mathrm{C}_0^L \cong M$. Let $J$ be any injective module; for $\ell < 0$ we have $\operatorname{Tor}_1^R(J, \mathrm{C}_\ell^L) = 0$ by Proposition (5.1.6), because the cokernel $\mathrm{C}_\ell^L = C_\ell$ is Gorenstein flat, and for $\ell \geq 0$ it follows by Lemma (4.1.7)(c) that

$$\operatorname{Tor}_1^R(J, \mathrm{C}_\ell^L) = \operatorname{Tor}_{1+\ell}^R(J, M) = 0.$$

Thus, $L$ is a complete flat resolution and hence, by the Lemma, a complete resolution by finite free modules. The proof is therefore complete when the short exact sequence (†) is established.

Since $M$ is Gorenstein flat there exists a complete flat resolution $F$ with $M \cong \mathrm{C}_0^F \cong \mathrm{Z}_{-1}^F$, cf. (A.1.7.3). That is, there is a short exact sequence

$$(\ddagger) \qquad 0 \to M \to F_{-1} \to \mathrm{C}_{-1}^F \to 0,$$

where $F_{-1}$ is flat and $\mathrm{C}_{-1}^F$ is Gorenstein flat, cf. Proposition (5.1.6). By Lazard's [46, Lemme 1.1] the map from $M$ into $F_{-1}$ factors through a finite free $R$-module $L_{-1}$, and the map $M \to L_{-1}$ is by necessity also injective. Thus, we have a short exact ladder

$$(\star) \qquad \begin{array}{ccccccccc} 0 & \longrightarrow & M & \longrightarrow & L_{-1} & \longrightarrow & C_{-1} & \longrightarrow & 0 \\ & & \big\| & & \big\downarrow & & \big\downarrow & & \\ 0 & \longrightarrow & M & \longrightarrow & F_{-1} & \longrightarrow & \mathrm{C}_{-1}^F & \longrightarrow & 0 \end{array}$$

To see that $C_{-1}$ is Gorenstein flat it is, by Corollary (5.1.9)(b), sufficient to prove that $\operatorname{Tor}_1^R(J, C_{-1}) = 0$ for every injective $R$-module $J$. This is easy: $\operatorname{Tor}_1^R(J, C_{-1}^F) = 0$ and $\operatorname{Tor}_1^R(J, L_{-1}) = 0$ for every injective module $J$, so we have a commutative diagram

$$\begin{array}{ccccccccc}
0 \to \operatorname{Tor}_1^R(J, C_{-1}) & \to & J \otimes_R M & \to & J \otimes_R L_{-1} & \to & J \otimes_R C_{-1} & \to & 0 \\
 & & \downarrow{=} & & \downarrow & & \downarrow & & \\
0 & \to & J \otimes_R M & \to & J \otimes_R F_{-1} & \to & J \otimes_R C_{-1}^F & \to & 0
\end{array}$$

and we can immediately see that the map $J \otimes_R M \to J \otimes_R L_{-1}$ is injective and, therefore, $\operatorname{Tor}_1^R(J, C_{-1}) = 0$ as desired. □

### Notes

The proof of Theorem (5.1.7) is based on an idea due to Enochs and Xu; it was communicated to the author by Foxby.

The Auslander class is defined for every local ring with a dualizing complex, but for non-Cohen–Macaulay rings the relation to Gorenstein flat modules is yet to be understood.

## 5.2 Gorenstein Flat Dimension

By Observation (5.1.2) every flat module is Gorenstein flat, and the definition of Gorenstein flat dimension, (5.2.3) below, makes sense over any Noetherian ring. However, as in the case of the Gorenstein projective dimension, we only know how to get a nice functorial description if we work over a Cohen–Macaulay local ring with a dualizing module.

(5.2.1) **Setup.** In this section $R$ is a **Cohen–Macaulay local ring with a dualizing module $D$.**

(5.2.2) **Definition.** We use the notation $\mathcal{C}^{\mathrm{GF}}(R)$ for the full subcategory (of $\mathcal{C}(R)$) of complexes of Gorenstein flat modules, and we use it with subscripts $\sqsubset$ and $\sqsupset$ (defined as usual cf. (2.3.1)).

(5.2.3) **Definition.** The *Gorenstein flat dimension*, $\operatorname{Gfd}_R X$, of $X \in \mathcal{C}_{(\sqsupset)}(R)$ is defined as

$$\operatorname{Gfd}_R X = \inf \{\sup \{\ell \in \mathbb{Z} \mid A_\ell \neq 0\} \mid X \simeq A \in \mathcal{C}_{\sqsupset}^{\mathrm{GF}}(R)\}.$$

Note that the set over which infimum is taken is non-empty: any complex $X \in \mathcal{C}_{(\sqsupset)}(R)$ has a projective resolution $X \xleftarrow{\simeq} P \in \mathcal{C}_{\sqsupset}^{\mathrm{P}}(R)$, and $\mathcal{C}_{\sqsupset}^{\mathrm{P}}(R) \subseteq \mathcal{C}_{\sqsupset}^{\mathrm{F}}(R) \subseteq \mathcal{C}_{\sqsupset}^{\mathrm{GF}}(R)$.

(5.2.4) **Observation.** We note the following facts about the Gorenstein flat dimension of $X \in \mathcal{C}_{(\sqsupset)}(R)$:

$$\operatorname{Gfd}_R X \in \{-\infty\} \cup \mathbb{Z} \cup \{\infty\};$$
$$\operatorname{fd}_R X \geq \operatorname{Gfd}_R X \geq \sup X; \quad \text{and}$$
$$\operatorname{Gfd}_R X = -\infty \Leftrightarrow X \simeq 0.$$

While the Definitions and the Observation above make perfect sense over any Noetherian ring, the proof (at least) of the next theorem relies heavily on the assumption that the base ring is local Cohen–Macaulay and has a dualizing module. For the proof we need a few auxiliary results; these have been deferred to the end of the section.

(5.2.5) **GFD Theorem.** *Let $X \in \mathcal{C}_{(\sqsupset)}(R)$ and $n \in \mathbb{Z}$. The following are equivalent:*

(*i*) *$X$ is equivalent to a complex $A \in \mathcal{C}_{\square}^{\mathrm{GF}}(R)$ concentrated in degrees at most $n$; and $A$ can be chosen with $A_\ell = 0$ for $\ell < \inf X$.*
(*ii*) *$\operatorname{Gfd}_R X \leq n$.*
(*iii*) *$X \in \mathcal{A}(R)$ and $n \geq \sup (U \otimes^{\mathbf{L}}_R X) - \sup U$ for all $U \not\simeq 0$ in $\mathcal{I}(R)$.*
(*iv*) *$X \in \mathcal{A}(R)$, $n \geq \sup X$, and $n \geq \sup (J \otimes^{\mathbf{L}}_R X)$ for all injective modules $J$.*
(*v*) *$n \geq \sup X$ and the module $\mathrm{C}_n^A$ is Gorenstein flat whenever $A \in \mathcal{C}_{\sqsupset}^{\mathrm{GF}}(R)$ is equivalent to $X$.*

*Proof.* It is immediate by Definition (5.2.3) that (*i*) implies (*ii*).

(*ii*) $\Rightarrow$ (*iii*): Choose a complex $A \in \mathcal{C}_{\square}^{\mathrm{GF}}(R)$ concentrated in degrees at most $n$ and equivalent to $X$. It follows by Proposition (3.1.14) that $A$, and thereby $X$, belongs to the Auslander class. Let $U \in \mathcal{I}(R)$ be homologically non-trivial, set $s = \sup U$, and choose by (A.5.1) a complex $J \simeq U$ in $\mathcal{C}_{\square}^{\mathrm{I}}(R)$ with $J_\ell = 0$ for $\ell > s$. By Proposition (5.2.17) the complex $J \otimes_R A$ represents $U \otimes^{\mathbf{L}}_R X$, in particular, $\sup (U \otimes^{\mathbf{L}}_R X) = \sup (J \otimes_R A)$. For $\ell > n + s$ and $p \in \mathbb{Z}$ either $p > s$ or $\ell - p \geq \ell - s > n$, so the module

$$(J \otimes_R A)_\ell = \coprod_{p \in \mathbb{Z}} J_p \otimes_R A_{\ell - p}$$

vanishes. In particular, $\mathrm{H}_\ell(J \otimes_R A) = 0$ for $\ell > n + s$, so $\sup (U \otimes^{\mathbf{L}}_R X) \leq n + s = n + \sup U$ as desired.

(*iii*) $\Rightarrow$ (*iv*): Since $D \in \mathcal{I}_0(R)$ we have

$$\sup X = \sup (D \otimes^{\mathbf{L}}_R X) \leq n,$$

cf. Lemma (3.4.3)(c).

(*iv*) $\Rightarrow$ (*v*): Choose a complex $A \in \mathcal{C}_{\sqsupset}^{\mathrm{GF}}(R)$ equivalent to $X$, and consider the short exact sequence of complexes $0 \to \sqsubset_{n-1} A \to \sqsubset_n A \to \Sigma^n \mathrm{C}_n^A \to 0$. By Proposition (3.1.14) the complex $\sqsubset_{n-1} A$ belongs to $\mathcal{A}(R)$, and since

$n \geq \sup X = \sup A$ we have $\sqsubset_n A \simeq A \simeq X \in \mathcal{A}(R)$ by (A.1.14.2). By Lemma (3.1.13) it now follows that $\mathrm{C}_n^A \in \mathcal{A}_0(R)$. For injective modules $J$ we have $\sup(J \otimes_R^{\mathbf{L}} \mathrm{C}_n^A) \leq \sup(J \otimes_R^{\mathbf{L}} X) - n \leq 0$ by Lemma (5.2.18), so it follows by Theorem (5.1.7) that $\mathrm{C}_n^A$ is Gorenstein flat.

(*v*) ⇒ (*i*): Choose by (A.3.2) a flat resolution $A \in \mathcal{C}_{\sqsupset}^{\mathrm{F}}(R) \subseteq \mathcal{C}_{\sqsupset}^{\mathrm{GF}}(R)$ of $X$ with $A_\ell = 0$ for $\ell < \inf X$. Since $n \geq \sup X = \sup A$ it follows by (A.1.14.2) that $X \simeq \sqsubset_n A$, and $\sqsubset_n A \in \mathcal{C}_{\square}^{\mathrm{GF}}(R)$ as $\mathrm{C}_n^A$ is Gorenstein flat. □

(5.2.6) **GFD Corollary.** *For a complex $X \in \mathcal{C}_{(\sqsupset)}(R)$ the next three conditions are equivalent.*

(*i*) *$X \in \mathcal{A}(R)$.*

(*ii*) *$\mathrm{Gfd}_R X < \infty$.*

(*iii*) *$X \in \mathcal{C}_{(\square)}(R)$ and $\mathrm{Gfd}_R X \leq \sup X + \dim R$.*

*Furthermore, if $X \in \mathcal{A}(R)$, then*

$$\begin{aligned}\mathrm{Gfd}_R X &= \sup\{\sup(U \otimes_R^{\mathbf{L}} X) - \sup U \mid U \in \mathcal{I}(R) \wedge U \not\simeq 0\}\\ &= \sup\{\sup(J \otimes_R^{\mathbf{L}} X) \mid J \in \mathcal{C}_0^{\mathrm{I}}(R)\}.\end{aligned}$$

*Proof.* It follows from the Theorem that (*ii*) implies (*i*), and (*iii*) is clearly stronger than (*ii*). For $X \in \mathcal{A}(R)$ and $J$ injective it follows by Lemma (3.4.13)(b) that

$$\sup(J \otimes_R^{\mathbf{L}} X) \leq \sup X + \dim R,$$

so by the equivalence of (*ii*) and (*iv*) in the Theorem we have $\mathrm{Gfd}_R X \leq \sup X + \dim R$ as wanted. This proves the equivalence of the three conditions.

For $X \in \mathcal{A}(R)$ the equalities now follow by the equivalence of (*ii*), (*iii*), and (*iv*) in the Theorem. □

(5.2.7) **Proposition.** *Let $X \in \mathcal{C}_{(\sqsupset)}(R)$. For every $\mathfrak{p} \in \operatorname{Spec} R$ there is an inequality:*

$$\mathrm{Gfd}_{R_\mathfrak{p}} X_\mathfrak{p} \leq \mathrm{Gfd}_R X.$$

*Proof.* If $X$ is equivalent to $A \in \mathcal{C}_{\sqsupset}^{\mathrm{GF}}(R)$, then $X_\mathfrak{p}$ is equivalent to $A_\mathfrak{p}$, and by Lemma (5.1.3) $A_\mathfrak{p}$ is a complex of Gorenstein flat $R_\mathfrak{p}$–modules. The inequality now follows by Definition (5.2.3). □

The next two propositions show that Gorenstein flat dimension is a refinement of flat dimension and a finer invariant than Gorenstein projective dimension. The second one also shows that the Gorenstein flat dimension agrees with the G–dimension (and, thereby, the Gorenstein projective one) for complexes with finite homology.

(5.2.8) **Proposition (GFD–FD Inequality).** *For every complex* $X \in \mathcal{C}_{(\sqsupset)}(R)$ *there is an inequality:*

$$\operatorname{Gfd}_R X \leq \operatorname{fd}_R X,$$

*and equality holds if* $\operatorname{fd}_R X < \infty$.

*Proof.* The inequality is, as we have already observed, immediate because flat modules are Gorenstein flat. Furthermore, equality holds if $X \simeq 0$, so we assume that $\operatorname{fd}_R X = f \in \mathbb{Z}$ and choose, by (A.5.6.1), an $R$-module $T$ such that $\sup(T \otimes^{\mathbf{L}}_R X) = f$. Also choose an injective module $J$ such that $T$ can be embedded in $J$. The short exact sequence of modules $0 \to T \to J \to C \to 0$ induces, cf. (A.4.17), a long exact sequence of homology modules:

$$\cdots \to \mathrm{H}_{f+1}(C \otimes^{\mathbf{L}}_R X) \to \mathrm{H}_f(T \otimes^{\mathbf{L}}_R X) \to \mathrm{H}_f(J \otimes^{\mathbf{L}}_R X) \to \cdots.$$

Since, by (A.5.6.1), $\mathrm{H}_{f+1}(C \otimes^{\mathbf{L}}_R X) = 0$ while $\mathrm{H}_f(T \otimes^{\mathbf{L}}_R X) \neq 0$, we conclude that also $\mathrm{H}_f(J \otimes^{\mathbf{L}}_R X)$ is non-zero. This proves, in view of the equalities in GFD Corollary (5.2.6), that $\operatorname{Gfd}_R X \geq f$, and hence equality holds. □

(5.2.9) **Proposition (GFD–GPD Inequality).** *For every* $X \in \mathcal{C}_{(\sqsupset)}(R)$ *there is an inequality:*

$$\operatorname{Gfd}_R X \leq \operatorname{Gpd}_R X,$$

*and the two dimensions are simultaneously finite; that is,*

$$\operatorname{Gfd}_R X < \infty \iff \operatorname{Gpd}_R X < \infty.$$

*Furthermore, for* $X \in \mathcal{C}^{(\mathrm{f})}_{(\sqsupset)}(R)$ *both dimensions agree with the G–dimension; that is,*

$$\operatorname{G-dim}_R X = \operatorname{Gfd}_R X = \operatorname{Gpd}_R X.$$

*Proof.* By Proposition (5.1.4) all Gorenstein projective modules are Gorenstein flat, so the inequality is immediate by Definitions (4.4.2) and (5.2.3). It follows by Corollaries (4.4.5) and (5.2.6) that the two dimensions are simultaneously finite, namely,

$$\operatorname{Gfd}_R X < \infty \Leftrightarrow X \in \mathcal{A}(R) \Leftrightarrow \operatorname{Gpd}_R X < \infty.$$

If $X \in \mathcal{C}^{(\mathrm{f})}_{(\sqsupset)}(R)$, then $\operatorname{G-dim}_R X = \operatorname{Gpd}_R X$ by the GD–GPD equality (4.4.6), and it follows by the above that the three dimensions are simultaneously finite. The equality $\operatorname{G-dim}_R X = \operatorname{Gfd}_R X$ now follows by the equalities in GFD Corollary (5.2.6) and ($\tau\mathrm{I}$) in Theorem (2.4.7). □

By GFD Corollary (5.2.6) the next theorem is just a rewrite of the $\mathcal{A}$ version (3.1.12).

(5.2.10) **Gorenstein Theorem, GFD Version.** *Let $R$ be a Cohen–Macaulay local ring with residue field $k$. If $R$ admits a dualizing module, then the following are equivalent:*

*(i)* *$R$ is Gorenstein.*

*(ii)* $\mathrm{Gfd}_R\, k < \infty$.

*(iii)* $\mathrm{Gfd}_R\, M < \infty$ *for all finite $R$–modules $M$.*

*(iv)* $\mathrm{Gfd}_R\, M < \infty$ *for all $R$–modules $M$.*

*(v)* $\mathrm{Gfd}_R\, X < \infty$ *for all complexes $X \in \mathcal{C}_{(\square)}(R)$.* □

In (5.2.11)–(5.2.15) we deal with Gorenstein flat dimension for modules: we rewrite (5.2.5) and (5.2.6) in classical terms of resolutions and Tor modules.

(5.2.11) **Definition.** A *Gorenstein flat resolution* of a module $M$ is defined the usual way, cf. (1.2.1). All modules have a flat resolution and hence a Gorenstein flat one.

(5.2.12) **Lemma.** *Let $M$ be an $R$–module. If $M$ is equivalent to $A \in \mathcal{C}_{\sqsupset}^{\mathrm{GF}}(R)$, then the truncated complex*

$$A_0\supset \;=\; \cdots \to A_\ell \to \cdots \to A_2 \to A_1 \to \mathrm{Z}_0^A \to 0$$

*is a Gorenstein flat resolution of $M$.*

*Proof.* The proof of Lemma (4.4.10) applies verbatim, only this time use Corollary (5.1.9) instead of (4.3.5). □

(5.2.13) **Remark.** It follows by the Lemma and Definition (5.2.3) that an $R$–module $M$ is Gorenstein flat if and only if $\mathrm{Gfd}_R\, M \leq 0$. That is,

$$M \text{ is Gorenstein flat} \quad\Longleftrightarrow\quad \mathrm{Gfd}_R\, M = 0 \;\vee\; M = 0.$$

(5.2.14) **GFD Theorem for Modules.** *Let $M$ be an $R$–module and $n \in \mathbb{N}_0$. The following are equivalent:*

*(i)* *$M$ has a Gorenstein flat resolution of length at most $n$. That is, there is an exact sequence of modules $0 \to A_n \to \cdots \to A_1 \to A_0 \to M \to 0$, where $A_0, A_1, \ldots, A_n$ are Gorenstein flat.*

*(ii)* $\mathrm{Gfd}_R\, M \leq n$.

*(iii)* $M \in \mathcal{A}_0(R)$ *and* $\mathrm{Tor}_m^R(T, M) = 0$ *for all $m > n$ and all $T \in \mathcal{I}_0(R)$.*

*(iv)* $M \in \mathcal{A}_0(R)$ *and* $\mathrm{Tor}_m^R(J, M) = 0$ *for all $m > n$ and all injective modules $J$.*

*(v)* *In any Gorenstein flat resolution of $M$,*

$$\cdots \to A_\ell \to A_{\ell-1} \to \cdots \to A_0 \to M \to 0,$$

*the kernel*[1] $K_n = \mathrm{Ker}(A_{n-1} \to A_{n-2})$ *is a Gorenstein flat module.*

[1] Appropriately interpreted for small $n$ as $K_0 = M$ and $K_1 = \mathrm{Ker}(A_0 \to M)$.

*Proof.* If the sequence $\cdots \to A_\ell \to A_{\ell-1} \to \cdots \to A_0 \to M \to 0$ is exact, then $M$ is equivalent to $A = \cdots \to A_\ell \to A_{\ell-1} \to \cdots \to A_0 \to 0$. The complex $A$ belongs to $\mathcal{C}_{\sqsupset}^{\mathrm{GF}}(R)$, and it has $\mathrm{C}_0^A \cong M$, $\mathrm{C}_1^A \cong \mathrm{Ker}(A_0 \to M)$, and $\mathrm{C}_\ell^A \cong \mathrm{Z}_{\ell-1}^A = \mathrm{Ker}(A_{\ell-1} \to A_{\ell-2})$ for $\ell \geq 2$. In view of the Lemma, the equivalence of the five conditions now follows from Theorem (5.2.5). □

(5.2.15) **GFD Corollary for Modules.** *For an $R$-module $M$ the next three conditions are equivalent.*

(*i*) $M \in \mathcal{A}_0(R)$.
(*ii*) $\mathrm{Gfd}_R\, M < \infty$.
(*iii*) $\mathrm{Gfd}_R\, M \leq \dim R$.

*Furthermore, if $M \in \mathcal{A}_0(R)$, then*

$$\begin{aligned}\mathrm{Gfd}_R\, M &= \sup\{m \in \mathbb{N}_0 \mid \exists\, T \in \mathcal{I}_0(R)\colon \mathrm{Tor}_m^R(T, M) \neq 0\}\\ &= \sup\{m \in \mathbb{N}_0 \mid \exists\, J \in \mathcal{C}_0^{\mathrm{I}}(R)\colon \mathrm{Tor}_m^R(J, M) \neq 0\}.\end{aligned}$$

*Proof.* Immediate from Corollary (5.2.6). □

The last results of this section are auxiliaries used in the proof of Theorem (5.2.5).

(5.2.16) **Lemma.** *If $A \in \mathcal{C}_{\sqsupset}^{\mathrm{GF}}(R)$ is homologically trivial and $J \in \mathcal{C}_{\square}^{\mathrm{I}}(R)$, then also the complex $J \otimes_R A$ is homologically trivial.*

*Proof.* If $J = 0$ the assertion is trivial, so we assume that $J$ is non-zero. We can also, without loss of generality, assume that $A_\ell = 0$ and $J_\ell = 0$ for $\ell < 0$. Set $u = \sup\{\ell \in \mathbb{Z} \mid J_\ell \neq 0\}$; we proceed by induction on $u$.

If $u = 0$ then $J$ is an injective module, and $\mathrm{Tor}_m^R(J, A_\ell) = 0$ for all $m > 0$ and all $\ell \in \mathbb{Z}$ by Theorem (5.1.7). Note that $\mathrm{C}_\ell^A = 0$ for $\ell \leq 0$; it follows by Lemma (4.1.7)(c) that

$$\mathrm{Tor}_1^R(J, \mathrm{C}_\ell^A) = \mathrm{Tor}_{1+\ell}^R(J, \mathrm{C}_0^A) = 0$$

for $\ell \geq 0$, so $J \otimes_R A$ is homologically trivial, again by (4.1.7)(c).

Let $u > 0$ and assume that $\widetilde{J} \otimes_R A$ is homologically trivial for all complexes $\widetilde{J} \in \mathcal{C}_{\square}^{\mathrm{I}}(R)$ concentrated in at most $u - 1$ degrees. The short exact sequence of complexes $0 \to \sqsubset_{u-1} J \to J \to \Sigma^u J_u \to 0$ is degree-wise split, cf. (A.1.17), so it stays exact after application of $- \otimes_R A$. Since the complexes $J_u \otimes_R A$ and $(\sqsubset_{u-1} J) \otimes_R A$ are homologically trivial by, respectively, the induction base an hypothesis, it follows that also $J \otimes_R A$ is homologically trivial. □

(5.2.17) **Proposition.** *If $X$ is equivalent to $A \in \mathcal{C}_{\sqsupset}^{\mathrm{GF}}(R)$ and $U \simeq J \in \mathcal{C}_{\square}^{\mathrm{I}}(R)$, then $U \otimes_R^{\mathbf{L}} X$ is represented by $J \otimes_R A$.*

*Proof.* Take a projective resolution $P \in \mathcal{C}_{\sqsupset}^{\mathrm{P}}(R)$ of $X$, then $U \otimes_R^{\mathbf{L}} X$ is represented by the complex $J \otimes_R P$. Since $P \simeq X \simeq A$ there is by (A.3.6) a quasi-isomorphism $\alpha \colon P \xrightarrow{\simeq} A$, and hence a morphism

$$J \otimes_R \alpha \colon \; J \otimes_R P \longrightarrow J \otimes_R A.$$

The mapping cone $\mathcal{M}(\alpha)$ is homologically trivial, and it follows by Corollary (5.1.9)(c) that it belongs to $\mathcal{C}_{\sqsupset}^{\mathrm{GF}}(R)$. By (A.2.4.4) we have

$$\mathcal{M}(J \otimes_R \alpha) \cong J \otimes_R \mathcal{M}(\alpha),$$

so it follows from the Lemma that the mapping cone $\mathcal{M}(J \otimes_R \alpha)$ is homologically trivial, and $J \otimes_R \alpha$ is, therefore, a quasi-isomorphism by (A.1.19). In particular, the two complexes $J \otimes_R A$ and $J \otimes_R P$ are equivalent, so also $J \otimes_R A$ represents $U \otimes_R^{\mathbf{L}} X$. □

(5.2.18) **Lemma.** *Let $J$ be an injective $R$–module. If $X \in \mathcal{C}_{(\square)}(R)$ is equivalent to $A \in \mathcal{C}_{\sqsupset}^{\mathrm{GF}}(R)$ and $n \geq \sup X$, then*

$$\mathrm{Tor}_m^R(J, \mathrm{C}_n^A) = \mathrm{H}_{m+n}(J \otimes_R^{\mathbf{L}} X)$$

*for $m > 0$. In particular, there is an inequality:*

$$\sup\,(J \otimes_R^{\mathbf{L}} \mathrm{C}_n^A) \leq \sup\,(J \otimes_R^{\mathbf{L}} X) - n.$$

*Proof.* Since $n \geq \sup X = \sup A$ we have $A_n \sqsupset \simeq \Sigma^n \mathrm{C}_n^A$, cf. (A.1.14.3), and since $J$ is injective it follows by the Proposition that $J \otimes_R^{\mathbf{L}} \mathrm{C}_n^A$ is represented by $J \otimes_R \Sigma^{-n}(A_n \sqsupset)$. For $m > 0$ the isomorphism class $\mathrm{Tor}_m^R(J, \mathrm{C}_n^A)$ is then represented by

$$\begin{aligned}
\mathrm{H}_m(J \otimes_R \Sigma^{-n}(A_n \sqsupset)) &= \mathrm{H}_m(\Sigma^{-n}(J \otimes_R (A_n \sqsupset))) \\
&= \mathrm{H}_{m+n}(J \otimes_R (A_n \sqsupset)) \\
&= \mathrm{H}_{m+n}((J \otimes_R A)_n \sqsupset) \\
&= \mathrm{H}_{m+n}(J \otimes_R A),
\end{aligned}$$

cf. (A.2.4.3), (A.1.3.1), and (A.1.20.1). From the Proposition it also follows that the complex $J \otimes_R A$ represents $J \otimes_R^{\mathbf{L}} X$, so $\mathrm{Tor}_m^R(J, \mathrm{C}_n^A) = \mathrm{H}_{m+n}(J \otimes_R^{\mathbf{L}} X)$ as wanted, and the inequality of suprema follows. □

## Notes

Theorems (5.2.5) and (5.2.14) are modeled on Cartan and Eilenberg's characterization of flat dimension [13, Exercise 6, p. 123]. The proofs copy the techniques used by Foxby in [33].

The equalities in GFD Corollary (5.2.15) are due to Enochs and Xu as announced in [39].

## 5.3 The Ultimate AB Formula

The restricted Tor–dimension is a functorial dimension[2] with a number of interesting properties: (1) it is finite for every homologically bounded complex, (2) it satisfies a formula of the Auslander–Buchsbaum (AB) type, and (3) it is a refinement of both flat and Gorenstein flat dimension. This means that the AB formula for restricted Tor–dimension includes, as special cases, the AB formula for flat dimension — originally proved by Chouinard — and an AB formula for Gorenstein flat dimension. This explains the, potentially jarring, title of the section.

We set up the basic properties of restricted Tor–dimension in this section, and in section 5.4 we compare it to the flat and Gorenstein flat dimensions.

(5.3.1) **Definition.** The *restricted Tor–dimension*, $\operatorname{Td}_R X$, of $X \in \mathcal{C}_{(\sqsupset)}(R)$ is defined as

$$\operatorname{Td}_R X = \sup \{\sup (T \otimes^{\mathbf{L}}_R X) \mid T \in \mathcal{F}_0(R)\}.$$

For an $R$–module $M$ the definition reads:

$$\operatorname{Td}_R M = \sup \{m \in \mathbb{N}_0 \mid \exists\, T \in \mathcal{F}_0(R) \colon \operatorname{Tor}^R_m(T, M) \neq 0\},$$

and this explains the name.

(5.3.2) **Proposition.** *If $X \in \mathcal{C}_{(\sqsupset)}(R)$, then*

(a) $$\operatorname{Td}_R X \in \{-\infty\} \cup \mathbb{Z} \cup \{\infty\},$$

*and there are inequalities:*

(b) $$\sup X \leq \operatorname{Td}_R X \leq \sup X + \dim R.$$

*In particular,*

(c) $$\operatorname{Td}_R X = -\infty \iff X \simeq 0,$$

*and if* $\dim R < \infty$, *then*

(d) $$\operatorname{Td}_R X < \infty \iff X \in \mathcal{C}_{(\square)}(R).$$

*Proof.* Part (a) is immediate by the definition, and so is the first inequality in (b):

$$\sup X = \sup (R \otimes^{\mathbf{L}}_R X) \leq \operatorname{Td}_R X.$$

For $T \in \mathcal{F}_0(R)$ we have

$$\sup (T \otimes^{\mathbf{L}}_R X) \leq \sup X + \operatorname{fd}_R T \leq \sup X + \dim R,$$

cf. (A.5.6.1), and this proves the second inequality in (b). The remaining two assertions follow from (b). $\square$

[2] By this we mean that it is defined solely in terms of derived functors, without reference to resolutions.

The next corollary is a special case of a result from the next section: in Proposition (5.4.8) we prove that restricted Tor–dimension is a refinement of Gorenstein flat dimension for complexes over Cohen–Macaulay local rings with a dualizing module.

(5.3.3) **Corollary.** *If $R$ is a Gorenstein local ring and $X \in \mathcal{C}_{(\sqsupset)}(R)$, then*

$$\mathrm{Td}_R X = \mathrm{Gfd}_R X.$$

*Proof.* By GFD Corollary (5.2.6), Gorenstein Theorem (5.2.10), and (d) in the Proposition we have

$$\mathrm{Gfd}_R X < \infty \Leftrightarrow X \in \mathcal{A}(R) \Leftrightarrow X \in \mathcal{C}_{(\square)}(R) \Leftrightarrow \mathrm{Td}_R X < \infty.$$

Since $R$ is Gorenstein we have $\mathcal{F}_0(R) = \mathcal{I}_0(R)$, cf. Theorem (3.3.4), and the equality now follows by GFD Corollary (5.2.6) and Definition (5.3.1). □

(5.3.4) **Proposition.** *Let $X \in \mathcal{C}_{(\sqsupset)}(R)$. For every $\mathfrak{p} \in \operatorname{Spec} R$ there is an inequality:*

$$\mathrm{Td}_{R_\mathfrak{p}} X_\mathfrak{p} \leq \mathrm{Td}_R X.$$

*Proof.* The ring $R_\mathfrak{p}$ is a flat $R$–algebra, so $\mathcal{F}_0(R_\mathfrak{p}) \subseteq \mathcal{F}_0(R)$. For $T \in \mathcal{F}_0(R_\mathfrak{p})$ we have

$$\sup (T \otimes^{\mathbf{L}}_{R_\mathfrak{p}} X_\mathfrak{p}) = \sup (T \otimes^{\mathbf{L}}_R X)_\mathfrak{p} \leq \sup (T \otimes^{\mathbf{L}}_R X),$$

and the desired inequality follows by Definition (5.3.1). □

(5.3.5) **Lemma.** *Let $X \in \mathcal{C}_{(\square)}(R)$. The following hold:*

(a) *If $U \in \mathcal{F}(R)$ and $U \otimes^{\mathbf{L}}_R X$ is homologically non-trivial, then $s = \sup (U \otimes^{\mathbf{L}}_R X) \in \mathbb{Z}$, and if $\mathfrak{p} \in \mathrm{Ass}_R(\mathrm{H}_s(U \otimes^{\mathbf{L}}_R X))$, then*

$$\sup (U \otimes^{\mathbf{L}}_R X) - \sup U \leq \operatorname{depth} R_\mathfrak{p} - \operatorname{depth}_{R_\mathfrak{p}} X_\mathfrak{p}.$$

(b) *If $T \in \mathcal{F}_0(R)$ is non-zero and $\mathfrak{p} \in \mathrm{Ass}_R T$, then*

$$\sup (T \otimes^{\mathbf{L}}_R X) \geq \operatorname{depth} R_\mathfrak{p} - \operatorname{depth}_{R_\mathfrak{p}} X_\mathfrak{p}.$$

*Proof.* (a): Assume that $U \otimes^{\mathbf{L}}_R X$ is homologically non-trivial and set $s = \sup (U \otimes^{\mathbf{L}}_R X)$; by (A.4.15.1) and (A.5.6.1) we then have

$$-\infty < \inf U + \inf X \leq s \leq \sup X + \mathrm{fd}_R U < \infty.$$

If $\mathfrak{p}$ is associated to the top homology module $\mathrm{H}_s(U \otimes^{\mathbf{L}}_R X)$, then

$$\begin{aligned}
\sup (U \otimes^{\mathbf{L}}_R X) - \sup U &= -\operatorname{depth}_{R_\mathfrak{p}} (U \otimes^{\mathbf{L}}_R X)_\mathfrak{p} - \sup U \\
&= -\operatorname{depth}_{R_\mathfrak{p}} (U_\mathfrak{p} \otimes^{\mathbf{L}}_{R_\mathfrak{p}} X_\mathfrak{p}) - \sup U \\
&= \operatorname{depth} R_\mathfrak{p} - \operatorname{depth}_{R_\mathfrak{p}} X_\mathfrak{p} - \operatorname{depth}_{R_\mathfrak{p}} U_\mathfrak{p} - \sup U \\
&\leq \operatorname{depth} R_\mathfrak{p} - \operatorname{depth}_{R_\mathfrak{p}} X_\mathfrak{p},
\end{aligned}$$

by (A.6.1.2), (A.6.7), and (A.6.1.1).

(b): If $\mathfrak{p} \in \operatorname{Ass}_R T$ then $\operatorname{depth}_{R_\mathfrak{p}} T_\mathfrak{p} = 0$, and by (A.6.1.1) and (A.6.7) we have

$$\begin{aligned}\sup(T \otimes^{\mathbf{L}}_R X) &\geq -\operatorname{depth}_{R_\mathfrak{p}}(T \otimes^{\mathbf{L}}_R X)_\mathfrak{p}\\ &= -\operatorname{depth}_{R_\mathfrak{p}}(T_\mathfrak{p} \otimes^{\mathbf{L}}_{R_\mathfrak{p}} X_\mathfrak{p})\\ &= \operatorname{depth} R_\mathfrak{p} - \operatorname{depth}_{R_\mathfrak{p}} X_\mathfrak{p} - \operatorname{depth}_{R_\mathfrak{p}} T_\mathfrak{p}\\ &= \operatorname{depth} R_\mathfrak{p} - \operatorname{depth}_{R_\mathfrak{p}} X_\mathfrak{p}. \quad \square\end{aligned}$$

We can now prove the "ultimate AB formula".

**(5.3.6) Theorem (AB Formula for TD).** *If $X \in \mathcal{C}_{(\square)}(R)$, then*

$$\operatorname{Td}_R X = \sup\{\operatorname{depth} R_\mathfrak{p} - \operatorname{depth}_{R_\mathfrak{p}} X_\mathfrak{p} \mid \mathfrak{p} \in \operatorname{Spec} R\}.$$

*Proof.* Let $T \in \mathcal{F}_0(R)$ be given. If $T \otimes^{\mathbf{L}}_R X$ is homologically trivial, then $\sup(T \otimes^{\mathbf{L}}_R X) = -\infty$, so the inequality

$$(\dagger) \qquad \sup(T \otimes^{\mathbf{L}}_R X) \leq \operatorname{depth} R_\mathfrak{p} - \operatorname{depth}_{R_\mathfrak{p}} X_\mathfrak{p}$$

holds for every $\mathfrak{p} \in \operatorname{Spec} R$. If $T \otimes^{\mathbf{L}}_R X$ is homologically non-trivial, we can use (a) in the Lemma: set $s = \sup(T \otimes^{\mathbf{L}}_R X)$, then (†) holds for prime ideals $\mathfrak{p}$ in $\operatorname{Ass}_R(\mathrm{H}_s(T \otimes^{\mathbf{L}}_R X))$, and this proves the inequality "$\leq$" in the formula.

To prove the opposite inequality "$\geq$", it is sufficient to see that for each prime ideal $\mathfrak{p}$, we can find module $T$ of finite flat dimension such that $\mathfrak{p} \in \operatorname{Ass}_R T$; then the inequality follows by part (b) in the Lemma. Let $\mathfrak{p} \in \operatorname{Spec} R$ be given and choose elements $x_1, \ldots, x_n$ in $\mathfrak{p}$ such that the fractions $^{x_1}/_1, \ldots, ^{x_n}/_1$ constitute a maximal $R_\mathfrak{p}$-sequence. The $R_\mathfrak{p}$-module $T = (R/(x_1, \ldots, x_n))_\mathfrak{p} \cong R_\mathfrak{p}/(^{x_1}/_1, \ldots, ^{x_n}/_1)$ has finite flat dimension, actually it belongs to $\mathcal{P}^{\mathrm{f}}_0(R_\mathfrak{p})$, and $\mathfrak{p}_\mathfrak{p} \in \operatorname{Ass}_{R_\mathfrak{p}} T$. It follows that $\mathfrak{p} \in \operatorname{Ass}_R T$, and since $R_\mathfrak{p}$ is a flat $R$-algebra, $T$ is also of finite flat dimension over $R$. $\square$

The rest of this section is devoted to Cohen–Macaulay rings. The important conclusion is that, over such rings, the restricted Tor–dimension can be tested by finite modules of finite projective dimension — even by special cyclic ones.

**(5.3.7) Lemma.** *Let $R$ be Cohen–Macaulay and let $\mathfrak{p}$ be a prime ideal in $R$. If $\boldsymbol{x} = x_1, \ldots, x_t$ is a maximal $R$–sequence in $\mathfrak{p}$, then $\mathfrak{p}$ is associated to the module $R/(\boldsymbol{x})$.*

*Proof.* Let $\mathfrak{p} \in \operatorname{Spec} R$ and let $\boldsymbol{x} = x_1, \ldots, x_t$ be a maximal $R$-sequence in $\mathfrak{p}$. Since $R$ is Cohen–Macaulay we have $t = \operatorname{grade}_R(\mathfrak{p}, R) = \operatorname{depth} R_\mathfrak{p}$, cf. [12, Theorem 2.1.3(b)], so the sequence of fractions $^{x_1}/_1, \ldots, ^{x_t}/_1$ in $\mathfrak{p}_\mathfrak{p}$ is a maximal $R_\mathfrak{p}$-sequence. That is, $\mathfrak{p}_\mathfrak{p}$ is associated to $R_\mathfrak{p}/(\boldsymbol{x})_\mathfrak{p} \cong (R/(\boldsymbol{x}))_\mathfrak{p}$ and, therefore, $\mathfrak{p} \in \operatorname{Ass}_R R/(\boldsymbol{x})$ as wanted. $\square$

(5.3.8) **Theorem.** *If $X \in \mathcal{C}_{(\square)}(R)$, then the numbers*

(D) $\quad \operatorname{Td}_R X,$

(TF) $\quad \sup\{\sup(U \otimes_R^{\mathbf{L}} X) - \sup U \mid U \in \mathcal{F}(R) \wedge U \not\simeq 0\},$ *and*

(TF$_0$) $\quad \sup\{\sup(T \otimes_R^{\mathbf{L}} X) \mid T \in \mathcal{F}_0(R)\}$

*are equal; and if R is Cohen–Macaulay, then they are equal to*

(Tx) $\quad \sup\{\sup(R/(\boldsymbol{x}) \otimes_R^{\mathbf{L}} X) \mid \boldsymbol{x} = x_1, \ldots, x_t$ is an $R$–sequence$\}$.

*Proof.* It is clear that (D) = (TF$_0$) $\leq$ (TF), cf. Definition (5.3.1). Let $U \not\simeq 0$ be a complex of finite flat dimension. If $s = \sup(U \otimes_R^{\mathbf{L}} X) \in \mathbb{Z}$, that is, $U \otimes_R^{\mathbf{L}} X$ is homologically non-trivial, then

$$(\dagger) \qquad \sup(U \otimes_R^{\mathbf{L}} X) - \sup U \leq \operatorname{depth} R_{\mathfrak{p}} - \operatorname{depth}_{R_{\mathfrak{p}}} X_{\mathfrak{p}}$$

for $\mathfrak{p} \in \operatorname{Ass}_R(\mathrm{H}_s(U \otimes_R^{\mathbf{L}} X))$, cf. Lemma (5.3.5)(a), and if $U \otimes_R^{\mathbf{L}} X \simeq 0$ then the inequality (†) holds for all $\mathfrak{p} \in \operatorname{Spec} R$. Now it follows by the AB formula (5.3.6) that (TF) $\leq$ (D).

If $\boldsymbol{x} = x_1, \ldots, x_t$ is an $R$–sequence, then $\operatorname{fd}_R R/(\boldsymbol{x}) = \operatorname{pd}_R R/(\boldsymbol{x}) = t$, so (Tx) $\leq$ (TF$_0$). On the other hand, let $\mathfrak{p} \in \operatorname{Spec} R$ be given, choose a maximal $R$–sequence $\boldsymbol{x} = x_1, \ldots, x_t$ in $\mathfrak{p}$, and set $T = R/(\boldsymbol{x})$. If $R$ is Cohen–Macaulay, then $\mathfrak{p} \in \operatorname{Ass}_R T$ by the Lemma, and it follows by Lemma (5.3.5)(b) and the AB formula (5.3.6) that (Tx) $\geq$ (D). This proves the equality. □

(5.3.9) **Corollary.** *If R is Cohen–Macaulay and M is an R–module, then the next three numbers are equal.*

(D) $\quad \operatorname{Td}_R M,$

(TF$_0$) $\quad \sup\{m \in \mathbb{N}_0 \mid \exists\, T \in \mathcal{F}_0(R) \colon \operatorname{Tor}_m^R(T, M) \neq 0\},$ *and*

(Tx) $\quad \sup\{m \in \mathbb{N}_0 \mid \operatorname{Tor}_m^R(R/(\boldsymbol{x}), M) \neq 0$ for some $R$–seq. $\boldsymbol{x} = x_1, \ldots, x_t\}$.

(5.3.10) **Theorem (AB Formula for TD, Local Finite Version).** *If R is a Cohen–Macaulay local ring, and $X \in \mathcal{C}_{(\square)}^{(\mathrm{f})}(R)$, then*

$$\operatorname{Td}_R X = \operatorname{depth} R - \operatorname{depth}_R X.$$

*Proof.* Let $\mathfrak{p} \in \operatorname{Spec} R$, by (A.6.2) we have

$$\begin{aligned}
\operatorname{depth} R_{\mathfrak{p}} - \operatorname{depth}_{R_{\mathfrak{p}}} X_{\mathfrak{p}} &\leq \operatorname{depth} R_{\mathfrak{p}} - (\operatorname{depth}_R X - \dim R/\mathfrak{p}) \\
&= \dim R_{\mathfrak{p}} + \dim R/\mathfrak{p} - \operatorname{depth}_R X \\
&\leq \dim R - \operatorname{depth}_R X \\
&= \operatorname{depth} R - \operatorname{depth}_R X.
\end{aligned}$$

The desired equality now follows by the AB formula (5.3.6). □

The next example shows that Theorem (5.3.10) need not hold if the ring is not Cohen–Macaulay. (In fact, it will only hold for Cohen–Macaulay rings; see the notes below.)

(5.3.11) **Example.** Let $(R, \mathfrak{m})$ be local with $\dim R = 1$ and $\operatorname{depth} R = 0$. Let $\mathfrak{p} \neq \mathfrak{m}$ be a prime ideal in $R$ and set $M = R/\mathfrak{p}$, then $\operatorname{depth}_R M > 0$ and

$$\operatorname{depth} R - \operatorname{depth}_R M < 0 \leq \operatorname{Td}_R M.$$

**Notes**

The AB formula for restricted Tor–dimension, Theorem (5.3.6), is an unpublished result of Foxby's. See also [19].

The test expression $(\tau_x)$ in Theorem (5.3.8) is valid over all rings $R$ of Cohen–Macaulay defect at most 1 (i.e., $\dim R_\mathfrak{p} - \operatorname{depth} R_\mathfrak{p} \leq 1$ for all $\mathfrak{p} \in \operatorname{Spec} R$); in fact, its validity characterizes local rings of Cohen–Macaulay defect at most 1. This is proved in [18, Part II] (see also [19]), and ibid it is established that the formula $\operatorname{Td}_R M = \operatorname{depth} R - \operatorname{depth}_R M$, cf. Theorem (5.3.10), holds for all finite modules over a local ring $R$ if and only if the ring is Cohen–Macaulay.

## 5.4 Comparing Tor–dimensions

The common feature of the flat, the Gorenstein flat, and the restricted Tor–dimension is that, under suitable circumstances, they can be computed by vanishing of certain Tor modules. This makes them easy to compare, and that is what we do in this section. Given the scope of this chapter, the most important outcome is Corollary (5.4.9): the coveted AB formula for Gorenstein flat dimension. But we start by proving that restricted Tor–dimension is a refinement of flat dimension.

(5.4.1) **Setup.** From (5.4.5) and on we assume that $R$ is a **Cohen–Macaulay local ring with a dualizing module $D$.**

(5.4.2) **Theorem (TD–FD Inequality).** *For every $X \in \mathcal{C}_{(\sqsupset)}(R)$ there is an inequality:*

$$\operatorname{Td}_R X \leq \operatorname{fd}_R X,$$

*and equality holds if $\operatorname{fd}_R X < \infty$.*

*Proof.* The inequality follows by Definition (5.3.1) and (A.5.6.1), and equality holds if $X$ is homologically trivial. If $\operatorname{fd}_R X = f \in \mathbb{Z}$, then $\beta_f^R(\mathfrak{p}, X) \neq 0$ for some $\mathfrak{p} \in \operatorname{Spec} R$, cf. (A.7.2); that is

$$(\dagger) \qquad \operatorname{H}_f(k(\mathfrak{p}) \otimes^{\mathbf{L}}_{R_\mathfrak{p}} X_\mathfrak{p}) \neq 0.$$

As in the proof of the AB formula (5.3.6) we choose a sequence $x_1, \dots, x_n$ of elements in $\mathfrak{p}$, such that the fractions ${}^{x_1}/_1, \dots, {}^{x_n}/_1$ constitute a maximal $R_\mathfrak{p}$–sequence. Then the module $T = R_\mathfrak{p}/({}^{x_1}/_1, \dots, {}^{x_n}/_1)$ has finite flat dimension over $R_\mathfrak{p}$, and the maximal ideal $\mathfrak{p}_\mathfrak{p}$ is associated to $T$, so the residue field $k(\mathfrak{p})$ of $R_\mathfrak{p}$ is isomorphic to a submodule of $T$. That is, there is an exact sequence of $R_\mathfrak{p}$–modules

$$0 \to k(\mathfrak{p}) \to T \to C \to 0;$$

and by (A.4.17) it induces a long exact sequence of homology modules

$$\cdots \to \mathrm{H}_{f+1}(C \otimes^{\mathbf{L}}_{R_\mathfrak{p}} X_\mathfrak{p}) \to \mathrm{H}_f(k(\mathfrak{p}) \otimes^{\mathbf{L}}_{R_\mathfrak{p}} X_\mathfrak{p}) \to \mathrm{H}_f(T \otimes^{\mathbf{L}}_{R_\mathfrak{p}} X_\mathfrak{p}) \to \cdots.$$

Since $\mathrm{H}_{f+1}(C \otimes^{\mathbf{L}}_R X) = 0$, cf. (A.5.6.1), also $\mathrm{H}_{f+1}(C \otimes^{\mathbf{L}}_{R_\mathfrak{p}} X_\mathfrak{p}) = 0$, and in view of (†) we conclude that

$$(\ddagger) \qquad \mathrm{H}_f(T \otimes^{\mathbf{L}}_{R_\mathfrak{p}} X_\mathfrak{p}) = \mathrm{H}_f(T \otimes^{\mathbf{L}}_R X)_\mathfrak{p} \neq 0.$$

The module $T$ has finite flat dimension over $R$, because $R_\mathfrak{p}$ is a flat $R$–algebra, so by (‡) we have

$$\mathrm{Td}_R X \geq \sup(T \otimes^{\mathbf{L}}_R X) \geq \sup(T \otimes^{\mathbf{L}}_R X)_\mathfrak{p} \geq f,$$

whence the desired equality holds. □

(5.4.3) **Corollary.** *If $X$ is an $R$–complex of finite flat dimension, then*

$$\mathrm{fd}_R X = \sup\{\operatorname{depth} R_\mathfrak{p} - \operatorname{depth}_{R_\mathfrak{p}} X_\mathfrak{p} \mid \mathfrak{p} \in \operatorname{Spec} R\}.$$

*Proof.* Immediate by the Theorem and the AB formula (5.3.6). □

Also the G–dimension is defined and computable over Noetherian rings in general, and restricted Tor–dimension is a refinement of G–dimension for complexes with finite homology:

(5.4.4) **Proposition (TD–GD Inequality).** *For every complex $X \in \mathcal{C}^{(\mathrm{f})}_{(\sqsupset)}(R)$ there is an inequality:*

$$\mathrm{Td}_R X \leq \mathrm{G\text{–}dim}_R X,$$

*and equality holds if* $\mathrm{G\text{–}dim}_R X < \infty$, *i.e.,* $X \in \mathcal{R}(R)$.

*Proof.* The inequality is trivial if the G–dimension of $X$ is infinite; and if $X \in \mathcal{R}(R)$, then equality holds by Definition (5.3.1) and ($TF_0$) in Theorem (2.4.7). □

In the rest of this section $R$ is a **Cohen–Macaulay local ring with a dualizing module** $D$. We will first use Foxby equivalence to establish a series of test expressions for Gorenstein flat dimension, then we can prove that the restricted Tor–dimension is a refinement and establish the AB formula for Gorenstein flat dimension.

**(5.4.5) Lemma.** *If $X \in \mathcal{A}(R)$ and $U \in \mathcal{F}(R)$, then*

$$\sup(U \otimes^{\mathbf{L}}_R X) = \sup((D \otimes^{\mathbf{L}}_R U) \otimes^{\mathbf{L}}_R X).$$

*Proof.* The first equality below follows as $X \in \mathcal{A}(R)$; it also uses commutativity (A.4.19). The second follows by tensor evaluation (A.4.23) as $U \in \mathcal{F}(R)$, the third by Lemma (3.4.3)(b), and the last one by commutativity and associativity (A.4.20).

$$\begin{aligned}\sup(U \otimes^{\mathbf{L}}_R X) &= \sup(\mathbf{R}\mathrm{Hom}_R(D, D \otimes^{\mathbf{L}}_R X) \otimes^{\mathbf{L}}_R U)\\ &= \sup(\mathbf{R}\mathrm{Hom}_R(D, (D \otimes^{\mathbf{L}}_R X) \otimes^{\mathbf{L}}_R U))\\ &= \sup((D \otimes^{\mathbf{L}}_R X) \otimes^{\mathbf{L}}_R U)\\ &= \sup((D \otimes^{\mathbf{L}}_R U) \otimes^{\mathbf{L}}_R X). \quad \square\end{aligned}$$

**(5.4.6) Theorem.** *If $X$ is a complex of finite Gorenstein flat dimension, i.e., $X \in \mathcal{A}(R)$, then the next six numbers are equal.*

- (D) $\mathrm{Gfd}_R X$,
- (TI) $\sup\{\sup(U \otimes^{\mathbf{L}}_R X) - \sup U \mid U \in \mathcal{I}(R) \wedge U \not\simeq 0\}$,
- (TF) $\sup\{\sup(U \otimes^{\mathbf{L}}_R X) - \sup U \mid U \in \mathcal{F}(R) \wedge U \not\simeq 0\}$,
- (Tx) $\sup\{\sup(R/(\boldsymbol{x}) \otimes^{\mathbf{L}}_R X) \mid \boldsymbol{x} = x_1, \ldots, x_t \text{ is an } R\text{-sequence}\}$,
- (TI$'_0$) $\sup\{\sup(T \otimes^{\mathbf{L}}_R X) \mid T \in \mathcal{I}^{\mathrm{f}}_0(R)\}$, *and*
- (TE) $\sup\{\sup(\mathrm{E}_R(R/\mathfrak{p}) \otimes^{\mathbf{L}}_R X) \mid \mathfrak{p} \in \mathrm{Spec}\, R\}$.

*Proof.* The numbers (D) and (TI) are equal by GFD Corollary (5.2.6). Every injective $R$–module is a direct sum of indecomposable injective modules, i.e., modules of the form $\mathrm{E}_R(R/\mathfrak{p})$, and the tensor product commutes with direct sums, so it follows, still by Corollary (5.2.6), that (D) = (TE). Furthermore, the numbers (Tx) and (TF) are equal by Theorem (5.3.8), so all in all we have

$$(\text{TI}'_0) \leq (\text{TI}) = (\text{D}) = (\text{TE}) \quad \text{and} \quad (\text{TF}) = (\text{Tx}).$$

This leaves us two implications to prove:

"(TE) $\leq$ (TF)": Let $E$ be an injective $R$–module, then, by Foxby equivalence (3.4.11), the module $T = \mathrm{Hom}_R(D,E)$ has finite flat dimension, and $E \cong D \otimes_R T$ represents $D \otimes^{\mathbf{L}}_R T$, cf. Theorems (3.4.9) and (3.4.6). It now follows by the Lemma that

$$\sup(T \otimes^{\mathbf{L}}_R X) = \sup(E \otimes^{\mathbf{L}}_R X),$$

and the desired inequality follows.

"(Tx) $\leq$ (TI$'_0$)": Let $\boldsymbol{x} = x_1, \ldots, x_t$ be an $R$–sequence and set $T = R/(\boldsymbol{x})$, then $T$ is a finite module of finite projective dimension, in particular, $T \in \mathcal{F}_0(R)$. By the Lemma we then have

$$\sup(T \otimes^{\mathbf{L}}_R X) = \sup((D \otimes^{\mathbf{L}}_R T) \otimes^{\mathbf{L}}_R X),$$

and $D \otimes_R^{\mathbf{L}} T$ is represented by $D \otimes_R T \in \mathcal{I}_0^{\mathrm{f}}(R)$, cf. Theorem (3.4.6) and the Foxby equivalence Theorem (3.4.11). This proves the desired inequality, and with that the six numbers are equal. □

(5.4.7) **Corollary.** *If $M$ is a module of finite Gorenstein flat dimension, i.e., $M \in \mathcal{A}_0(R)$, then the next six numbers are equal.*

(D) $\mathrm{Gfd}_R M$,

(TI₀) $\sup\{m \in \mathbb{N}_0 \mid \exists\, T \in \mathcal{I}_0(R) \colon \mathrm{Tor}_m^R(T, M) \neq 0\}$,

(TF₀) $\sup\{m \in \mathbb{N}_0 \mid \exists\, T \in \mathcal{F}_0(R) \colon \mathrm{Tor}_m^R(T, M) \neq 0\}$,

(Tx) $\sup\{m \in \mathbb{N}_0 \mid \mathrm{Tor}_m^R(R/(\boldsymbol{x}), M) \neq 0 \text{ for some } R\text{–seq. } \boldsymbol{x} = x_1, \ldots, x_t\}$,

(TI′₀) $\sup\{m \in \mathbb{N}_0 \mid \exists\, T \in \mathcal{I}_0^{\mathrm{f}}(R) \colon \mathrm{Tor}_m^R(T, M) \neq 0\}$, *and*

(TE) $\sup\{m \in \mathbb{N}_0 \mid \exists\, \mathfrak{p} \in \mathrm{Spec}\, R \colon \mathrm{Tor}_m^R(\mathrm{E}_R(R/\mathfrak{p}), M) \neq 0\}$. □

(5.4.8) **Proposition (TD–GFD Inequality).** *For every complex $X \in \mathcal{C}_{(\sqsupset)}(R)$ there is an inequality:*

$$\mathrm{Td}_R X \leq \mathrm{Gfd}_R X,$$

*and equality holds if $\mathrm{Gfd}_R X < \infty$, i.e., $X \in \mathcal{A}(R)$.*

*Proof.* The inequality is trivial if the Gorenstein flat dimension of $X$ is infinite; and if $X \in \mathcal{A}(R)$, then equality holds by (TF) in Theorems (5.3.8) and (5.4.6). □

(5.4.9) **Corollary (AB Formula for GFD).** *If $X$ is a complex of finite Gorenstein flat dimension, i.e., $X \in \mathcal{A}(R)$, then*

$$\mathrm{Gfd}_R X = \sup\{\mathrm{depth}\, R_\mathfrak{p} - \mathrm{depth}_{R_\mathfrak{p}} X_\mathfrak{p} \mid \mathfrak{p} \in \mathrm{Spec}\, R\}.$$

*Proof.* Immediate by the Proposition and the AB formula (5.3.6). □

## Notes

The AB formula for flat dimension, Corollary (5.4.3), was originally proved for modules by Chouinard [14, Corollary 1.2] and later extended to complexes by Foxby.

The test expression (TF) in Theorem (5.4.6) and, thereby, the AB formula for Gorenstein flat dimension, Corollary (5.4.9), are due to Foxby [39, Section 4].

# Chapter 6

# G–injectivity

The central notion in this chapter is 'Gorenstein injective modules' as introduced by Enochs and Jenda in [25][1]. The first two sections follow the familiar pattern from chapters 4 and 5: first we introduce Gorenstein injective modules (over general Noetherian rings), next we prove that (over local Cohen–Macaulay rings) they are distinguished modules in an Auslander category, and then a neat theory for Gorenstein injective dimension unfolds. In section 6.3 we study duality between G–flatness and G–injectivity, and in the final section 6.4 we collect additional stability results, mostly in the form of exercises.

## 6.1 Gorenstein Injective Modules

We introduce Gorenstein injective modules — a notion that includes the usual injective modules — and we characterize Gorenstein injective modules over Cohen Macaulay rings as distinguished modules in the Bass class. This view is due to Enochs, Jenda, and Xu [32].

(6.1.1) **Definitions.** Let $I \in \mathcal{C}^{\mathrm{I}}(R)$ be homologically trivial. We say that $I$ is a *complete injective resolution* if and only if the complex $\mathrm{Hom}_R(J, I)$ is homologically trivial for every injective $R$–module $J$.

A module $N$ is said to be *Gorenstein injective* if and only if there exists a complete injective resolution $I$ with $\mathrm{Z}_0^I \cong N$.

(6.1.2) **Observation.** Every injective module is Gorenstein injective: let $I'$ be injective, then the complex $I = 0 \to I' \xrightarrow{=} I' \to 0$, concentrated in degrees 1 and 0, is a complete injective resolution with $\mathrm{Z}_0^I \cong I'$.

(6.1.3) **Remark.** If $N$ is a Gorenstein injective $R$–module and $\mathfrak{p}$ is a prime ideal in $R$, then it is not obvious from the definition that $N_\mathfrak{p}$ is a Gorenstein injective

[1] Gorenstein injective modules over Gorenstein rings were studied by the same authors in an earlier paper [22].

$R_{\mathfrak{p}}$–module. It is, however, so (at least) if $R$ is a Cohen–Macaulay local ring with a dualizing module; we prove this in Proposition (6.2.13).

(6.1.4) **Lemma.** *Let $N$ be an $R$–module and assume that $\mathrm{Ext}_R^m(J,N) = 0$ for all $m > 0$ and all injective modules $J$. If $T$ is a module of finite injective dimension, then $\mathrm{Ext}_R^m(T,N) = 0$ for $m > 0$.*

*Proof.* Let

$$J = 0 \to J_0 \to J_{-1} \to \cdots \to J_{-v} \to 0$$

be an injective resolution of $T$, then $\inf J = 0$, $\mathrm{Z}_0^J \cong T$, and $\mathrm{Z}_{-v}^J = J_{-v}$. For $m > 0$ we then have

$$\mathrm{Ext}_R^m(T,N) = \mathrm{Ext}_R^{m+v}(J_{-v},N)$$

by Lemma (4.1.6)(a) and, therefore, $\mathrm{Ext}_R^m(T,N) = 0$ for $m > 0$. □

(6.1.5) **Proposition.** *If $I \in \mathcal{C}^{\mathrm{I}}(R)$ is homologically trivial, then the following are equivalent:*

(*i*) *$I$ is a complete injective resolution.*

(*ii*) *All the kernels $\mathrm{Z}_\ell^I$, $\ell \in \mathbb{Z}$, are Gorenstein injective modules.*

(*iii*) *$\mathrm{Hom}_R(T,I)$ is homologically trivial for every module $T \in \mathcal{I}_0(R)$.*

*In particular: if $N$ is Gorenstein injective and $T \in \mathcal{I}_0(R)$, then $\mathrm{Ext}_R^m(T,N) = 0$ for $m > 0$.*

*Proof.* It is clear from the definitions in (6.1.1) that (*i*)⇒(*ii*) and (*iii*)⇒(*i*). If all the kernels in $I$ are Gorenstein injective, then, by (6.1.1) and Lemma (4.1.6)(c), we have $\mathrm{Ext}_R^m(J,\mathrm{Z}_\ell^I) = 0$ for all $m > 0$, all $\ell \in \mathbb{Z}$, and all injective modules $J$. For every $\ell \in \mathbb{Z}$ and $T \in \mathcal{I}_0(R)$ it now follows by Lemma (6.1.4) that $\mathrm{Ext}_R^m(T,\mathrm{Z}_\ell^I) = 0$ for $m > 0$. This proves the last assertion, and it follows, again by Lemma (4.1.6)(c), that $\mathrm{Hom}_R(T,I)$ is homologically trivial, so (*ii*) implies (*iii*). □

The last assertion in (6.1.5) can be interpreted as saying that, as far as modules of finite injective dimension are concerned, Gorenstein injective modules behave as injectives.

The key ingredient in the proof of the main result of the section is Enochs' notion of injective precovers; we start by recalling the definition.

(6.1.6) ***Injective Precovers.*** Let $N$ be an $R$–module. A homomorphism $\eta\colon I \to N$, where $I$ is an injective $R$–module, is said to be an ***injective precover*** of $N$ if and only if the sequence

$$\mathrm{Hom}_R(I',I) \xrightarrow{\mathrm{Hom}_R(I',\eta)} \mathrm{Hom}_R(I',N) \longrightarrow 0$$

is exact for every injective $R$–module $I'$. That is, if $I'$ is injective and $v\colon I' \to N$ is a homomorphism, then there exists a $v' \in \mathrm{Hom}_R(I', I)$ such that $v = \eta v'$.

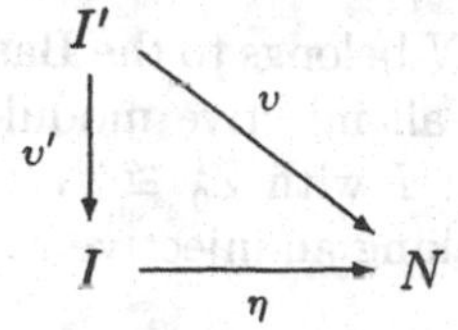

Every module over a Noetherian ring has an injective precover, cf. [21, Proposition 2.2].

(6.1.7) **Theorem.** *Let $R$ be a Cohen–Macaulay local ring with a dualizing module. For an $R$–module $N$ the next three conditions are then equivalent.*

(*i*) *$N$ is Gorenstein injective.*

(*ii*) *$N \in \mathcal{B}_0(R)$ and $\mathrm{Ext}_R^m(J, N) = 0$ for all $m > 0$ and all injective modules $J$.*

(*iii*) *$N \in \mathcal{B}_0(R)$ and $\mathrm{Ext}_R^m(T, N) = 0$ for all $m > 0$ and all $T \in \mathcal{I}_0(R)$.*

*Proof.* The third condition is stronger than the second; this leaves us two implications to prove.

(*i*) $\Rightarrow$ (*iii*): It was proved in Proposition (6.1.5) that $\mathrm{Ext}_R^m(T, N) = 0$ for all $m > 0$ and all $T \in \mathcal{I}_0(R)$. The dualizing module $D$ has finite injective dimension so, in particular, $\mathrm{Ext}_R^m(D, N) = 0$ for $m > 0$. That is, $N$ meets the first condition in Theorem (3.4.9), and we now prove that it also meets conditions (2) and (3). Let $I$ be a complete injective resolution with $\mathrm{Z}_0^I \cong N$. It follows by Proposition (6.1.5) that the complex $\mathrm{Hom}_R(D, I)$ is homologically trivial, and by (A.1.7.3) and (b) in Lemma (4.1.6) we have

$$(\ddagger) \qquad \mathrm{C}_1^{\mathrm{Hom}_R(D,I)} \cong \mathrm{Z}_0^{\mathrm{Hom}_R(D,I)} \cong \mathrm{Hom}_R(D, \mathrm{Z}_0^I) \cong \mathrm{Hom}_R(D, N).$$

Also the complex $D \otimes_R \mathrm{Hom}_R(D, I)$ is homologically trivial; this follows because it is isomorphic to the complete injective resolution $I$: the isomorphism is the natural one, $\xi_I^D$, where the $\ell$-th component $(\xi_I^D)_\ell = \xi_{I_\ell}^D$ is invertible as $I_\ell \in \mathcal{B}_0(R)$. For the same reason, for each $\ell \in \mathbb{Z}$ we have $\mathrm{Tor}_m^R(D, \mathrm{Hom}_R(D, I_\ell)) = 0$ for $m > 0$, and $\mathrm{Hom}_R(D, I_\ell) = \mathrm{Hom}_R(D, I)_\ell$, so by Lemma (4.1.7)(c) it follows that $\mathrm{Tor}_m^R(D, \mathrm{C}_\ell^{\mathrm{Hom}_R(D,I)}) = 0$ for all $\ell \in \mathbb{Z}$ and $m > 0$. In particular, $\mathrm{Tor}_m^R(D, \mathrm{Hom}_R(D, N)) = 0$ for $m > 0$, cf. ($\ddagger$), so $N$ satisfies also the second condition in (3.4.9). In view of ($\ddagger$) it follows by Lemma (4.1.7)(b) that

$$\mathrm{C}_1^{D \otimes_R \mathrm{Hom}_R(D,I)} \cong D \otimes_R \mathrm{Hom}_R(D, N),$$

and $\mathrm{C}_1^I \cong N$, cf. (A.1.7.3), so we have an exact ladder

$$\begin{array}{ccccccccc}
\cdots \to & D \otimes_R \mathrm{Hom}_R(D, I_2) & \to & D \otimes_R \mathrm{Hom}_R(D, I_1) & \to & D \otimes_R \mathrm{Hom}_R(D, N) & \to 0 \\
 & \cong \big\downarrow \xi_{I_2}^D & & \cong \big\downarrow \xi_{I_1}^D & & \big\downarrow \xi_N^D & \\
\cdots \to & I_2 & \to & I_1 & \to & N & \to 0
\end{array}$$

and the five lemma applies to show that the canonical map $\xi_N^D$ is an isomorphism. With this, also the third condition in Theorem (3.4.9) is satisfied, and it follows that $N \in \mathcal{B}_0(R)$.

*(ii)*$\Rightarrow$*(i)*: We assume that $N$ belongs to the Bass class and has $\mathrm{Ext}_R^m(J,N) = 0$ for all integers $m > 0$ and all injective modules $J$. We want to construct a complete injective resolution $I$ with $\mathrm{Z}_0^I \cong N$. The right half of a complex $I \in \mathcal{C}^{\mathrm{I}}(R)$ we get for free by taking an injective resolution of $N$:

$$0 \to I_0 \to I_{-1} \to \cdots \to I_\ell \to \cdots.$$

To establish the left half of $I$, it is sufficient to prove the existence of a short exact sequence

$$(\star) \qquad 0 \to Z_1 \to I_1 \to N \to 0$$

where $I_1$ is injective and $Z_1$ is a module with the same properties as $N$. Then the left half can be constructed recursively: the $n$-th step supplies an injective module $I_n$ (and an obvious differential) and a module $Z_n \in \mathcal{B}_0(R)$ with $\mathrm{Ext}_R^m(J, Z_n) = 0$ for $m > 0$ and $J$ injective. A complex $I$ established this way is homologically trivial and has $\mathrm{Z}_0^I \cong N$. Let $J$ be an injective $R$-module; for $\ell \leq 0$ we have $\mathrm{Ext}_R^1(J, \mathrm{Z}_\ell^I) = \mathrm{Ext}_R^{1-\ell}(J,N) = 0$ by Lemma (4.1.6)(c) and the assumptions on $N$, and for $\ell > 0$ we have $\mathrm{Ext}_R^1(J, \mathrm{Z}_\ell^I) = 0$ because $\mathrm{Z}_\ell^I = Z_\ell$ is a module with the same properties as $N$. Thus, $I$ will be a complete injective resolution, and the Theorem is, therefore, proved when we have established the short exact sequence $(\star)$.

First, choose a projective module $P$ such that $\mathrm{Hom}_R(D,N)$ is a homomorphic image of $P$, and apply $D \otimes_R -$ to the sequence $P \to \mathrm{Hom}_R(D,N) \to 0$. This yields an exact sequence

$$(*) \qquad T \xrightarrow{\gamma} N \to 0,$$

where we have used that $D \otimes_R \mathrm{Hom}_R(D,N) \cong N$ as $N \in \mathcal{B}_0(R)$, and we have set $T = D \otimes_R P$. Next, choose an injective module $I'$ such that $T$ can be embedded in $I'$, and consider the short exact sequence

$$(\dagger\dagger) \qquad 0 \to T \xrightarrow{\iota} I' \to C \to 0.$$

Applying $\mathrm{Hom}_R(-,N)$ to $(\dagger\dagger)$ we get an exact sequence

$$\mathrm{Hom}_R(I',N) \xrightarrow{\mathrm{Hom}_R(\iota,N)} \mathrm{Hom}_R(T,N) \to \mathrm{Ext}_R^1(C,N).$$

Since $I'$ is injective and $T \in \mathcal{I}_0(R)$, by Foxby equivalence (3.4.11), also $C \in \mathcal{I}_0(R)$ and, therefore, $\mathrm{Ext}_R^1(C,N) = 0$ by Lemma (6.1.4) and the assumptions on $N$. Consequently, the composition map $\mathrm{Hom}_R(\iota,N)$ is surjective, so there exists a homomorphism $v \in \mathrm{Hom}_R(I',N)$ such that $\gamma = v\iota$, and since $\gamma$ is surjective so is $v$. Now, take an injective precover $\eta\colon I \to N$, cf. (6.1.6). Since $I'$ is injective there is a homomorphism $v' \in \mathrm{Hom}_R(I',I)$ such that $v = \eta v'$, and since $v$ is surjective also $\eta$ must be surjective. Set $Z_1 = \operatorname{Ker}\eta$, then we have a short exact sequence

$$(\ddagger\ddagger) \qquad 0 \to Z_1 \to I_1 \xrightarrow{\eta} N \to 0.$$

What now remains to be proved is that $Z_1$ has the same properties as $N$. Both $N$ and the injective module $I_1$ belong to the Bass class, so by Corollary (3.4.10)(a) it follows from (‡‡) that also $Z_1 \in \mathcal{B}_0(R)$. Let $J$ be injective; for $m > 0$ we have $\operatorname{Ext}_R^m(J, I_1) = 0 = \operatorname{Ext}_R^m(J, N)$, so it follows from the long exact sequence of Ext modules associated to (‡‡) that $\operatorname{Ext}_R^m(J, Z_1) = 0$ for $m > 1$. Now, consider the right-exact sequence

$$\operatorname{Hom}_R(J, I_1) \xrightarrow{\operatorname{Hom}_R(J,\eta)} \operatorname{Hom}_R(J, N) \to \operatorname{Ext}_R^1(J, Z_1) \to 0.$$

The induced map $\operatorname{Hom}_R(J, \eta)$ is surjective because $I_1$ is an injective precover of $N$, so also $\operatorname{Ext}_R^1(J, Z_1) = 0$. This concludes the proof. □

The Bass class is defined for every local ring with a dualizing complex, but for non-Cohen–Macaulay rings the relation to Gorenstein injective modules is yet to be uncovered.

The next result is [25, Theorem 2.13]. A straightforward proof, similar to that of Corollary (4.3.5), applies when the base ring is local Cohen–Macaulay with a dualizing module.

(6.1.8) **Corollary.** *Let $0 \to N' \to N \to N'' \to 0$ be a short exact sequence of $R$-modules. The following hold:*

(a) *If $N'$ is Gorenstein injective, then $N$ is Gorenstein injective if and only if $N''$ is so.*

(b) *If $N$ and $N''$ are Gorenstein injective, then $N'$ is Gorenstein injective if and only if $\operatorname{Ext}_R^1(J, N') = 0$ for all injective modules $J$.*

(c) *If the sequence splits, then $N$ is Gorenstein injective if and only if both $N'$ and $N''$ are so.* □

(6.1.9) **Definition.** We use the notation $\mathcal{C}^{\mathrm{GI}}(R)$ for the full subcategory (of $\mathcal{C}(R)$) of complexes of Gorenstein injective modules, and we use it with subscripts □ and ⊏ (defined as usual cf. (2.3.1)).

In the rest of this section, that is, in (6.1.10)–(6.1.12) we assume that $R$ is a **Cohen–Macaulay local ring with a dualizing module**. These last three results are auxiliaries needed for the proof of the main theorem in section 6.2.

(6.1.10) **Lemma.** *If $B \in \mathcal{C}_{\sqsubset}^{\mathrm{GI}}(R)$ is homologically trivial and $J \in \mathcal{C}_{\square}^{\mathrm{I}}(R)$, then also the complex $\operatorname{Hom}_R(J, B)$ is homologically trivial.*

*Proof.* If $J = 0$ the assertion is trivial, so we assume that $J$ is non-zero. We can also, without loss of generality, assume that $B_\ell = 0$ for $\ell > 0$ and $J_\ell = 0$ for $\ell < 0$. Set $u = \sup\{\ell \in \mathbb{Z} \mid J_\ell \neq 0\}$; we proceed by induction on $u$.

If $u = 0$ then $J$ is an injective module, and $\mathrm{Ext}_R^m(J, B_\ell) = 0$ for all $m > 0$ and all $\ell \in \mathbb{Z}$, cf. Theorem (6.1.7). Note that $\mathrm{Z}_\ell^B = 0$ for $\ell \geq 0$; it follows by Lemma (4.1.6)(c) that

$$\mathrm{Ext}_R^1(J, \mathrm{Z}_\ell^B) = \mathrm{Ext}_R^{1-\ell}(J, \mathrm{Z}_0^B) = 0$$

for $\ell \leq 0$, so $\mathrm{Hom}_R(J, B)$ is homologically trivial, again by (4.1.6)(c).

Let $u > 0$ and assume that $\mathrm{Hom}_R(\widetilde{J}, B)$ is homologically trivial for all complexes $\widetilde{J} \in \mathcal{C}_\square^{\mathrm{I}}(R)$ concentrated in at most $u - 1$ degrees. The short exact sequence of complexes $0 \to \sqsubset_{u-1} J \to J \to \Sigma^u J_u \to 0$ is degree-wise split, cf. (A.1.17), so it stays exact after application of $\mathrm{Hom}_R(-, B)$. As the complexes $\mathrm{Hom}_R(J_u, B)$ and $\mathrm{Hom}_R(\sqsubset_{u-1} J, B)$ are homologically trivial by, respectively, the induction base an hypothesis, it follows that also $\mathrm{Hom}_R(J, B)$ is homologically trivial. □

(6.1.11) **Proposition.** *If $Y$ is equivalent to $B \in \mathcal{C}_\sqsubset^{\mathrm{gI}}(R)$ and $U \simeq J \in \mathcal{C}_\square^{\mathrm{I}}(R)$, then $\mathbf{R}\mathrm{Hom}_R(U, Y)$ is represented by $\mathrm{Hom}_R(J, B)$.*

*Proof.* Take an injective resolution $I \in \mathcal{C}_\sqsubset^{\mathrm{I}}(R)$ of $Y$, then $\mathbf{R}\mathrm{Hom}_R(U, Y)$ is represented by the complex $\mathrm{Hom}_R(J, I)$. Since $B \simeq Y \simeq I$ there is by (A.3.5) a quasi-isomorphism $\beta\colon B \xrightarrow{\simeq} I$, and hence a morphism

$$\mathrm{Hom}_R(J, \beta) : \mathrm{Hom}_R(J, B) \longrightarrow \mathrm{Hom}_R(J, I).$$

The mapping cone $\mathcal{M}(\beta)$ is homologically trivial, and it follows by Corollary (6.1.8)(c) that it belongs to $\mathcal{C}_\sqsubset^{\mathrm{gI}}(R)$. By (A.2.1.2) we have

$$\mathcal{M}(\mathrm{Hom}_R(J, \beta)) = \mathrm{Hom}_R(J, \mathcal{M}(\beta)),$$

so it follows from the Lemma that the mapping cone $\mathcal{M}(\mathrm{Hom}_R(J, \beta))$ is homologically trivial, and $\mathrm{Hom}_R(J, \beta)$ is, therefore, a quasi-isomorphism, cf. (A.1.19). In particular, the two complexes $\mathrm{Hom}_R(J, B)$ and $\mathrm{Hom}_R(J, I)$ are equivalent, so also $\mathrm{Hom}_R(J, B)$ represents $\mathbf{R}\mathrm{Hom}_R(U, Y)$. □

(6.1.12) **Lemma.** *Let $J$ be an injective $R$–module. If $Y \in \mathcal{C}_{(\square)}(R)$ is equivalent to $B \in \mathcal{C}_\sqsubset^{\mathrm{gI}}(R)$ and $n \geq -\inf Y$, then*

$$\mathrm{Ext}_R^m(J, \mathrm{Z}_{-n}^B) = \mathrm{H}_{-(m+n)}(\mathbf{R}\mathrm{Hom}_R(J, Y))$$

*for $m > 0$. In particular, there is an inequality:*

$$\inf\,(\mathbf{R}\mathrm{Hom}_R(J, \mathrm{Z}_{-n}^B)) \geq \inf\,(\mathbf{R}\mathrm{Hom}_R(J, Y)) + n.$$

*Proof.* Since $-n \leq \inf Y = \inf B$ we have $\sqsubset_{-n} B \simeq \Sigma^{-n} \mathrm{Z}^B_{-n}$, cf. (A.1.14.1), and since $J$ is injective it follows by the Proposition that $\mathbf{R}\mathrm{Hom}_R(J, \mathrm{Z}^B_{-n})$ is represented by $\mathrm{Hom}_R(J, \Sigma^n \sqsubset_{-n} B)$. For $m > 0$ the isomorphism class $\mathrm{Ext}^m_R(J, \mathrm{Z}^B_{-n})$ is now represented by

$$\begin{aligned} \mathrm{H}_{-m}(\mathrm{Hom}_R(J, \Sigma^n \sqsubset_{-n} B)) &= \mathrm{H}_{-m}(\Sigma^n \mathrm{Hom}_R(J, \sqsubset_{-n} B)) \\ &= \mathrm{H}_{-(m+n)}(\mathrm{Hom}_R(J, \sqsubset_{-n} B)) \\ &= \mathrm{H}_{-(m+n)}(\sqsubset_{-n} \mathrm{Hom}_R(J, B)) \\ &= \mathrm{H}_{-(m+n)}(\mathrm{Hom}_R(J, B)), \end{aligned}$$

cf. (A.2.1.1), (A.1.3.1), and (A.1.20.1). It also follows by the Proposition that the complex $\mathrm{Hom}_R(J, B)$ represents $\mathbf{R}\mathrm{Hom}_R(J, Y)$, so $\mathrm{Ext}^m_R(J, \mathrm{Z}^B_{-n}) = \mathrm{H}_{-(m+n)}(\mathbf{R}\mathrm{Hom}_R(J, Y))$ as wanted, and the inequality of infima follows. □

## 6.2 Gorenstein Injective Dimension

By Observation (6.1.2) every injective module is Gorenstein injective, and the definition of Gorenstein injective dimension, (6.2.2) below, makes sense over any Noetherian ring. However, as for Gorenstein projective and flat dimensions, we only know how to get a nice functorial description if we work over a Cohen–Macaulay local ring with a dualizing module.

(6.2.1) **Setup.** In this section $R$ **is a Cohen–Macaulay local ring with a dualizing module** $D$.

(6.2.2) **Definition.** The *Gorenstein injective dimension*, $\mathrm{Gid}_R\, Y$, of a complex $Y \in \mathcal{C}_{(\sqsubset)}(R)$ is defined as

$$\mathrm{Gid}_R\, Y = \inf \{ \sup \{ \ell \in \mathbb{Z} \mid B_{-\ell} \neq 0 \} \mid Y \simeq B \in \mathcal{C}^{\mathrm{GI}}_{\sqsubset}(R) \}.$$

Note that the set over which infimum is taken is non-empty: any complex $Y \in \mathcal{C}_{(\sqsubset)}(R)$ has an injective resolution $Y \xrightarrow{\simeq} I \in \mathcal{C}^{\mathrm{I}}_{\sqsubset}(R)$, and $\mathcal{C}^{\mathrm{I}}_{\sqsubset}(R) \subseteq \mathcal{C}^{\mathrm{GI}}_{\sqsubset}(R)$.

(6.2.3) **Observation.** We note the following facts about the Gorenstein injective dimension of $Y \in \mathcal{C}_{(\sqsubset)}(R)$:

$$\begin{gathered} \mathrm{Gid}_R\, Y \in \{-\infty\} \cup \mathbb{Z} \cup \{\infty\}; \\ \mathrm{id}_R\, Y \geq \mathrm{Gid}_R\, Y \geq -\inf Y; \text{ and} \\ \mathrm{Gid}_R\, Y = -\infty \Leftrightarrow Y \simeq 0. \end{gathered}$$

While the Definition and the Observation above make perfect sense over any Noetherian ring, the proof (at least) of the next theorem relies heavily on the assumption that the base ring is local Cohen–Macaulay and has a dualizing module.

(6.2.4) **GID Theorem.** *Let $Y \in \mathcal{C}_{(\sqsubset)}(R)$ and $n \in \mathbb{Z}$. The following are equivalent:*

*(i) $Y$ is equivalent to a complex $B \in \mathcal{C}_{\square}^{\mathrm{GI}}(R)$ concentrated in degrees at least $-n$; and $B$ can be chosen with $B_\ell = 0$ for $\ell > \sup Y$.*

*(ii) $\mathrm{Gid}_R\, Y \leq n$.*

*(iii) $Y \in \mathcal{B}(R)$ and $n \geq -\sup U - \inf(\mathbf{R}\mathrm{Hom}_R(U,Y))$ for all $U \not\simeq 0$ in $\mathcal{I}(R)$.*

*(iv) $Y \in \mathcal{B}(R)$, $n \geq -\inf Y$, and $n \geq -\inf(\mathbf{R}\mathrm{Hom}_R(J,Y))$ for all injective modules $J$.*

*(v) $n \geq -\inf Y$ and the module $\mathrm{Z}_{-n}^{B}$ is Gorenstein injective whenever $B \in \mathcal{C}_{\sqsubset}^{\mathrm{GI}}(R)$ is equivalent to $Y$.*

*Proof.* It is immediate by Definition (6.2.2) that *(i)* implies *(ii)*.

*(ii)* $\Rightarrow$ *(iii)*: Choose a complex $B \in \mathcal{C}_{\square}^{\mathrm{GI}}(R)$ concentrated in degrees at least $-n$ and equivalent to $Y$. It follows by Proposition (3.2.13) that $B$, and thereby $Y$, belongs to the Bass class. Let $U \in \mathcal{I}(R)$ be homologically non-trivial, set $s = \sup U$, and choose by (A.5.1) a complex $J \simeq U$ in $\mathcal{C}_{\square}^{\mathrm{I}}(R)$ with $J_\ell = 0$ for $\ell > s$. By Proposition (6.1.11) the complex $\mathrm{Hom}_R(J,B)$ represents $\mathbf{R}\mathrm{Hom}_R(U,Y)$, in particular, $\inf(\mathbf{R}\mathrm{Hom}_R(U,Y)) = \inf(\mathrm{Hom}_R(J,B))$. For $\ell < -s-n$ and $p \in \mathbb{Z}$ either $p > s$ or $p + \ell \leq s + \ell < -n$, so the module

$$\mathrm{Hom}_R(J,B)_\ell = \prod_{p\in\mathbb{Z}} \mathrm{Hom}_R(J_p, B_{p+\ell})$$

vanishes. In particular, $\mathrm{H}_\ell(\mathrm{Hom}_R(J,B)) = 0$ for $\ell < -s-n$ and, therefore, $\inf(\mathbf{R}\mathrm{Hom}_R(U,Y)) \geq -s-n = -\sup U - n$, as desired.

*(iii)* $\Rightarrow$ *(iv)*: Since $D \in \mathcal{I}_0(R)$ we have

$$-\inf Y = -\inf(\mathbf{R}\mathrm{Hom}_R(D,Y)) \leq n,$$

cf. Lemma (3.4.3)(d).

*(iv)* $\Rightarrow$ *(v)*: Choose a complex $B \in \mathcal{C}_{\sqsubset}^{\mathrm{GI}}(R)$ equivalent to $Y$, and consider the short exact sequence of complexes $0 \to \Sigma^{-n}\mathrm{Z}_{-n}^{B} \to B_{-n}\supset \to B_{1-n}\sqsupset \to 0$. By Proposition (3.2.13) the complex $B_{1-n}\sqsupset$ belongs to $\mathcal{B}(R)$, and since $-n \leq \inf Y = \inf B$ we have $B_{-n}\supset \simeq B \simeq Y \in \mathcal{B}(R)$, cf. (A.1.14.4). By Lemma (3.2.12) it now follows that $\mathrm{Z}_{-n}^{B} \in \mathcal{B}_0(R)$. For injective modules $J$ we have $-\inf(\mathbf{R}\mathrm{Hom}_R(J,\mathrm{Z}_{-n}^{B})) \leq -\inf(\mathbf{R}\mathrm{Hom}_R(J,Y)) - n \leq 0$ by Lemma (6.1.12), so it follows by Theorem (6.1.7) that $\mathrm{Z}_{-n}^{B}$ is Gorenstein injective.

*(v)*$\Rightarrow$*(i)*: Choose by (A.3.2) an injective resolution $B \in \mathcal{C}_{\sqsubset}^{\mathrm{I}}(R) \subseteq \mathcal{C}_{\sqsubset}^{\mathrm{GI}}(R)$ of $Y$ with $B_\ell = 0$ for $\ell > \sup Y$. Since $-n \leq \inf Y = \inf B$ it follows from (A.1.14.4) that $Y \simeq B_{-n}\supset$, and $B_{-n}\supset \in \mathcal{C}_{\square}^{\mathrm{GI}}(R)$ as $\mathrm{Z}_{-n}^{B}$ is Gorenstein injective. $\square$

(6.2.5) **GID Corollary.** *For a complex $Y \in \mathcal{C}_{(\sqsubset)}(R)$ the next three conditions are equivalent.*

(*i*) $Y \in \mathcal{B}(R)$.

(*ii*) $\operatorname{Gid}_R Y < \infty$.

(*iii*) $Y \in \mathcal{C}_{(\square)}(R)$ *and* $\operatorname{Gid}_R Y \leq -\inf Y + \dim R$.

*Furthermore, if* $Y \in \mathcal{B}(R)$, *then*

$$\begin{aligned}\operatorname{Gid}_R Y &= \sup\{-\sup U - \inf(\mathbf{R}\mathrm{Hom}_R(U,Y)) \mid U \in \mathcal{I}(R) \wedge U \not\simeq 0\}\\ &= \sup\{-\inf(\mathbf{R}\mathrm{Hom}_R(J,Y)) \mid J \in \mathcal{C}_0^{\mathrm{I}}(R)\}.\end{aligned}$$

*Proof.* It follows by the Theorem that (*ii*) implies (*i*), and (*iii*) is clearly stronger than (*ii*). For $Y \in \mathcal{B}(R)$ and $J$ injective it follows by Lemma (3.4.13)(c) that

$$-\inf(\mathbf{R}\mathrm{Hom}_R(J,Y)) \leq -\inf Y + \dim R,$$

so by the equivalence of (*ii*) and (*iv*) in the Theorem we have $\operatorname{Gid}_R Y \leq -\inf Y + \dim R$ as wanted. This proves the equivalence of the three conditions.

For $Y \in \mathcal{B}(R)$ the equalities now follow by the equivalence of (*ii*), (*iii*), and (*iv*) in the Theorem. □

The next proposition shows that Gorenstein injective dimension is a refinement of injective dimension.

(6.2.6) **Proposition (GID–ID Inequality).** *For every complex $Y \in \mathcal{C}_{(\sqsubset)}(R)$ there is an inequality:*

$$\operatorname{Gid}_R Y \leq \operatorname{id}_R Y,$$

*and equality holds if* $\operatorname{id}_R Y < \infty$.

*Proof.* The inequality is, as we have already observed, immediate because injective modules are Gorenstein injective. Furthermore, equality holds if $Y$ is homologically trivial, so we assume that $\operatorname{id}_R Y = j \in \mathbb{Z}$ and choose, by (A.5.2.1), an $R$–module $T$ such that $j = -\inf(\mathbf{R}\mathrm{Hom}_R(T,Y))$. Also choose a injective module $J$ such that $T$ can be embedded in $J$. The short exact sequence of modules $0 \to T \to J \to C \to 0$ induces, cf. (A.4.8), a long exact sequence of homology modules:

$$\cdots \to \mathrm{H}_{-j}(\mathbf{R}\mathrm{Hom}_R(J,Y)) \to \mathrm{H}_{-j}(\mathbf{R}\mathrm{Hom}_R(T,Y)) \to \mathrm{H}_{-(j+1)}(\mathbf{R}\mathrm{Hom}_R(C,Y)) \to \cdots$$

Since, by (A.5.2.1), $\mathrm{H}_{-(j+1)}(\mathbf{R}\mathrm{Hom}_R(C,Y)) = 0$ while $\mathrm{H}_{-j}(\mathbf{R}\mathrm{Hom}_R(T,Y)) \neq 0$, we conclude that also $\mathrm{H}_{-j}(\mathbf{R}\mathrm{Hom}_R(J,Y))$ is non-zero. This proves, in view of GID Corollary (6.2.5), that $\operatorname{Gid}_R Y \geq j$, and hence equality holds. □

By GID Corollary (6.2.5) the next theorem is just a rewrite of the $\mathcal{B}$ version (3.2.10).

(6.2.7) **Gorenstein Theorem, GID Version.** *Let $R$ be a Cohen–Macaulay local ring with residue field $k$. If $R$ admits a dualizing module, then the following are equivalent:*

(*i*) *$R$ is Gorenstein.*

(*ii*) $\operatorname{Gid}_R k < \infty$.

(*iii*) $\operatorname{Gid}_R N < \infty$ *for all finite $R$–modules $N$.*

(*iv*) $\operatorname{Gid}_R N < \infty$ *for all $R$–modules $N$.*

(*v*) $\operatorname{Gid}_R Y < \infty$ *for all complexes $Y \in \mathcal{C}_{(\sqsubset)}(R)$.* □

In (6.2.8)–(6.2.12) we consider Gorenstein injective dimension for modules: we rewrite (6.2.4) and (6.2.5) in classical terms of resolutions and Ext modules.

(6.2.8) **Definition.** Let $N$ be an $R$-module. A *Gorenstein injective resolution* of $N$ is a complex of Gorenstein injective $R$–modules,

$$B = 0 \to B_0 \to B_{-1} \to \cdots \to B_\ell \to \cdots,$$

with homology concentrated in degree zero and $\mathrm{H}_0(B) = \mathrm{Z}_0^B \cong N$. That is, there is a homomorphism $\iota\colon N \to B_0$ such that the sequence

$$0 \to N \xrightarrow{\iota} B_0 \to B_{-1} \to \cdots \to B_\ell \to \cdots$$

is exact.

Every module has an injective resolution and hence a Gorenstein injective one.

(6.2.9) **Lemma.** *Let $N$ be an $R$-module. If $N$ is equivalent to $B \in \mathcal{C}_{\sqsubset}^{\mathrm{GI}}(R)$, then the truncated complex*

$$\sqsubset_0 B = 0 \to \mathrm{C}_0^B \to B_{-1} \to B_{-2} \to \cdots \to B_\ell \to \cdots$$

*is a Gorenstein injective resolution of $N$.*

*Proof.* Suppose $N$ is equivalent to $B \in \mathcal{C}_{\sqsubset}^{\mathrm{GI}}(R)$, then $\sup B = 0$, so $\sqsubset_0 B \simeq B \simeq N$ by (A.1.14.2), and we have an exact sequence of modules:

$$(\dagger) \qquad 0 \to N \to \mathrm{C}_0^B \to B_{-1} \to B_{-2} \to \cdots \to B_\ell \to \cdots.$$

Set $u = \sup\{\ell \in \mathbb{Z} \mid B_\ell \neq 0\}$, then also the sequence

$$0 \to B_u \to B_{u-1} \to \cdots \to B_0 \to \mathrm{C}_0^B \to 0$$

is exact. All the modules $B_u, \ldots, B_0$ are Gorenstein injective, so it follows by repeated applications of Corollary (6.1.8)(a) that $\mathrm{C}_0^B$ is Gorenstein injective, and therefore $\sqsubset_0 B$ is a Gorenstein injective resolution of $N$, cf. ($\dagger$). □

(6.2.10) **Remark.** It follows by the Lemma and Definition (6.2.2) that an $R$-module $N$ is Gorenstein injective if and only if $\operatorname{Gid}_R N \leq 0$. That is,

$$N \text{ is Gorenstein injective} \iff \operatorname{Gid}_R N = 0 \vee N = 0.$$

(6.2.11) **GID Theorem for Modules.** *Let $N$ be an $R$-module and $n \in \mathbb{N}_0$. The following are equivalent:*

(*i*) *$N$ has a Gorenstein injective resolution of length at most $n$. That is, there is an exact sequence of modules $0 \to N \to B_0 \to B_{-1} \to \cdots \to B_{-n} \to 0$, where $B_0, B_{-1}, \ldots, B_{-n}$ are Gorenstein injective.*

(*ii*) $\operatorname{Gid}_R N \leq n$.

(*iii*) *$N \in \mathcal{B}_0(R)$ and $\operatorname{Ext}_R^m(T,N) = 0$ for all $m > n$ and all $T \in \mathcal{I}_0(R)$.*

(*iv*) *$N \in \mathcal{B}_0(R)$ and $\operatorname{Ext}_R^m(J,N) = 0$ for all $m > n$ and all injective modules $J$.*

(*v*) *In any Gorenstein injective resolution of $N$,*

$$0 \to N \to B_0 \to B_{-1} \to \cdots \to B_\ell \to \cdots$$

*the cokernel*[2] *$W_{-n} = \operatorname{Coker}(B_{-n+2} \to B_{-n+1})$ is a Gorenstein injective module.*

*Proof.* If the sequence $0 \to N \to B_0 \to B_{-1} \to \cdots \to B_\ell \to \cdots$ is exact, then $N$ is equivalent to $B = 0 \to B_0 \to B_{-1} \to \cdots \to B_\ell \to \cdots$. The complex $B$ belongs to $\mathcal{C}_\sqsubset^{\mathrm{GI}}(R)$, and it has $\mathrm{Z}_0^B \cong N$, $\mathrm{Z}_{-1}^B \cong \operatorname{Coker}(N \to B_0)$, and $\mathrm{Z}_{-\ell}^B \cong \mathrm{C}_{-\ell+1}^B = \operatorname{Coker}(B_{-\ell+2} \to B_{-\ell+1})$ for $\ell \geq 2$. In view of the Lemma the equivalence of the five conditions now follows from Theorem (6.2.4). □

(6.2.12) **GID Corollary for Modules.** *For an $R$-module $N$ the next three conditions are equivalent.*

(*i*) $N \in \mathcal{B}_0(R)$.

(*ii*) $\operatorname{Gid}_R N < \infty$.

(*iii*) $\operatorname{Gid}_R N \leq \dim R$.

*Furthermore, if $N \in \mathcal{B}_0(R)$, then*

$$\begin{aligned}\operatorname{Gid}_R N &= \sup\{m \in \mathbb{N}_0 \mid \exists\, T \in \mathcal{I}_0(R) : \operatorname{Ext}_R^m(T,N) \neq 0\} \\ &= \sup\{m \in \mathbb{N}_0 \mid \exists\, J \in \mathcal{C}_0^{\mathrm{I}}(R) : \operatorname{Ext}_R^m(J,N) \neq 0\}.\end{aligned}$$

*Proof.* Immediate from Corollary (6.2.5). □

The next proposition shows that the Gorenstein injective dimension cannot grow under localization. In particular, it follows that $N_\mathfrak{p}$ is Gorenstein injective over $R_\mathfrak{p}$ if $N$ is Gorenstein injective over $R$ and, as we remarked in (6.1.3), this is not immediate from the definition.

[2] Appropriately interpreted for small $n$ as $W_0 = N$ and $W_{-1} = \operatorname{Coker}(N \to B_0)$.

(6.2.13) **Proposition.** *Let* $Y \in \mathcal{C}_{(\sqsubset)}(R)$. *For every* $\mathfrak{p} \in \operatorname{Spec} R$ *there is an inequality:*

$$\operatorname{Gid}_{R_\mathfrak{p}} Y_\mathfrak{p} \leq \operatorname{Gid}_R Y.$$

*Proof.* If $Y$ is equivalent to $B \in \mathcal{C}^{\mathrm{gI}}_{\sqsubset}(R)$, then $Y_\mathfrak{p}$ is equivalent to $B_\mathfrak{p}$. It is, therefore, sufficient to prove that a localized module $N_\mathfrak{p}$ is Gorenstein injective over $R_\mathfrak{p}$ if $N$ is Gorenstein injective over $R$.

Let $N$ be a Gorenstein injective $R$–module, and set $d = \dim R_\mathfrak{p}$. It follows from the definitions in (6.1.1) that there is an exact sequence

$$(\dagger) \qquad 0 \to K \to I_d \to \cdots \to I_2 \to I_1 \to N \to 0,$$

where the modules $I_d, \ldots, I_1$ are injective. Since $N$ and the injective modules all belong to the Bass class, it follows by repeated applications of Corollary (3.4.10)(a) that also $K \in \mathcal{B}_0(R)$. Localizing at $\mathfrak{p}$ we get an exact sequence

$$(\ddagger) \qquad 0 \to K_\mathfrak{p} \to (I_d)_\mathfrak{p} \to \cdots \to (I_2)_\mathfrak{p} \to (I_1)_\mathfrak{p} \to N_\mathfrak{p} \to 0,$$

where the modules $(I_\ell)_\mathfrak{p}$ are injective over $R_\mathfrak{p}$, while $N_\mathfrak{p}$ and $K_\mathfrak{p}$ belong to $\mathcal{B}(R_\mathfrak{p})$, cf. Observation (3.2.7). From GID Corollary (6.2.12) it follows that $\operatorname{Gid}_{R_\mathfrak{p}} K_\mathfrak{p} \leq d$, and since ($\ddagger$) is exact it follows by GID Theorem (6.2.11) that $N_\mathfrak{p}$ is Gorenstein injective. □

We will now use Foxby equivalence to prove a formula for Gorenstein injective dimension like that of Bass' for injective dimension (see page 13).

(6.2.14) **Lemma.** *If* $Y \in \mathcal{B}(R)$ *and* $U \in \mathcal{P}^{(\mathrm{f})}(R)$, *then*

$$\inf\left(\mathbf{R}\mathrm{Hom}_R(U,Y)\right) = \inf\left(\mathbf{R}\mathrm{Hom}_R(D \otimes^{\mathbf{L}}_R U, Y)\right).$$

*Proof.* The first equality in the calculation below follows as $Y \in \mathcal{B}(R)$; it also uses commutativity (A.4.19). The second equality follows by tensor evaluation (A.4.23) as $U \in \mathcal{P}^{(\mathrm{f})}(R)$, the third by Lemma (3.4.3)(a), the fourth by adjointness (A.4.21), and the last one by commutativity.

$$\begin{aligned}
\inf\left(\mathbf{R}\mathrm{Hom}_R(U,Y)\right) &= \inf\left(\mathbf{R}\mathrm{Hom}_R(U, \mathbf{R}\mathrm{Hom}_R(D,Y) \otimes^{\mathbf{L}}_R D)\right) \\
&= \inf\left(\mathbf{R}\mathrm{Hom}_R(U, \mathbf{R}\mathrm{Hom}_R(D,Y)) \otimes^{\mathbf{L}}_R D\right) \\
&= \inf\left(\mathbf{R}\mathrm{Hom}_R(U, \mathbf{R}\mathrm{Hom}_R(D,Y))\right) \\
&= \inf\left(\mathbf{R}\mathrm{Hom}_R(U \otimes^{\mathbf{L}}_R D, Y)\right) \\
&= \inf\left(\mathbf{R}\mathrm{Hom}_R(D \otimes^{\mathbf{L}}_R U, Y)\right). \quad \square
\end{aligned}$$

(6.2.15) **Theorem (Bass Formula for GID).** *If $Y$ is a complex with finite homology and finite Gorenstein injective dimension, i.e., $Y \in \mathcal{B}^{(\mathrm{f})}(R)$, then*

$$\mathrm{Gid}_R\, Y = \operatorname{depth} R - \inf Y.$$

*In particular,*

$$\mathrm{Gid}_R\, N = \operatorname{depth} R$$

*for finite modules $N \neq 0$ of finite Gorenstein injective dimension.*

*Proof.* By GID Corollary (6.2.5) we have

$$\mathrm{Gid}_R\, Y \le \dim R - \inf Y = \operatorname{depth} R - \inf Y$$

as $R$ is Cohen–Macaulay. To prove the opposite inequality, let $x_1, \ldots, x_d$ be a maximal $R$-sequence and set $T = R/(x_1, \ldots, x_d)$, then $T$ belongs to $\mathcal{P}_0^{\mathrm{f}}(R)$ and has $\mathrm{pd}_R\, T = \operatorname{depth} R$. By Theorem (3.4.6) the module $D \otimes_R T$ represents $D \otimes_R^{\mathbf{L}} T$, and $D \otimes_R T \in \mathcal{I}_0(R)$ by Foxby equivalence (3.4.11). Now the inequality in demand follows by (A.7.8), Lemma (6.2.14), and GID Corollary (6.2.5):

$$\begin{aligned}\operatorname{depth} R - \inf Y &= \mathrm{pd}_R\, T - \inf Y \\ &= -\inf\,(\mathbf{R}\mathrm{Hom}_R(T, Y)) \\ &= -\inf\,(\mathbf{R}\mathrm{Hom}_R(D \otimes_R T, Y)) \\ &\le \mathrm{Gid}_R\, Y. \quad \square\end{aligned}$$

(6.2.16) **Remark.** We do not know if the existence of a finite $R$-module $N \neq 0$ of finite Gorenstein injective dimension has any implications for the ring. By the celebrated Bass conjecture a local ring must be Cohen–Macaulay to accommodate a non-trivial finite module of finite injective dimension, so the question seems to be: If among the non-trivial finite modules there is one of finite Gorenstein injective dimension, is there then also one of finite injective dimension?

### Notes

The Bass formula for Gorenstein injective dimension, Theorem (6.2.15), was proved over Gorenstein rings by Enochs and Jenda [26, Theorem 4.3], see also [30, Corollary 4.11].

If $Y \in \mathcal{C}_{(\square)}(R)$ has finite injective dimension, then

$$\mathrm{id}_R\, Y = \sup\,\{\operatorname{depth} R_{\mathfrak{p}} - \mathrm{width}_{R_{\mathfrak{p}}}\, Y_{\mathfrak{p}} \mid \mathfrak{p} \in \operatorname{Spec} R\}.$$

This formula was proved for modules by Chouinard [14, Corollary 3.1] and extended to complexes by Yassemi [63, Theorem 2.10]; it holds over Noetherian rings in general.

One must ask if a similar formula holds for the Gorenstein injective dimension. That is, (to be modest) if $R$ is a Cohen–Macaulay local ring with a dualizing module, is then

$$\operatorname{Gid}_R Y = \sup\{\operatorname{depth} R_\mathfrak{p} - \operatorname{width}_{R_\mathfrak{p}} Y_\mathfrak{p} \mid \mathfrak{p} \in \operatorname{Spec} R\}$$

for all complexes $Y$ in the Bass class? For complexes with finite homology the answer is positive: by (A.6.3.2) and Cohen–Macaulayness of $R$ we have

$$\begin{aligned} \sup\{\operatorname{depth} R_\mathfrak{p} - \operatorname{width}_{R_\mathfrak{p}} Y_\mathfrak{p} \mid \mathfrak{p} \in \operatorname{Spec} R\} &= \sup\{\dim R_\mathfrak{p} - \inf Y_\mathfrak{p} \mid \mathfrak{p} \in \operatorname{Spec} R\} \\ &= \dim R - \inf Y \\ &= \operatorname{depth} R - \inf Y, \end{aligned}$$

so the desired formula was established in Theorem (6.2.15). In general, however, the answer is not yet known. Enochs and Jenda have in [30] established a couple of special cases where $\operatorname{Gid}_R N = \operatorname{depth} R - \operatorname{width}_R N$ for a non-finite module $N$; we return briefly to this point in Observation (6.3.7).

## 6.3 G–injective versus G–flat Dimension

In this section we establish the "G–parallels" of Ishikawa's formulas for flat and injective dimension.

(6.3.1) **Setup.** In this section $R$ is a **Cohen–Macaulay local ring with a dualizing module** $D$.

We start by rewriting Gorenstein Theorems (3.3.5) and (3.4.12) in terms of finiteness of Gorenstein dimensions.

(6.3.2) **Gorenstein Theorem, GFD/GID Version.** *Let $R$ be a Cohen–Macaulay local ring. If $D$ is a dualizing module for $R$, then the following are equivalent:*

- *(i)* *$R$ is Gorenstein.*
- *(ii)* $\operatorname{Gid}_R R < \infty$.
- *(ii')* $\operatorname{Gfd}_R D < \infty$.
- *(iii)* $\operatorname{Gid}_R N < \infty$ *and* $\operatorname{fd}_R N < \infty$ *for some $R$–module $N$ of finite depth.*
- *(iii')* $\operatorname{id}_R M < \infty$ *and* $\operatorname{Gfd}_R M < \infty$ *for some $R$–module $M$ of finite depth.*
- *(iv)* *A homologically bounded complex $X$ has finite Gorenstein flat dimension if and only if it has finite Gorenstein injective dimension; that is,* $\operatorname{Gfd}_R X < \infty \Leftrightarrow \operatorname{Gid}_R X < \infty$.

*Proof.* In view of GID Corollary (6.2.5) and GFD Corollary (5.2.6) the theorem is just a reformulation of the the special complexes version (3.3.5) and the special modules version (3.4.12). □

The next theorem is a parallel to Ishikawa's [42, Theorem 1.4].

(6.3.3) **Theorem.** *Let $E$ be an injective $R$–module. For every $X \in \mathcal{C}_{(\sqsupset)}(R)$ there is an inequality:*

$$\operatorname{Gid}_R(\operatorname{Hom}_R(X,E)) \leq \operatorname{Gfd}_R X,$$

*and equality holds if $E$ is faithfully injective.*

*Proof.* The inequality is trivial if $X$ is not of finite Gorenstein flat dimension, so we assume that $X \in \mathcal{A}(R)$. Then, by Lemma (3.2.9)(a), $\operatorname{Hom}_R(X,E)$ belongs to the Bass class, and for every injective module $J$ we have

$$(\dagger)\qquad \begin{aligned} -\inf(\mathbf{R}\operatorname{Hom}_R(J,\operatorname{Hom}_R(X,E))) &= -\inf(\mathbf{R}\operatorname{Hom}_R(J\otimes^{\mathbf{L}}_R X,E))\\ &\leq \sup(J\otimes^{\mathbf{L}}_R X), \end{aligned}$$

by adjointness (A.4.21) and (A.5.2.1). The desired equality now follows by GID Corollary (6.2.5) and GFD Corollary (5.2.6).

If $E$ is faithfully injective then, again by Lemma (3.2.9)(a), $X$ belongs to $\mathcal{A}(R)$ if and only if $\operatorname{Hom}_R(X,E)$ is in $\mathcal{B}(R)$; that is, the two dimensions are simultaneously finite. Furthermore, equality holds in (†), cf. (A.4.10), so the desired equality follows by Corollaries (6.2.5) and (5.2.6). □

As an immediate corollary we obtain a special case of Theorem (6.4.2):

(6.3.4) **Corollary.** *An $R$–module $M$ is Gorenstein flat if and only if $\operatorname{Hom}_R(M,E)$ is Gorenstein injective for every injective $R$–module $E$.* □

The next result is supposed to be the dual of Theorem (6.3.3) — i.e., the G–parallel of Ishikawa's [42, Theorem 1.5] — but, alas, it is not quite so. In (6.3.8) and (6.3.9) we work out a couple of special cases where the duality works as it should or, rather, as one could hope that it would.

(6.3.5) **Proposition.** *Let $E$ be an injective $R$–module. For every $Y \in \mathcal{C}_{(\sqsubset)}(R)$ there is an inequality:*

$$\operatorname{Gfd}_R(\operatorname{Hom}_R(Y,E)) \leq \operatorname{Gid}_R Y,$$

*and if $E$ is faithfully injective, then the two dimensions are simultaneously finite; that is,*

$$\operatorname{Gfd}_R(\operatorname{Hom}_R(Y,E)) < \infty \iff \operatorname{Gid}_R Y < \infty.$$

*Proof.* The inequality is trivial if $Y$ is not of finite Gorenstein injective dimension, so we assume that $Y \in \mathcal{B}(R)$. Then $\operatorname{Hom}_R(Y,E)$ belongs to the Auslander class, cf. Lemma (3.2.9)(b), and for every finite $R$–module $T$ we have

$$\begin{aligned} \sup(T\otimes^{\mathbf{L}}_R \operatorname{Hom}_R(Y,E)) &= \sup(\mathbf{R}\operatorname{Hom}_R(\mathbf{R}\operatorname{Hom}_R(T,Y),E))\\ &\leq -\inf(\mathbf{R}\operatorname{Hom}_R(T,Y)) \end{aligned}$$

by Hom evaluation (A.4.24) and (A.4.6.1). The desired inequality now follows by GID Corollary (6.2.5) and ($TI_0'$) in Theorem (5.4.6).

If $E$ is faithfully injective, then, again by Lemma (3.2.9)(b), we have $\mathrm{Hom}_R(Y,E) \in \mathcal{A}(R)$ if and only if $Y \in \mathcal{B}(R)$; that is, the two dimensions are simultaneously finite. □

(6.3.6) **Corollary.** *If $N$ is a Gorenstein injective $R$–module, then $\mathrm{Hom}_R(N,E)$ is Gorenstein flat for every injective $R$–module $E$.* □

(6.3.7) **Observation.** Let $E = \mathrm{E}_R(k)$ be the injective hull of the residue field. For any $R$–module $N$ it then follows by (A.6.4) that $\mathrm{width}_R N = \mathrm{depth}_R(\mathrm{Hom}_R(N,E))$, this is [63, Lemma 2.2], so $\mathrm{width}_R N \leq \dim R = \mathrm{depth}\, R$. Now, if $N \neq 0$ is Gorenstein injective, then $\mathrm{Hom}_R(N,E) \neq 0$ is Gorenstein flat by the Corollary, so

$$\begin{aligned} 0 = \mathrm{Gfd}_R(\mathrm{Hom}_R(N,E)) &\geq \mathrm{depth}\, R - \mathrm{depth}_R(\mathrm{Hom}_R(N,E)) \\ &= \mathrm{depth}\, R - \mathrm{width}_R N \end{aligned}$$

by the AB formula (5.4.9). Thus,

$$\mathrm{width}_R N = \mathrm{depth}\, R \tag{6.3.7.1}$$

for every Gorenstein injective $R$–module $N$.

(6.3.8) **Proposition.** *Let $E$ be a faithfully injective $R$–module. For every complex $Y \in \mathcal{C}_{(\sqsubset)}^{(\mathrm{f})}(R)$ there is an equality:*

$$\mathrm{Gfd}_R(\mathrm{Hom}_R(Y,E)) = \mathrm{Gid}_R\, Y.$$

*Proof.* By Proposition (6.3.5) it is sufficient to prove that $\mathrm{Gid}_R Y \leq \mathrm{Gfd}_R(\mathrm{Hom}_R(Y,E))$ for $Y \in \mathcal{B}^{(\mathrm{f})}(R)$. Let $x_1,\ldots,x_d$ be a maximal $R$–sequence and set $T = R/(x_1,\ldots,x_d)$, then it follows by Theorem (5.4.6), (A.4.10), (A.7.8), and Theorem (6.2.15) that

$$\begin{aligned} \mathrm{Gfd}_R(\mathrm{Hom}_R(Y,E)) &\geq \sup\,(R/(x_1,\ldots,x_d) \otimes^{\mathbf{L}}_R \mathrm{Hom}_R(Y,E)) \\ &= \sup\,(\mathbf{R}\mathrm{Hom}_R(\mathbf{R}\mathrm{Hom}_R(R/(x_1,\ldots,x_d),Y),E)) \\ &= -\inf\,(\mathbf{R}\mathrm{Hom}_R(R/(x_1,\ldots,x_d),Y)) \\ &= \mathrm{pd}_R\, R/(x_1,\ldots,x_d) - \inf Y \\ &= \mathrm{depth}\, R - \inf Y \\ &= \mathrm{Gid}_R\, Y. \quad \square \end{aligned}$$

(6.3.9) **Observation.** Let $X \in \mathcal{C}_{(\sqsupset)}(R)$, and let $E$ and $E'$ be faithfully injective $R$–modules. For $T \in \mathcal{P}_0^{\mathrm{f}}(R)$ we have

$$\begin{aligned}\sup(T \otimes^{\mathbf{L}}_R X) &= -\inf(\mathbf{R}\mathrm{Hom}_R(T \otimes^{\mathbf{L}}_R X, E')) \\ &= -\inf(\mathbf{R}\mathrm{Hom}_R(T, \mathbf{R}\mathrm{Hom}_R(X, E'))) \\ &= \sup(\mathbf{R}\mathrm{Hom}_R(\mathbf{R}\mathrm{Hom}_R(T, \mathrm{Hom}_R(X, E')), E)) \\ &= \sup(T \otimes^{\mathbf{L}}_R \mathrm{Hom}_R(\mathrm{Hom}_R(X, E'), E))\end{aligned}$$

by adjointness (A.4.21), (A.4.10), and tensor evaluation (A.4.23). By ($T_x$) in Theorem (5.4.6) and Lemma (3.2.9) it now follows that

$$(\dagger) \qquad \mathrm{Gfd}_R X = \mathrm{Gfd}_R(\mathrm{Hom}_R(\mathrm{Hom}_R(X, E'), E)).$$

Set $Y = \mathrm{Hom}_R(X, E')$, then $Y \in \mathcal{C}_{(\sqsubset)}(R)$, and by (†), Proposition (6.3.5), and Theorem (6.3.3) we have

$$\mathrm{Gfd}_R X = \mathrm{Gfd}_R(\mathrm{Hom}_R(Y, E)) \leq \mathrm{Gid}_R Y \leq \mathrm{Gfd}_R X.$$

That is, if $Y \in \mathcal{C}_{(\sqsubset)}(R)$ is equivalent to a complex $\mathrm{Hom}_R(X, E')$, where $X \in \mathcal{A}(R)$ and $E'$ is faithfully injective, then

$$\mathrm{Gfd}_R(\mathrm{Hom}_R(Y, E)) = \mathrm{Gid}_R Y$$

for every faithfully injective $R$–module $E$.

(6.3.10) **Theorem.** *If $Y$ is a complex of finite Gorenstein injective dimension, i.e., $Y \in \mathcal{B}(R)$, then the following numbers are equal:*

(D) $\mathrm{Gid}_R Y$,

(EI) $\sup\{-\sup U - \inf(\mathbf{R}\mathrm{Hom}_R(U, Y)) \mid U \in \mathcal{I}(R) \wedge U \not\simeq 0\}$, *and*

(EE) $\sup\{-\inf(\mathbf{R}\mathrm{Hom}_R(\mathrm{E}_R(R/\mathfrak{p}), Y)) \mid \mathfrak{p} \in \mathrm{Spec}\, R\}$.

*Furthermore, if $Y \simeq \mathrm{Hom}_R(X, E)$, where $X \in \mathcal{A}(R)$ and $E$ is a faithfully injective $R$–module, then also the next three numbers are equal, and equal to those above.*

(EI$_0^{\prime}$) $\sup\{-\inf(\mathbf{R}\mathrm{Hom}_R(T, Y)) \mid T \in \mathcal{I}_0^{\mathrm{f}}(R)\}$,

(EF) $\sup\{-\sup U - \inf(\mathbf{R}\mathrm{Hom}_R(U, Y)) \mid U \in \mathcal{F}(R) \wedge U \not\simeq 0\}$, *and*

(Ex) $\sup\{-\inf(\mathbf{R}\mathrm{Hom}_R(R/(\boldsymbol{x}), Y)) \mid \boldsymbol{x} = x_1, \ldots, x_t$ is an $R$–sequence$\}$.

*Proof.* It was shown in GID Corollary (6.2.5) that the numbers (D) and (EI) are equal. Furthermore, every injective $R$–module is a direct sum of indecomposable injectives, i.e., modules of the form $\mathrm{E}_R(R/\mathfrak{p})$, so in view of the functorial isomorphism

$$\mathrm{Hom}_R(\coprod_p E_p, -) \cong \prod_p \mathrm{Hom}_R(E_p, -),$$

it follows, still by Corollary (6.2.5), that ($D$) = ($EE$).

Now, assume that $Y \simeq \operatorname{Hom}_R(X,E)$, where $X \in \mathcal{A}(R)$ and $E$ is a faithfully injective $R$–module. For $U \in \mathcal{C}_{(\square)}(R)$ we then have

$$
\begin{aligned}
-\inf(\mathbf{R}\mathrm{Hom}_R(U,Y)) &= -\inf(\mathbf{R}\mathrm{Hom}_R(U,\operatorname{Hom}_R(X,E)))\\
(\dagger)\qquad &= -\inf(\mathbf{R}\mathrm{Hom}_R(U\otimes^{\mathbf{L}}_R X,E))\\
&= \sup(U\otimes^{\mathbf{L}}_R X)
\end{aligned}
$$

by adjointness (A.4.21) and (A.4.10). Since ($D$) = $\operatorname{Gfd}_R X$ by Theorem (6.3.3), it is sufficient to prove that the numbers ($EI'_0$), ($EF$), and ($Ex$) are equal to $\operatorname{Gfd}_R X$; and in view of (†) this is immediate from Theorem (5.4.6). □

(6.3.11) **Corollary.** *If $N$ is a module of finite Gorenstein injective dimension, i.e., $N \in \mathcal{B}_0(R)$, then the following numbers are equal:*

($D$) $\operatorname{Gid}_R N$,

($EI_0$) $\sup\{m \in \mathbb{N}_0 \mid \exists\, T \in \mathcal{I}_0(R) \colon \operatorname{Ext}^m_R(T,N) \neq 0\}$, *and*

($EE$) $\sup\{m \in \mathbb{N}_0 \mid \exists\, \mathfrak{p} \in \operatorname{Spec} R \colon \operatorname{Ext}^m_R(\mathrm{E}_R(R/\mathfrak{p}),N) \neq 0\}$.

*Furthermore, if $N \cong \operatorname{Hom}_R(M,E)$, where $M \in \mathcal{A}_0(R)$ and $E$ is a faithfully injective $R$–module, then also the next three numbers are equal, and equal to those above.*

($EI'_0$) $\sup\{m \in \mathbb{N}_0 \mid \exists\, T \in \mathcal{I}^{\mathrm{f}}_0(R) \colon \operatorname{Ext}^m_R(T,N) \neq 0\}$,

($EF_0$) $\sup\{m \in \mathbb{N}_0 \mid \exists\, T \in \mathcal{F}_0(R) \colon \operatorname{Ext}^m_R(T,N) \neq 0\}$, *and*

($Ex$) $\sup\{m \in \mathbb{N}_0 \mid \operatorname{Ext}^m_R(R/(\boldsymbol{x}),N) \neq 0$ for some $R$–seq. $\boldsymbol{x} = x_1,\dots,x_t\}$.

### Notes

The equality (6.3.7.1) was proved by Enochs and Jenda [30, Lemma 4.1] under slightly different conditions.

Theorem (6.3.2) — the GFD/GID version of the Gorenstein Theorem — strengthens the PD/ID version (see page 6), and it is natural to ask if an even stronger version exists: does the existence of an $R$–module (or complex) of finite depth, finite Gorenstein flat dimension, and finite Gorenstein injective dimension imply that $R$ is Gorenstein? The answer is not known (to the author).

## 6.4 Exercises in Stability

In the previous section we used the functorial characterizations of Gorenstein flat and injective dimensions to prove a couple of stability results. While this approach is fast, it also has a serious drawback: it only works over certain Cohen–Macaulay rings. Some of the results, however, hold over general Noetherian rings; and in this section we show how to prove them by working with resolutions.

The main theorem of this section is (6.4.2): a module is Gorenstein flat if and only if the dual with respect to every injective module is Gorenstein injective; it is the general version of Corollary (6.3.4).

While detailed proofs are provided for the first three results, the rest of the section can be taken as a series of exercises; the proofs are, at any rate, reduced to hints. This is particularly true for the final (6.4.13) which is only interesting from a "derived category point of view" and should be proved by within this framework.

(6.4.1) **Proposition.** *Let $E$ be an injective $R$-module. If $F \in \mathcal{C}^{\mathrm{F}}(R)$ is a complete flat resolution, then* $\mathrm{Hom}_R(F, E)$ *is a complete injective resolution; and the converse holds if $E$ is faithfully injective.*

*Proof.* If $F \in \mathcal{C}^{\mathrm{F}}(R)$ and $E$ is injective, then $\mathrm{Hom}_R(F, E)$ is a complex of injective modules. Furthermore, if $F$ is homologically trivial, then so is $\mathrm{Hom}_R(F, E)$; and the converse holds if $E$ is faithfully injective. For every (injective) module $J$ we have

$$\mathrm{Hom}_R(J \otimes_R F, E) \cong \mathrm{Hom}_R(J, \mathrm{Hom}_R(F, E))$$

by adjointness (A.2.8), so if $J \otimes_R F$ is homologically trivial, then so is $\mathrm{Hom}_R(J, \mathrm{Hom}_R(F, E))$; and, again, the converse holds if $E$ is faithfully injective. □

(6.4.2) **Theorem.** *The following are equivalent for an $R$-module $M$:*

(*i*) *$M$ is Gorenstein flat.*

(*ii*) $\mathrm{Hom}_R(M, E)$ *is Gorenstein injective for some faithfully injective $R$-module $E$.*

(*iii*) $\mathrm{Hom}_R(M, E)$ *is Gorenstein injective for every injective $R$-module $E$.*

*Proof.* It is evident that (*iii*) implies (*ii*); this leaves us two implications to prove.

(*i*) ⇒ (*iii*): Let $F$ be a complete flat resolution with $\mathrm{C}_0^F \cong M$, and let $E$ be injective. Then, by the Proposition, $\mathrm{Hom}_R(F, E)$ is a complete injective resolution, and $\mathrm{Z}_0^{\mathrm{Hom}_R(F,E)} \cong \mathrm{Hom}_R(M, E)$ by Lemma (4.1.1)(b), so $\mathrm{Hom}_R(M, E)$ is Gorenstein injective as wanted.

(*ii*) ⇒ (*i*): We assume that $E$ is a faithfully injective $R$-module such that $\mathrm{Hom}_R(M, E)$ is Gorenstein injective, and we set out to construct a complete flat resolution $F$ with $\mathrm{C}_0^F \cong M$. If we can construct a short exact sequence

$$(\ddagger) \qquad 0 \to M \to F_{-1} \to C_{-1} \to 0,$$

where $F_{-1}$ is flat and $C_{-1}$ is a module with the same property as $M$ (that is, $\mathrm{Hom}_R(C_{-1}, E)$ is Gorenstein injective), then the right half of a complex $F \in \mathcal{C}^{\mathrm{F}}(R)$ can be constructed recursively. The left half of $F$ we get for free by taking a flat resolution of $M$, and a complex $F$ established this way is homologically trivial with $\mathrm{C}_0^F \cong M$. Consider the homologically trivial complex $\mathrm{Hom}_R(F,E)$ of injective modules. By Lemma (4.1.1)(b) we have

$$(\star) \qquad \mathrm{Z}_\ell^{\mathrm{Hom}_R(F,E)} \cong \mathrm{Hom}_R(\mathrm{C}_{-\ell}^F, E),$$

so for $\ell > 0$ the kernel $\mathrm{Z}_\ell^{\mathrm{Hom}_R(F,E)}$ is a Gorenstein injective module, because $\mathrm{C}_{-\ell}^F = C_{-\ell}$ is a module with the same property as $M$. Let $J$ be an injective module; for $\ell > 0$ we then have

$$\mathrm{Ext}_R^1(J, \mathrm{Z}_\ell^{\mathrm{Hom}_R(F,E)}) = 0,$$

cf. Proposition (6.1.5), and for $\ell \leq 0$ we have

$$\begin{aligned}\mathrm{Ext}_R^1(J, \mathrm{Z}_\ell^{\mathrm{Hom}_R(F,E)}) &= \mathrm{Ext}_R^{1-\ell}(J, \mathrm{Z}_0^{\mathrm{Hom}_R(F,E)}) \\ &= \mathrm{Ext}_R^{1-\ell}(J, \mathrm{Hom}_R(M,E)) = 0\end{aligned}$$

by Lemma (4.1.6)(c), (4.1.1)(b), and the assumption on $M$. Thus, it follows by (4.1.6)(c) that $\mathrm{Hom}_R(J, \mathrm{Hom}_R(F,E))$ is homologically trivial for every injective module $J$; that is, $\mathrm{Hom}_R(F,E)$ is a complete injective resolution and, therefore, $F$ is a complete flat resolution by Proposition (6.4.1). To prove the theorem it is now sufficient to construct the short exact sequence (‡).

The module $M^\vee = \mathrm{Hom}_R(M,E)$ is Gorenstein injective by assumption, so by definition we have a short exact sequence

$$0 \to Z \to I \xrightarrow{\partial} M^\vee \to 0,$$

where $I$ is injective. Applying the exact functor $-^\vee = \mathrm{Hom}_R(-,E)$, we get another short exact sequence

$$0 \to M^{\vee\vee} \xrightarrow{\partial^\vee} I^\vee \to Z^\vee \to 0.$$

The canonical map $\delta_M^E \colon M \to M^{\vee\vee}$ is injective because $E$ is faithfully injective, so we have an injective map $\nu = \partial^\vee \delta_M^E$ from $M$ into the flat module $I^\vee$. Let $\phi\colon M \to F_{-1}$ be a flat preenvelope of $M$, then $\phi$ is injective by Lemma (4.3.3), so with $C_{-1} = \mathrm{Coker}\,\phi$ we have an exact sequence

$$(*) \qquad 0 \to M \xrightarrow{\phi} F_{-1} \to C_{-1} \to 0.$$

We now want to prove that $\mathrm{Hom}_R(C_{-1},E)$ is Gorenstein injective. From $(*)$ we get a short exact sequence

$$0 \to \mathrm{Hom}_R(C_{-1},E) \to \mathrm{Hom}_R(F_{-1},E) \xrightarrow{\mathrm{Hom}_R(\phi,E)} \mathrm{Hom}_R(M,E) \to 0;$$

where the module $\mathrm{Hom}_R(F_{-1},E)$ is injective and $\mathrm{Hom}_R(M,E)$ is Gorenstein injective by assumption. To prove that also $\mathrm{Hom}_R(C_{-1},E)$ is Gorenstein injective it is, by Corollary (6.1.8), sufficient to see that $\mathrm{Ext}^1_R(J, \mathrm{Hom}_R(C_{-1}, E)) = 0$ for all injective modules $J$. Let $J$ be an injective module, $\mathrm{Ext}^1_R(J, \mathrm{Hom}_R(C_{-1}, E))$ vanishes if and only if the map

$$\mathrm{Hom}_R(J, \mathrm{Hom}_R(\phi, E)) \colon \mathrm{Hom}_R(J, \mathrm{Hom}_R(F_{-1}, E)) \to \mathrm{Hom}_R(J, \mathrm{Hom}_R(M, E))$$

is surjective ($\mathrm{Ext}^1_R(J, \mathrm{Hom}_R(F_{-1}, E)) = 0$ because $\mathrm{Hom}_R(F_{-1}, E)$ is injective), so we consider the commutative diagram

$$\begin{array}{ccc} \mathrm{Hom}_R(J, \mathrm{Hom}_R(F_{-1}, E)) & \xrightarrow{\mathrm{Hom}_R(J, \mathrm{Hom}_R(\phi, E))} & \mathrm{Hom}_R(J, \mathrm{Hom}_R(M, E)) \\ \cong \big\downarrow \varsigma_{JF_{-1}E} & & \cong \big\downarrow \varsigma_{JME} \\ \mathrm{Hom}_R(F_{-1}, \mathrm{Hom}_R(J, E)) & \xrightarrow{\mathrm{Hom}_R(\phi, \mathrm{Hom}_R(J, E))} & \mathrm{Hom}_R(M, \mathrm{Hom}_R(J, E)) \end{array}$$

The module $\mathrm{Hom}_R(J, E)$ is flat, and $\phi$ is a flat preenvelope of $M$, so the map $\mathrm{Hom}_R(\phi, \mathrm{Hom}_R(J, E))$ is surjective, cf. (4.3.2), and hence so is $\mathrm{Hom}_R(J, \mathrm{Hom}_R(\phi, E))$. This concludes the proof. $\square$

(6.4.3) **Theorem.** *Let $X \in \mathcal{C}_{(\sqsupset)}(R)$; if $U$ is a complex of finite injective dimension, i.e., $U \in \mathcal{I}(R)$, then*

$$\mathrm{Gid}_R(\mathbf{R}\mathrm{Hom}_R(X, U)) \leq \mathrm{Gfd}_R X + \mathrm{id}_R U.$$

*Proof.* We can assume that $U$ is homologically non-trivial, otherwise the inequality is trivial; and we set $s = \sup U$ and $i = \mathrm{id}_R U$. The inequality is also trivial if $X$ is homologically trivial or not of finite Gorenstein flat dimension, so we assume that $X \not\simeq 0$ and set $g = \mathrm{Gfd}_R X \in \mathbb{Z}$. We can now choose a complex $A \in \mathcal{C}^{\mathrm{GF}}_{\square}(R)$ which is equivalent to $X$ and has $A_\ell = 0$ for $\ell > g$; we set $v = \inf \{\ell \in \mathbb{Z} \mid A_\ell \neq 0\}$. By (A.5.1) $U$ is equivalent to a complex $I$ of injective modules concentrated in degrees $s, \ldots, -i$. Now, $\mathbf{R}\mathrm{Hom}_R(X, U)$ is represented by the complex $\mathrm{Hom}_R(A, I)$ with

$$(\dagger) \qquad \mathrm{Hom}_R(A, I)_\ell = \prod_{p \in \mathbb{Z}} \mathrm{Hom}_R(A_p, I_{p+\ell}) = \bigoplus_{p=v}^{g} \mathrm{Hom}_R(A_p, I_{p+\ell}).$$

The modules $\mathrm{Hom}_R(A_p, I_{p+\ell})$ are Gorenstein injective by Theorem (6.4.2), and finite sums of Gorenstein injective modules are Gorenstein injective, cf. Corollary (6.1.8)(c), so $\mathrm{Hom}_R(A, I) \in \mathcal{C}^{\mathrm{GI}}(R)$. Furthermore, it is easy to see that $\mathrm{Hom}_R(A, I)$ is bounded: by ($\dagger$) we have $\mathrm{Hom}_R(A, I)_\ell = 0$ for $\ell > s - v$; and if $\ell < -(i+g)$, then either $p > g$ or $p + \ell \leq g + \ell < -i$, so also for $\ell < -(i+g)$ is $\mathrm{Hom}_R(A, I)_\ell = 0$. That is, $\mathrm{Hom}_R(A, I)$ is a bounded complex of Gorenstein injective modules concentrated in degrees at least $-(i+g)$ and, therefore, $\mathrm{Gid}_R(\mathbf{R}\mathrm{Hom}_R(X, U)) \leq i + g = \mathrm{Gfd}_R X + \mathrm{id}_R U$ as wanted. $\square$

(6.4.4) **Proposition.** *If $F \in \mathcal{C}^{\mathrm{F}}(R)$ is a complete flat resolution, then so is $F \otimes_R F'$ for every flat $R$–module $F'$.*

*In particular: if $M$ is Gorenstein flat, then $M \otimes_R F'$ is the same for every flat module $F'$.*

*Proof.* Use the definitions and associativity. □

(6.4.5) **Theorem.** *Let $X \in \mathcal{C}_{(\sqsupset)}(R)$; if $U$ is a complex of finite flat dimension, i.e., $U \in \mathcal{F}(R)$, then*

$$\mathrm{Gfd}_R(X \otimes^{\mathbf{L}}_R U) \leq \mathrm{Gfd}_R X + \mathrm{fd}_R U.$$

*Proof.* Apply the technique from the proof of Theorem (6.4.3); only this time use Proposition (6.4.4) and Corollary (5.1.9). □

Note that (6.4.5) generalizes (2.3.17)(b).

(6.4.6) **Proposition.** *If $P \in \mathcal{C}^{\mathrm{P}}(R)$ is a complete projective resolution, then so is $\mathrm{Hom}_R(P', P)$ for every finite projective $R$–module $P'$.*

*In particular: if $M$ is Gorenstein projective, then $\mathrm{Hom}_R(P', M)$ is the same for every finite projective module $P'$.*

*Proof.* Use the definitions and Hom evaluation. □

The next result generalizes (2.3.17)(a).

(6.4.7) **Theorem.** *Let $X \in \mathcal{C}_{(\sqsupset)}(R)$; if $U$ is a complex with finite homology and finite projective dimension, i.e., $U \in \mathcal{P}^{(\mathrm{f})}(R)$, then*

$$\mathrm{Gpd}_R(\mathbf{R}\mathrm{Hom}_R(U, X)) \leq \mathrm{Gpd}_R X - \inf U.$$

*Proof.* Apply the technique from the proof of Theorem (6.4.3); only this time use Proposition (6.4.6) and Corollary (4.3.5). □

(6.4.8) **Proposition.** *If $I \in \mathcal{C}^{\mathrm{I}}(R)$ is a complete injective resolution, then so is $\mathrm{Hom}_R(P', I)$ for every finite projective $R$–module $P'$.*

*In particular: if $N$ is Gorenstein injective, then $\mathrm{Hom}_R(P', N)$ is the same for every finite projective module $P'$.*

*Proof.* Use the definitions and swap. □

(6.4.9) **Theorem.** *Let $Y \in \mathcal{C}_{(\sqsubset)}(R)$; if $U$ is a complex with finite homology and finite projective dimension, i.e., $U \in \mathcal{P}^{(\mathrm{f})}(R)$, then*

$$\mathrm{Gid}_R(\mathbf{R}\mathrm{Hom}_R(U, Y)) \leq \mathrm{Gid}_R Y + \mathrm{pd}_R U.$$

*Proof.* Apply the technique from the proof of Theorem (6.4.3); only this time use Proposition (6.4.8) and Corollary (6.1.8). □

(6.4.10) **Theorem.** *Let $R$ be a Cohen–Macaulay local ring with a dualizing module, and let $Y \in \mathcal{C}_{(\sqsubset)}(R)$. If $U$ is a complex of finite injective dimension, i.e., $U \in \mathcal{I}(R)$, then*

$$\mathrm{Gfd}_R(\mathbf{R}\mathrm{Hom}_R(Y,U)) \leq \mathrm{Gid}_R\, Y + \sup U.$$

*Proof.* Apply the technique from the proof of (6.4.3), but use Corollary (6.3.6) instead of Theorem (6.4.2) and Corollary (5.1.9) instead of (6.1.8). See also (6.4.12) below. □

(6.4.11) **Remark.** We do not know if it is, at all, true that an $R$–module $N$ is Gorenstein injective if and only if $\mathrm{Hom}_R(N,E)$ is Gorenstein flat for every injective $R$–module $E$; not even if $R$ is local Cohen–Macaulay with a dualizing module, cf. Corollary (6.3.6). In particular, it is not obvious that the technique from the proof of Theorem (6.4.2) can be used to solve the problem. One of the obstructions seems to be that it is not clear whether $\mathrm{Hom}_R(\eta,E)\colon \mathrm{Hom}_R(N,E) \longrightarrow \mathrm{Hom}_R(I,E)$ is a flat preenvelope of $\mathrm{Hom}_R(N,E)$ whenever $\eta\colon I \to N$ is an injective precover of $N$ and $E$ is injective. The dual is, however, true: if $\phi\colon M \to F$ is a flat preenvelope of $M$, then $\mathrm{Hom}_R(\phi,E)\colon \mathrm{Hom}_R(F,E) \longrightarrow \mathrm{Hom}_R(M,E)$ is an injective precover of $\mathrm{Hom}_R(M,E)$ for every injective module $E$. This can be deduced from the closing argument in the proof of (6.4.2).

(6.4.12) **Remark.** A different proof of Theorem (6.4.10) is available: it is possible (and using derived category methods it is even easy) to prove that $\mathbf{R}\mathrm{Hom}_R(Y,U)$ belongs to the Auslander class when $Y \in \mathcal{B}(R)$ and $U \in \mathcal{I}(R)$. For every $T \in \mathcal{P}_0^{\mathrm{f}}(R)$ we have

$$\begin{aligned}
\sup(T \otimes_R \mathbf{R}\mathrm{Hom}_R(Y,U)) &= \sup(\mathbf{R}\mathrm{Hom}_R(\mathbf{R}\mathrm{Hom}_R(T,Y),U))\\
&\leq \sup U - \inf(\mathbf{R}\mathrm{Hom}_R(T,Y))\\
&= -\inf(\mathbf{R}\mathrm{Hom}_R(D \otimes^{\mathbf{L}}_R T,Y)) + \sup U\\
&= -\inf(\mathbf{R}\mathrm{Hom}_R(D \otimes_R T,Y)) + \sup U\\
&\leq \mathrm{Gid}_R\, Y + \sup U
\end{aligned}$$

by Hom evaluation (A.4.24), (A.4.6.1), Lemma (6.2.14), Theorems (3.4.6) and (3.4.11), and GID Corollary (6.2.5), so

$$\mathrm{Gfd}_R(\mathbf{R}\mathrm{Hom}_R(Y,U)) \leq \mathrm{Gid}_R\, Y + \sup U$$

by Theorem (5.4.6).

The reader is invited to apply similar methods reestablish special cases of Theorems (6.4.3), (6.4.5), (6.4.7), and (6.4.9) in the following form:

**(6.4.13) Theorem (Stability of Auslander Categories).** *Let $R$ be a Cohen–Macaulay local ring with a dualizing module. The following hold for complexes $A \in \mathcal{A}(R)$, $B \in \mathcal{B}(R)$, $F \in \mathcal{F}(R)$, $I \in \mathcal{I}(R)$, and $P \in \mathcal{P}^{(f)}(R)$:*

- $\mathbf{R}\mathrm{Hom}_R(A, I) \in \mathcal{B}(R)$ *with*
$$\mathrm{Gid}_R(\mathbf{R}\mathrm{Hom}_R(A, I)) \leq \mathrm{Gfd}_R\, A + \mathrm{id}_R\, I;$$
- $\mathbf{R}\mathrm{Hom}_R(B, I) \in \mathcal{A}(R)$ *with*
$$\mathrm{Gfd}_R(\mathbf{R}\mathrm{Hom}_R(B, I)) \leq \mathrm{Gid}_R\, B + \sup I;$$
- $A \otimes^{\mathbf{L}}_R F \in \mathcal{A}(R)$ *with*
$$\mathrm{Gfd}_R(A \otimes^{\mathbf{L}}_R F) \leq \mathrm{Gfd}_R\, A + \mathrm{fd}_R\, F;$$
- $\mathbf{R}\mathrm{Hom}_R(P, A) \in \mathcal{A}(R)$ *with*
$$\mathrm{Gpd}_R(\mathbf{R}\mathrm{Hom}_R(P, A)) \leq \mathrm{Gpd}_R\, A - \inf P; \quad \textit{and}$$
- $\mathbf{R}\mathrm{Hom}_R(P, B) \in \mathcal{B}(R)$ *with*
$$\mathrm{Gid}_R(\mathbf{R}\mathrm{Hom}_R(P, B)) \leq \mathrm{Gid}_R\, B + \mathrm{pd}_R\, P.$$

## Notes

A special case of Theorem (6.4.2) follows from [27, Lemma 3.4]: a module $M$ over a Gorenstein ring is Gorenstein flat if and only if the Pontryagin dual, $\mathrm{Hom}_{\mathbb{Z}}(M, \mathbb{Q}/\mathbb{Z})$, is Gorenstein injective. Special cases of some of the other stability results can also be found in [27].

The stability results for Auslander categories, Theorem (6.4.13), are selected special cases taken from a series of unpublished results by Foxby.

# Appendix

# Hyperhomology

This appendix offers a crash course in hyperhomological algebra. The aim is, first of all, to provide an easy reference for readers who are used to plain modules and do not feel entirely at home among complexes. Of course, this aim could also be accomplished by referring consistently to one or two textbooks on the subject — if only such existed. While there are numerous introductions to the technical side of hyperhomology, including the construction of derived categories, a thorough and coherent introduction to homological dimensions (not to mention depth, Krull dimension, and other invariants from commutative algebra) for complexes has yet to be published. The appearance of an introductory text like that may not be imminent, but this appendix is based on Foxby's notes [33], and their eventual publication will render it obsolete.

In general we omit references and proofs for standard definitions and results: they can be found almost everywhere in the literature. Specific references are, however, given for the more special results on homological dimensions and other invariants for complexes; the primary sources are [35, 36] (Foxby) and [7] (Avramov and Foxby).

The author's favorite sources for background stuff are

- [49] (Matsumura) and [12] (Bruns and Herzog) for commutative algebra,
- [13] (Cartan & Eilenberg) and [60] (Weibel) for homological algebra, and
- [47] (MacLane) for categories.

But also [20] (Eisenbud) has definitions, usually accompanied by illustrative examples and exercises, of many basic and advanced notions in commutative algebra and homological algebra.

**The blanket assumptions** (see page 9) **are still in force**; in particular, $R$ is always assumed to be a **non-trivial, commutative, and Noetherian ring**. Needless to say, most results in this appendix hold in a far more general setting, but to keep it simple we only state what we need.

## A.1 Basic Definitions and Notation

The definitions given in this section are all standard, and the same can be said about most of the notation.

Further details (in particular about the central notion of equivalence of complexes) can be extracted from the first chapters in [41] and [43].

(A.1.1) ***Complexes***. An *$R$-complex* $X$ is a sequence of $R$-modules $X_\ell$ and $R$-linear maps $\partial_\ell^X$, $\ell \in \mathbb{Z}$,

$$X = \cdots \longrightarrow X_{\ell+1} \xrightarrow{\partial_{\ell+1}^X} X_\ell \xrightarrow{\partial_\ell^X} X_{\ell-1} \longrightarrow \cdots$$

The module $X_\ell$ is called the *module in degree $\ell$*, and the map $\partial_\ell^X : X_\ell \to X_{\ell-1}$ is the *$\ell$-th differential*; composition of two consecutive differentials always yields the zero-map, i.e., $\partial_\ell^X \partial_{\ell+1}^X = 0$ for all $\ell \in \mathbb{Z}$. The *degree* of an element $x$ is denoted by $|x|$, i.e.,

$$|x| = \ell \quad \Longleftrightarrow \quad x \in X_\ell.$$

A complex $X$ is said to be *concentrated in degrees $u, \ldots, v$* if $X_\ell = 0$ for $\ell > u$ and $\ell < v$. A complex $X$ concentrated in degree zero is identified with the module $X_0$, and a module $M$ is thought of as the complex

$$M = 0 \to M \to 0,$$

with $M$ in degree zero. Of course, the complex $M$ has 0 in all degrees except (possibly) degree zero, i.e., $M = \cdots \to 0 \to 0 \to M \to 0 \to 0 \to \cdots$, but we never write superfluous zeros. In line with this the zero-complex is denoted by 0.

(A.1.2) ***Homology***. For an $R$-complex $X$ and $\ell \in \mathbb{Z}$ the following notation is used:

$$\begin{aligned} \mathrm{Z}_\ell^X &= \operatorname{Ker} \partial_\ell^X; \\ \mathrm{B}_\ell^X &= \operatorname{Im} \partial_{\ell+1}^X; \quad \text{and} \\ \mathrm{C}_\ell^X &= \operatorname{Coker} \partial_{\ell+1}^X. \end{aligned}$$

Both $\mathrm{B}_\ell^X$ and $\mathrm{Z}_\ell^X$ are submodules of $X_\ell$, and $\mathrm{B}_\ell^X \subseteq \mathrm{Z}_\ell^X$ because $\partial_\ell^X \partial_{\ell+1}^X = 0$. The residue class module

$$\mathrm{H}_\ell(X) = \mathrm{Z}_\ell^X \,/\, \mathrm{B}_\ell^X$$

is called the *homology module in degree $\ell$*, and the *homology complex* $\mathrm{H}(X)$ is defined by setting

$$\mathrm{H}(X)_\ell = \mathrm{H}_\ell(X) \quad \text{and} \quad \partial_\ell^{\mathrm{H}(X)} = 0$$

for $\ell \in \mathbb{Z}$.

Note that $\mathrm{H}(\mathrm{H}(X)) = \mathrm{H}(X)$ for any complex $X$, and $\mathrm{H}(M) = M$ when $M$ is a module.

A complex is said to be ***homologically trivial*** when $\mathrm{H}(X) = 0$. Thus, a homologically trivial $R$–complex is what is classically called an exact sequence of $R$–modules.

(A.1.3) ***Shift.*** If $m$ is an integer and $X$ is a complex, then $\Sigma^m X$ denotes the complex $X$ *shifted* $m$ degrees (to the left); it is given by

$$(\Sigma^m X)_\ell = X_{\ell-m} \quad \text{and} \quad \partial_\ell^{\Sigma^m X} = (-1)^m \partial_{\ell-m}^X$$

for $\ell \in \mathbb{Z}$. Note that

$$\mathrm{H}_\ell(\Sigma^m X) = \mathrm{H}_{\ell-m}(X). \tag{A.1.3.1}$$

For example, if $M$ is a module, then the complex $\Sigma^m M$ has $M$ in degree $m$ and 0 elsewhere.

(A.1.4) ***Morphisms.*** A *morphism* $\alpha\colon X \to Y$ of $R$–complexes is a family $\alpha = (\alpha_\ell)_{\ell\in\mathbb{Z}}$ of $R$–linear maps $\alpha_\ell\colon X_\ell \to Y_\ell$ making the diagram

$$\begin{array}{ccccccccc}
\cdots & \longrightarrow & X_{\ell+1} & \xrightarrow{\partial_{\ell+1}^X} & X_\ell & \xrightarrow{\partial_\ell^X} & X_{\ell-1} & \longrightarrow & \cdots \\
 & & \downarrow{\scriptstyle \alpha_{\ell+1}} & & \downarrow{\scriptstyle \alpha_\ell} & & \downarrow{\scriptstyle \alpha_{\ell-1}} & & \\
\cdots & \longrightarrow & Y_{\ell+1} & \xrightarrow[\partial_{\ell+1}^Y]{} & Y_\ell & \xrightarrow[\partial_\ell^Y]{} & Y_{\ell-1} & \longrightarrow & \cdots
\end{array}$$

commutative. That is, $\partial_\ell^Y \alpha_\ell = \alpha_{\ell-1}\partial_\ell^X$ for all $\ell \in \mathbb{Z}$.

For an element $r \in R$ and an $R$–complex $X$ the morphism $r_X\colon X \to X$ is the *homothety* given by multiplication by $r$. In line with this we denote the identity morphism on $X$ by $1_X$.

Any morphism $\alpha\colon X \to Y$ induces a morphism $\mathrm{H}(\alpha)\colon \mathrm{H}(X) \to \mathrm{H}(Y)$ in homology. The homology H is a functor in the category of all $R$–complexes and all morphisms of $R$–complexes.

(A.1.5) ***Isomorphisms.*** A morphism $\alpha\colon X \to Y$ of $R$–complexes is said to be an *isomorphism* when there exists a morphism $\alpha^{-1}\colon Y \to X$ such that $\alpha\alpha^{-1} = 1_Y$ and $\alpha^{-1}\alpha = 1_X$. Isomorphisms are indicated by the symbol $\cong$ next to their arrows, and two complexes $X$ and $Y$ are *isomorphic*, $X \cong Y$ in symbols, if and only if there exists an isomorphism $X \xrightarrow{\cong} Y$.

Note that $\alpha\colon X \to Y$ is an isomorphism of $R$–complexes if and only if all the maps $\alpha_\ell\colon X_\ell \to Y_\ell$ are isomorphisms of $R$–modules. In particular, two modules are isomorphic as complexes if and only if they are so as modules.

If $\alpha\colon X \to Y$ is an isomorphism, then so is the induced morphism in homology, and $\mathrm{H}(\alpha)^{-1} = \mathrm{H}(\alpha^{-1})$.

(A.1.6) ***Quasi-isomorphisms.*** A morphism $\alpha\colon X \to Y$ is said to be a *quasi-isomorphism* if the induced morphism $\mathrm{H}(\alpha)\colon \mathrm{H}(X) \to \mathrm{H}(Y)$ is an isomorphism. Quasi-isomorphisms are also called homology isomorphisms, and they are indicated by the symbol $\simeq$ next to their arrows.

Note that all isomorphisms are quasi-isomorphisms. A morphism of complexes concentrated in degree zero (that is, a homomorphism of modules) is a quasi-isomorphism if and only if it is an isomorphism.

(A.1.7) ***Homological Position and Size.*** The numbers *supremum, infimum,* and *amplitude*:

$$\begin{aligned} \sup X &= \sup\{\ell \in \mathbb{Z} \mid \mathrm{H}_\ell(X) \neq 0\}, \\ \inf X &= \inf\{\ell \in \mathbb{Z} \mid \mathrm{H}_\ell(X) \neq 0\}, \quad \text{and} \\ \operatorname{amp} X &= \sup X - \inf X \end{aligned}$$

capture the homological position and size of the complex $X$. By the conventions for supremum and infimum of the empty set it follows that $\sup X = -\infty$ and $\inf X = \infty$ if $X$ is homologically trivial; otherwise we have

$$\infty \geq \sup X \geq \inf X \geq -\infty.$$

Let $X$ be any complex; it is immediate that

(A.1.7.1) $$\mathrm{Z}_\ell^X = \mathrm{B}_\ell^X \quad \text{for } \ell < \inf X,$$

and

(A.1.7.2) $$\mathrm{B}_\ell^X \cong X_{\ell+1}/\mathrm{Z}_{\ell+1}^X = \mathrm{C}_{\ell+1}^X \quad \text{for } \ell \geq \sup X.$$

In particular, if $X$ is homologically trivial, then

(A.1.7.3) $$\mathrm{Z}_\ell^X = \mathrm{B}_\ell^X \cong \mathrm{C}_{\ell+1}^X$$

for all $\ell \in \mathbb{Z}$.

(A.1.8) ***The Category of R–complexes.*** We use the notation $\mathcal{C}(R)$ for the category of all $R$–complexes and all morphism of $R$–complexes. Recall that a full subcategory $\mathcal{S}$ of $\mathcal{C}(R)$ is defined by specifying its objects, the arrows in $\mathcal{S}$ are simply all morphisms between the specified objects. We shall consider a number of full subcategories $\mathcal{S}$ of $\mathcal{C}(R)$ (the first ones are introduced below); of course the notation $X \in \mathcal{S}$ means that $X$ is an object in $\mathcal{S}$, and for two full subcategories the notation $\mathcal{S}_1 \subseteq \mathcal{S}_2$ means that every object in $\mathcal{S}_1$ is also an object in $\mathcal{S}_2$.

(A.1.9) ***Categories of (Homologically) Bounded Complexes.*** An $R$–complex $X$ is said to be *bounded to the left* if there is an integer $u$ such that $X_\ell = 0$ for all $\ell > u$; similarly $X$ is *bounded to the right* if there is an integer $v$ such that

$X_\ell = 0$ for all $\ell < v$. A complex which is bounded to the right as well as to the left is said to be *bounded*.

We define the full subcategories $\mathcal{C}_{\sqsubset}(R)$, $\mathcal{C}_{\sqsupset}(R)$, $\mathcal{C}_{\square}(R)$, and $\mathcal{C}_0(R)$ of $\mathcal{C}(R)$ by specifying their objects as follows:

- $\mathcal{C}_{\sqsubset}(R)$: complexes bounded to the left;
- $\mathcal{C}_{\sqsupset}(R)$: complexes bounded to the right;
- $\mathcal{C}_{\square}(R)$: bounded complexes; and
- $\mathcal{C}_0(R)$: modules (considered as complexes concentrated in degree 0).

An $R$–complex $X$ is said to be ***homologically bounded*** (*to the left/right*) when the homology complex $\mathrm{H}(X)$ is bounded (to the left/right). We also consider the following full subcategories of $\mathcal{C}(R)$:

- $\mathcal{C}_{(\sqsubset)}(R)$: complexes homologically bounded to the left;
- $\mathcal{C}_{(\sqsupset)}(R)$: complexes homologically bounded to the right;
- $\mathcal{C}_{(\square)}(R)$: homologically bounded complexes; and
- $\mathcal{C}_{(0)}(R)$: complexes with homology concentrated in degree zero.

Note that these last four subcategories can be characterized as follows:

$$X \in \mathcal{C}_{(\sqsubset)}(R) \iff \sup X < \infty;$$
$$X \in \mathcal{C}_{(\sqsupset)}(R) \iff \inf X > -\infty;$$
$$X \in \mathcal{C}_{(\square)}(R) \iff \operatorname{amp} X < \infty; \quad \text{and}$$
$$X \in \mathcal{C}_{(0)}(R) \iff \sup X \le \inf X.$$

(A.1.10) ***A Menagerie of Categories.*** We also consider the following full subcategories of $\mathcal{C}(R)$:

- $\mathcal{C}^{\mathrm{f}}(R)$: complexes of finite modules;
- $\mathcal{C}^{(\mathrm{f})}(R)$: complexes with finite homology modules;
- $\mathcal{C}^{\mathrm{I}}(R)$: complexes of injective modules;
- $\mathcal{C}^{\mathrm{F}}(R)$: complexes of flat modules;
- $\mathcal{C}^{\mathrm{P}}(R)$: complexes of projective modules;
- $\mathcal{C}^{\mathrm{fP}}(R)$: complexes of finite projective modules; and
- $\mathcal{C}^{\mathrm{L}}(R)$: complexes of finite free modules.

Superscripts and subscripts are freely mixed to produce new notation. E.g.,

- $\mathcal{C}^{(\mathrm{f})}_{(\square)}(R)$: homologically bounded complexes with finite homology modules (for short, complexes with *finite homology*);
- $\mathcal{C}^{\mathrm{f}}_0(R)$: finite modules;
- $\mathcal{C}^{\mathrm{L}}_{\sqsupset}(R)$: bounded to the right complexes of finite free modules; and
- $\mathcal{C}^{\mathrm{I}}_0(R)$: injective modules.

The principle behind this notation is that subscripts indicate boundedness conditions on the complexes, while superscripts indicate conditions on the modules of the complexes. We use (or rather allow) any combination

$$\mathcal{C}^{\&}_{\sharp}(R) = \mathcal{C}_{\sharp}(R) \cap \mathcal{C}^{\&}(R)$$

of subscripts ($\sqsubset$, $\sqsupset$, $\square$, or 0) and superscripts (f, I, F, P, fP, or L).

Sub- and superscripts in parentheses indicate the corresponding conditions on the homology complexes. That is,

$$X \in \mathcal{C}^{(\&)}_{(\sharp)}(R) \iff \mathrm{H}(X) \in \mathcal{C}^{\&}_{\sharp}(R).$$

(A.1.11) ***Equivalence***. Equivalence of $R$–complexes can be defined as follows: two complexes $X$ and $Y$ are *equivalent*, we write $X \simeq Y$, if and only if there exists a third complex $Z$ and two quasi-isomorphisms: $X \xrightarrow{\simeq} Z \xleftarrow{\simeq} Y$. In particular, the existence of a quasi-isomorphism $X \xrightarrow{\simeq} Y$ implies that $X$ and $Y$ are equivalent. Equivalence $\simeq$ is an equivalence relation in $\mathcal{C}(R)$.

The following implications always hold:

$$X \cong Y \implies X \simeq Y \implies \mathrm{H}(X) \cong \mathrm{H}(Y),$$

but the reverse implications do not hold in general. For $R$–modules $M$ and $N$, however, we have

$$M \cong N \iff M \simeq N;$$

that is, equivalent objects in $\mathcal{C}_0(R)$ are "just" isomorphic modules.

Also note that a complex is homologically trivial if and only if it is equivalent to the zero-complex. That is,

$$X \simeq 0 \iff \mathrm{H}(X) = 0.$$

The symbol $\sim$ denotes equivalence up to a shift. That is,

$$X \sim Y \iff X \simeq \Sigma^m Y \quad \text{for some } m \in \mathbb{Z}.$$

(A.1.12) ***Notation for Equivalence Classes***. Let $\boldsymbol{X}$ be an equivalence class of $R$–complexes; all representatives of $\boldsymbol{X}$ have isomorphic homology modules, so it makes sense to say, e.g., that $\boldsymbol{X}$ has finite homology: meaning, of course, that every representative of $\boldsymbol{X}$ belongs to $\mathcal{C}^{(\mathrm{f})}_{(\square)}(R)$. Generally speaking, notation of the form $\boldsymbol{X} \in \mathcal{C}^{(\&)}_{(\sharp)}(R)$, cf. (A.1.10), will make sense but $\boldsymbol{X} \in \mathcal{C}^{\&}_{\sharp}(R)$ will not!

Sensible notation should be interpreted like this:

$$\begin{aligned} \boldsymbol{X} \in \mathcal{C}^{(\&)}_{(\sharp)}(R) &\iff X \in \mathcal{C}^{(\&)}_{(\sharp)}(R) \quad \text{for some representative } X \text{ of } \boldsymbol{X} \\ &\iff X \in \mathcal{C}^{(\&)}_{(\sharp)}(R) \quad \text{for every representative } X \text{ of } \boldsymbol{X}. \end{aligned}$$

We sometimes identify a complex with its equivalence class, and — abusing the notation slightly — we may write, e.g., $X' = \boldsymbol{X}$ intending, of course, that the complex $X'$ represents the equivalence class $\boldsymbol{X}$.

(A.1.13) **Remarks.** The *derived category* of the category of $R$–modules — usually denoted by $\mathcal{D}(R)$ — is the localization of $\mathcal{C}(R)$ at the class of all quasi-isomorphisms. That is, the quasi-isomorphisms are formally inverted and become the isomorphisms in $\mathcal{D}(R)$; the notation $X \simeq Y$ then means that $X$ and $Y$ are isomorphic objects in $\mathcal{D}(R)$.

The classical references for derived categories are [59] (Verdier) and [41] (Hartshorne/Grothendieck), but most readers will probably find chapter 10 in [60] (Weibel) more accessible.

Now, we don't use derived categories, but [41] is still well worth perusing in order to develop a gut feeling for the preliminary notion of equivalent complexes — and so is [43] (Iversen).

(A.1.14) ***Truncations.*** Let $X$ be an $R$–complex and let $u, v$ be integers. The *hard left-truncation*, $\sqsubset_u X$, of $X$ at $u$ and the *hard right-truncation*, $X_v\sqsupset$, of $X$ at $v$ are given by:

$$\sqsubset_u X = 0 \longrightarrow X_u \xrightarrow{\partial_u^X} X_{u-1} \xrightarrow{\partial_{u-1}^X} X_{u-2} \xrightarrow{\partial_{u-2}^X} \cdots \quad \text{and}$$

$$X_v\sqsupset = \cdots \xrightarrow{\partial_{v+3}^X} X_{v+2} \xrightarrow{\partial_{v+2}^X} X_{v+1} \xrightarrow{\partial_{v+1}^X} X_v \longrightarrow 0.$$

The *soft left-truncation*, $\subset_u X$, of $X$ at $u$ and the *soft right-truncation*, $X_v\supset$, of $X$ at $v$ are given by:

$$\subset_u X = 0 \longrightarrow \mathrm{C}_u^X \xrightarrow{\bar{\partial}_u^X} X_{u-1} \xrightarrow{\partial_{u-1}^X} X_{u-2} \xrightarrow{\partial_{u-2}^X} \cdots \quad \text{and}$$

$$X_v\supset = \cdots \xrightarrow{\partial_{v+3}^X} X_{v+2} \xrightarrow{\partial_{v+2}^X} X_{v+1} \xrightarrow{\partial_{v+1}^X} \mathrm{Z}_v^X \longrightarrow 0.$$

The differential $\bar{\partial}_u^X$ is the induced map on residue classes.

It is easy to see that

$$\mathrm{H}_\ell(\sqsubset_u X) = \begin{cases} 0 & \text{for } \ell > u, \\ \mathrm{Z}_u^X & \text{for } \ell = u, \quad \text{and} \\ \mathrm{H}_\ell(X) & \text{for } \ell < u. \end{cases}$$

For $u \leq \inf X$ the natural inclusion of $\Sigma^u \mathrm{Z}_u^X$ into $\sqsubset_u X$,

$$\begin{array}{ccccccc} 0 & \longrightarrow & \mathrm{Z}_u^X & \longrightarrow & 0 & & \\ \downarrow & & \downarrow & & \downarrow & & \\ 0 & \longrightarrow & X_u & \longrightarrow & X_{u-1} & \longrightarrow & \cdots \end{array}$$

is, therefore, a quasi-isomorphism. In particular,

$$\sqsubset_u X \simeq \Sigma^u \mathrm{Z}_u^X \quad \text{for } u \leq \inf X. \tag{A.1.14.1}$$

It is also evident that

$$\mathrm{H}_\ell(\subset_u X) = \begin{cases} 0 & \text{for } \ell > u, \text{ and} \\ \mathrm{H}_\ell(X) & \text{for } \ell \leq u; \end{cases}$$

so the canonical morphism $X \to \subset_u X$,

$$\begin{array}{ccccccccc} \cdots & \longrightarrow & X_{u+1} & \longrightarrow & X_u & \longrightarrow & X_{u-1} & \longrightarrow & \cdots \\ & & \downarrow & & \downarrow & & \downarrow & & \\ & & 0 & \longrightarrow & \mathrm{C}_u^X & \longrightarrow & X_{u-1} & \longrightarrow & \cdots \end{array}$$

is a quasi-isomorphism if and only if $u \geq \sup X$. In particular,

(A.1.14.2) $$\subset_u X \simeq X \quad \text{for } u \geq \sup X.$$

It is equally easy to see that

(A.1.14.3) $$X_v\sqsupset \simeq \Sigma^v \mathrm{C}_v^X \quad \text{for } v \geq \sup X; \text{ and}$$
(A.1.14.4) $$X_v\supset \simeq X \quad \text{for } v \leq \inf X.$$

(A.1.15) **Remark.** It follows by (A.1.14.2) and (A.1.14.4) that any $R$–complex $X$ is equivalent to a complex $X'$ with $X'_\ell = 0$ for $\ell > \sup X$ and $\ell < \inf X$. In particular, it follows that

(A.1.15.1) $$X \in \mathcal{C}_{(0)}(R) \iff X \simeq \mathrm{H}_0(X).$$

(A.1.16) ***Short Exact Sequences.*** Consider three $R$–complexes $X, Y$, and $Z$, and morphisms $\alpha\colon X \to Y$ and $\beta\colon Y \to Z$. We say that

(A.1.16.1) $$0 \to X \xrightarrow{\alpha} Y \xrightarrow{\beta} Z \to 0$$

is a *short exact sequence of $R$–complexes* if it is exact in each degree. That is,

$$0 \to X_\ell \xrightarrow{\alpha_\ell} Y_\ell \xrightarrow{\beta_\ell} Z_\ell \to 0$$

is a short exact sequence of $R$–modules for each $\ell \in \mathbb{Z}$.

A short exact sequence like (A.1.16.1) induces a long exact sequence of homology modules

$$\cdots \xrightarrow{\mathrm{H}_{\ell+1}(\beta)} \mathrm{H}_{\ell+1}(Z) \xrightarrow{\Delta_{\ell+1}} \mathrm{H}_\ell(X) \xrightarrow{\mathrm{H}_\ell(\alpha)} \mathrm{H}_\ell(Y) \xrightarrow{\mathrm{H}_\ell(\beta)} \mathrm{H}_\ell(Z) \xrightarrow{\Delta_\ell} \cdots$$

Note that homological triviality of two complexes in a short exact sequence implies homological triviality of the third.

(A.1.17) **Remark.** For any $X \in \mathcal{C}(R)$ and $n \in \mathbb{Z}$ the diagram

$$
\begin{array}{ccccccccccc}
 & & & & 0 & \longrightarrow & X_{n-1} & \longrightarrow & X_{n-2} & \longrightarrow & \cdots \\
 & & & & \downarrow & & \downarrow & & \downarrow & & \\
\cdots & \longrightarrow & X_{n+1} & \longrightarrow & X_n & \longrightarrow & X_{n-1} & \longrightarrow & X_{n-2} & \longrightarrow & \cdots \\
 & & \downarrow & & \downarrow & & \downarrow & & & & \\
\cdots & \longrightarrow & X_{n+1} & \longrightarrow & X_n & \longrightarrow & 0 & & & &
\end{array}
$$

is commutative, so we have a short exact sequence of complexes:

$$0 \longrightarrow \sqsubset_{n-1}X \longrightarrow X \longrightarrow X_n\sqsupset \longrightarrow 0,$$

which is split (even trivial) in each degree.

(A.1.18) ***Mapping Cones.*** To a morphism $\alpha\colon X \to Y$ we associate a complex, $\mathcal{M}(\alpha)$, called the *mapping cone* of $\alpha$. It is given by

$$
\begin{aligned}
\mathcal{M}(\alpha)_\ell &= Y_\ell \oplus X_{\ell-1} \quad \text{and} \\
\partial_\ell^{\mathcal{M}(\alpha)}(y_\ell, x_{\ell-1}) &= (\partial_\ell^Y(y_\ell) + \alpha_{\ell-1}(x_{\ell-1}), -\partial_{\ell-1}^X(x_{\ell-1})).
\end{aligned}
$$

(A.1.19) **Lemma.** *A morphism $\alpha\colon X \to Y$ of $R$–complexes is a quasi-isomorphism if and only if the mapping cone $\mathcal{M}(\alpha)$ is homologically trivial.*

*Proof.* It is easy to check that the inclusion $Y \to \mathcal{M}(\alpha)$ and the (degree-wise projection) $\mathcal{M}(\alpha) \to \Sigma^1 X$ are morphisms and make up a short exact sequence of complexes

$$0 \to Y \to \mathcal{M}(\alpha) \to \Sigma^1 X \to 0.$$

In the induced long exact sequence of homology modules

$$\cdots \to \mathrm{H}_{\ell+1}(\mathcal{M}(\alpha)) \to \mathrm{H}_{\ell+1}(\Sigma^1 X) \xrightarrow{\Delta_{\ell+1}} \mathrm{H}_\ell(Y) \to \mathrm{H}_\ell(\mathcal{M}(\alpha)) \to \cdots$$

we have $\mathrm{H}_{\ell+1}(\Sigma^1 X) = \mathrm{H}_\ell(X)$, cf. (A.1.3.1), and the connecting map $\Delta_{\ell+1}$ is just the induced map $\mathrm{H}_\ell(\alpha)\colon \mathrm{H}_\ell(X) \to \mathrm{H}_\ell(Y)$. The assertion is now immediate. □

(A.1.20) ***Induced Functors.*** Any additive module functor $\mathrm{T}\colon \mathcal{C}_0(R) \to \mathcal{C}_0(R')$ induces a functor $\mathrm{T}\colon \mathcal{C}(R) \to \mathcal{C}(R')$ on complexes.

Let $X \in \mathcal{C}(R)$. If the functor T is covariant, then $\mathrm{T}(X) \in \mathcal{C}(R')$ is given by

(A.1.20.1) $$\mathrm{T}(X)_\ell = \mathrm{T}(X_\ell) \quad \text{and} \quad \partial_\ell^{\mathrm{T}(X)} = \mathrm{T}(\partial_\ell^X);$$

and if T is contravariant, then $\mathrm{T}(X)$ is given by

(A.1.20.2) $$\mathrm{T}(X)_\ell = \mathrm{T}(X_{-\ell}) \quad \text{and} \quad \partial_\ell^{\mathrm{T}(X)} = \mathrm{T}(\partial_{-\ell+1}^X).$$

If T is exact, then $\mathrm{H}(\mathrm{T}(X)) = \mathrm{T}(\mathrm{H}(X))$ for every $X \in \mathcal{C}(R)$, and T preserves quasi-isomorphisms and equivalences. For example, the functor $- \otimes_R R_\mathfrak{p}$ (localization at $\mathfrak{p}$) is exact for every prime ideal $\mathfrak{p} \in \operatorname{Spec} R$.

If the module functor T is faithful (i.e., T is "injective on homomorphisms"), then, in particular, $M = 0$ if $\mathrm{T}(M) = 0$, so $\mathrm{T}(M) \neq 0 \Leftrightarrow M \neq 0$ as T is additive. Thus, if T is faithful, exact, and covariant, then the induced functor on complexes preserves suprema and infima:

$$\sup(\mathrm{T}(X)) = \sup X \quad \text{and} \quad \inf(\mathrm{T}(X)) = \inf X. \tag{A.1.20.3}$$

If T is faithful, exact, and contravariant, then the induced functor "swaps" suprema and infima:

$$\sup(\mathrm{T}(X)) = -\inf X \quad \text{and} \quad \inf(\mathrm{T}(X)) = -\sup X. \tag{A.1.20.4}$$

## A.2 Standard Functors and Morphisms

Ever since the highly influential book [13] by Cartan and Eilenberg appeared, the functors Hom and tensor product — and the associated standard homomorphisms — have formed the core of almost any course in homological algebra. In this section we review Hom, tensor product, and standard morphisms for complexes.

(A.2.1) ***Homomorphisms.*** For $R$–complexes $X$ and $Y$ we define the *homomorphism complex* $\operatorname{Hom}_R(X,Y) \in \mathcal{C}(R)$ as follows:

$$\operatorname{Hom}_R(X,Y)_\ell = \prod_{p\in\mathbb{Z}} \operatorname{Hom}_R(X_p, Y_{p+\ell})$$

and when $\psi = (\psi_p)_{p\in\mathbb{Z}}$ belongs to $\operatorname{Hom}_R(X,Y)_\ell$ the family $\partial_\ell^{\operatorname{Hom}_R(X,Y)}(\psi)$ in $\operatorname{Hom}_R(X,Y)_{\ell-1}$ has $p$-th component

$$\partial_\ell^{\operatorname{Hom}_R(X,Y)}(\psi)_p = \partial_{p+\ell}^Y \psi_p - (-1)^\ell \psi_{p-1}\partial_p^X.$$

If $V$ and $W$ are fixed $R$–complexes, then $\operatorname{Hom}_R(V,-)$ and $\operatorname{Hom}_R(-,W)$ are functors in $\mathcal{C}(R)$. No ambiguity arises when one or both involved complexes are modules. If $M \in \mathcal{C}_0(R)$ and $X \in \mathcal{C}(R)$, then the homomorphism complexes $\operatorname{Hom}_R(M,X)$ and $\operatorname{Hom}_R(X,M)$ agree with the complexes yielded by applying, respectively, $\operatorname{Hom}_R(M,-)$ and $\operatorname{Hom}_R(-,M)$ to $X$. In particular, for $M,N \in \mathcal{C}_0(R)$ the homomorphism complex $\operatorname{Hom}_R(M,N)$ is concentrated in degree zero, where it is the module $\operatorname{Hom}_R(M,N)$.

The covariant functor $\operatorname{Hom}_R(V,-)$ commutes with shift and mapping cones:

$$\operatorname{Hom}_R(V,\Sigma^m Y) = \Sigma^m \operatorname{Hom}_R(V,Y); \quad \text{and} \tag{A.2.1.1}$$

$$\mathcal{M}(\operatorname{Hom}_R(V,\alpha)) = \operatorname{Hom}_R(V,\mathcal{M}(\alpha)). \tag{A.2.1.2}$$

For the contravariant functor $\operatorname{Hom}_R(-,W)$ we have the following:

(A.2.1.3) $$\operatorname{Hom}_R(\Sigma^m X, W) \cong \Sigma^{-m}\operatorname{Hom}_R(X,W); \quad \text{and}$$

(A.2.1.4) $$\mathcal{M}(\operatorname{Hom}_R(\alpha, W)) \cong \Sigma^1 \operatorname{Hom}_R(\mathcal{M}(\alpha), W).$$

When $X \in \mathcal{C}_{\sqsupset}(R)$ and $Y \in \mathcal{C}_{\sqsubset}(R)$ all the products $\prod_{p\in\mathbb{Z}} \operatorname{Hom}_R(X_p, Y_{p+\ell})$ are finite; the next two lemmas are, therefore, direct consequences of the similar results for modules.

**(A.2.2) Lemma.** *If $X \in \mathcal{C}^{\mathrm{f}}_{\sqsupset}(R)$ and $Y \in \mathcal{C}^{\mathrm{f}}_{\sqsubset}(R)$, then $\operatorname{Hom}_R(X,Y) \in \mathcal{C}^{\mathrm{f}}_{\sqsubset}(R)$.*

**(A.2.3) Lemma.** *If $X \in \mathcal{C}^{\mathrm{f}}_{\sqsupset}(R)$, $Y \in \mathcal{C}_{\sqsubset}(R)$, and $\mathfrak{p}$ is a prime ideal in $R$, then there is an isomorphism of $R_\mathfrak{p}$–complexes:*

$$\operatorname{Hom}_R(X,Y)_\mathfrak{p} \cong \operatorname{Hom}_{R_\mathfrak{p}}(X_\mathfrak{p}, Y_\mathfrak{p}).$$

(A.2.4) ***Tensor Products.*** For $R$–complexes $X$ and $Y$ we define the *tensor product complex* $X \otimes_R Y \in \mathcal{C}(R)$ as follows:

$$(X \otimes_R Y)_\ell = \coprod_{p\in\mathbb{Z}} X_p \otimes_R Y_{\ell-p}$$

and the $\ell$-th differential $\partial_\ell^{X\otimes_R Y}$ is given on a generator $x_p \otimes y_{\ell-p}$ in $(X \otimes_R Y)_\ell$ by

$$\partial_\ell^{X\otimes_R Y}(x_p \otimes y_{\ell-p}) = \partial_p^X(x_p) \otimes y_{\ell-p} + (-1)^p x_p \otimes \partial_{\ell-p}^Y(y_{\ell-p}),$$

which is an element in $(X \otimes_R Y)_{\ell-1}$.

The tensor product is commutative: for complexes $X$ and $Y$ the *commutativity* isomorphism

(A.2.4.1) $$\tau_{XY} : X \otimes_R Y \xrightarrow{\cong} Y \otimes_R X,$$

with the map in degree $\ell$ given on generators by

(A.2.4.2) $$\tau_{XY\ell}(x_p \otimes y_{\ell-p}) = (-1)^{p(\ell-p)} y_{\ell-p} \otimes x_p,$$

is natural in $X$ and $Y$.

If $V$ is a fixed $R$–complex, then $V \otimes_R -$ (and thereby $- \otimes_R V$) is a functor in $\mathcal{C}(R)$. No ambiguity arises when one or both involved complexes are modules. If $M \in \mathcal{C}_0(R)$ and $X \in \mathcal{C}(R)$, then the tensor product complex $M \otimes_R X$ agrees with the complex yielded by applying $M \otimes_R -$ to $X$. In particular, for $M, N \in \mathcal{C}_0(R)$ the tensor product complex $M \otimes_R N$ is concentrated in degree zero, where it is the module $M \otimes_R N$.

The tensor product functor $V \otimes_R -$ commutes with shift and mapping cones:

(A.2.4.3) $$V \otimes_R (\Sigma^m Y) = \Sigma^m (V \otimes_R Y); \quad \text{and}$$

(A.2.4.4) $$\mathcal{M}(V \otimes_R \alpha) \cong V \otimes_R \mathcal{M}(\alpha).$$

The first lemma below is a direct consequence of the corresponding result for modules; and so is the second, because all the sums $\coprod_{p\in\mathbb{Z}} X_p \otimes_R Y_{\ell-p}$ are finite when $X$ and $Y$ are bounded to the right.

(A.2.5) **Lemma.** *If $X$ and $Y$ are $R$–complexes, and $\mathfrak{p}$ is a prime ideal in $R$, then there is an isomorphism of $R_\mathfrak{p}$–complexes:*

$$(X \otimes_R Y)_\mathfrak{p} \cong X_\mathfrak{p} \otimes_{R_\mathfrak{p}} Y_\mathfrak{p}.$$

(A.2.6) **Lemma.** *If $X$ and $Y$ belong to $\mathcal{C}^{\mathrm{f}}_{\sqsupset}(R)$, then also $X \otimes_R Y \in \mathcal{C}^{\mathrm{f}}_{\sqsupset}(R)$.*

The five **standard homomorphisms of modules** (see page 11) **induce** five **natural morphisms of complexes**; these are described below in (A.2.7)–(A.2.11).

In the rest of this section $S$ denotes an $R$–algebra, and in most applications $S$ will be $R$ itself.

(A.2.7) **Associativity.** *Let $Z, Y \in \mathcal{C}(S)$ and $X \in \mathcal{C}(R)$. Then $Z \otimes_S Y \in \mathcal{C}(R)$ and $Y \otimes_R X \in \mathcal{C}(S)$, and*

$$\sigma_{ZYX} : (Z \otimes_S Y) \otimes_R X \xrightarrow{\cong} Z \otimes_S (Y \otimes_R X)$$

*is a natural isomorphism of $S$–complexes.*

(A.2.8) **Adjointness.** *Let $Z, Y \in \mathcal{C}(S)$ and $X \in \mathcal{C}(R)$. Then $Z \otimes_S Y \in \mathcal{C}(R)$ and $\mathrm{Hom}_R(Y, X) \in \mathcal{C}(S)$, and*

$$\rho_{ZYX} : \mathrm{Hom}_R(Z \otimes_S Y, X) \xrightarrow{\cong} \mathrm{Hom}_S(Z, \mathrm{Hom}_R(Y, X))$$

*is a natural isomorphism of $S$–complexes.*

(A.2.9) **Swap.** *Let $Z, Y \in \mathcal{C}(S)$ and $X \in \mathcal{C}(R)$. Then $\mathrm{Hom}_R(X, Y) \in \mathcal{C}(S)$ and $\mathrm{Hom}_S(Z, Y) \in \mathcal{C}(R)$, and*

$$\varsigma_{ZXY} : \mathrm{Hom}_S(Z, \mathrm{Hom}_R(X, Y)) \xrightarrow{\cong} \mathrm{Hom}_R(X, \mathrm{Hom}_S(Z, Y))$$

*is a natural isomorphism of $S$–complexes.*

(A.2.10) **Tensor Evaluation.** *Let $Z, Y \in \mathcal{C}(S)$ and $X$ belong to $\mathcal{C}(R)$. Then $\mathrm{Hom}_S(Z, Y) \in \mathcal{C}(R)$ and $Y \otimes_R X \in \mathcal{C}(S)$, and*

$$\omega_{ZYX} : \mathrm{Hom}_S(Z, Y) \otimes_R X \longrightarrow \mathrm{Hom}_S(Z, Y \otimes_R X)$$

*is a natural morphism of $S$–complexes. The morphism is invertible under each of the next two extra conditions:*

- *$Z \in \mathcal{C}^{\mathrm{fP}}_{\square}(S)$, $Y \in \mathcal{C}_{\sqsupset}(S)$, and $X \in \mathcal{C}_{\sqsupset}(R)$; or*
- *$Z \in \mathcal{C}^{\mathrm{fP}}_{\sqsupset}(S)$, $Y \in \mathcal{C}_{\sqsubset}(S)$, and $X \in \mathcal{C}_{\square}(R)$.*

(A.2.11) **Hom Evaluation.** *Let* $Z, Y \in \mathcal{C}(S)$ *and* $X$ *belong to* $\mathcal{C}(R)$. *Then* $\mathrm{Hom}_S(Z,Y) \in \mathcal{C}(R)$ *and* $\mathrm{Hom}_R(Y,X) \in \mathcal{C}(S)$, *and*

$$\theta_{ZYX} : Z \otimes_S \mathrm{Hom}_R(Y,X) \longrightarrow \mathrm{Hom}_R(\mathrm{Hom}_S(Z,Y),X)$$

*is a natural morphism of S–complexes. The morphism is invertible under each of the next two extra conditions:*

- $Z \in \mathcal{C}_{\square}^{\mathrm{fP}}(S)$, $Y \in \mathcal{C}_{\sqsupset}(S)$, *and* $X \in \mathcal{C}_{\sqsubset}(R)$; *or*
- $Z \in \mathcal{C}_{\sqsupset}^{\mathrm{fP}}(S)$, $Y \in \mathcal{C}_{\sqsubset}(S)$, *and* $X \in \mathcal{C}_{\square}(R)$.

(A.2.12) *Proof of (A.2.7)–(A.2.11).* Basically, the morphisms are defined by applying the corresponding homomorphisms of modules in each degree. However, by the "universal sign rule" a sign $(-1)^{pq}$ is introduced whenever two elements of degrees, respectively, $p$ and $q$ are interchanged. In short (but suggestive) notation we can write the definitions as follows:

$$
\begin{array}{lrcl}
(\sigma_{ZYX}) & (z \otimes y) \otimes x & \longmapsto & z \otimes (y \otimes x); \\
(\rho_{ZYX}) & \psi & \longmapsto & [z \mapsto [y \mapsto \psi(z \otimes y)]]; \\
(\varsigma_{ZXY}) & \psi & \longmapsto & [x \mapsto [z \mapsto (-1)^{|x||z|}\psi(z)(x)]]; \\
(\omega_{ZYX}) & \psi \otimes x & \longmapsto & [z \mapsto (-1)^{|x||z|}\psi(z) \otimes x]; \quad \text{and} \\
(\theta_{ZYX}) & z \otimes \psi & \longmapsto & [\vartheta \mapsto (-1)^{|z|(|\psi|+|\vartheta|)}\psi\vartheta(z)].
\end{array}
$$

Of course, it must be verified that the modules have the right form, so that these definitions make sense; they do. It must also be checked that the degree-wise maps commute with the differentials; they do. Finally, it must be verified that the extra conditions listed in (A.2.10) and (A.2.11) ensure that the degree-wise maps are invertible. This boils down to the direct sums and products (making up the modules in the Hom and tensor product complexes) being finite in each degree; and they are. Details are given in [33, Chapter 5].

# A.3 Resolutions

To do hyperhomological algebra we must first establish the existence of resolutions. It should be emphasized right away that we are aiming for a **relative homological algebra**: we will not resolve objects in $\mathcal{C}(R)$ by projective, injective, or flat objects in that category, but rather by complexes of projective, injective, or flat modules.

(A.3.1) **Definitions.** We define resolutions for (appropriately bounded) complexes as follows:

(I) An *injective resolution* of a complex $Y \in \mathcal{C}_{(\sqsubset)}(R)$ is a quasi-isomorphism $\iota \colon Y \xrightarrow{\simeq} I$ where $I \in \mathcal{C}_{\sqsubset}^{\mathrm{I}}(R)$.

(F) A ***flat resolution*** of a complex $X \in \mathcal{C}_{(\sqsupset)}(R)$ is a quasi-isomorphism $\varphi\colon F \xrightarrow{\simeq} X$ where $F \in \mathcal{C}^{\mathrm{F}}_{\sqsupset}(R)$.

(P) A ***projective resolution*** of a complex $X \in \mathcal{C}_{(\sqsupset)}(R)$ is a quasi-isomorphism $\pi\colon P \xrightarrow{\simeq} X$ where $P \in \mathcal{C}^{\mathrm{P}}_{\sqsupset}(R)$.

(L) A ***resolution by finite free modules*** of a complex $Z \in \mathcal{C}^{(\mathrm{f})}_{(\sqsupset)}(R)$ is a quasi-isomorphism $\lambda\colon L \xrightarrow{\simeq} Z$ where $L \in \mathcal{C}^{\mathrm{L}}_{\sqsupset}(R)$.

In (A.3.11) we shall see how these definitions relate to the usual concepts of injective, flat, and projective resolutions of modules.

(A.3.2) **Theorem (Existence of Resolutions).** *The following hold:*

(I) *Every complex $Y \in \mathcal{C}_{(\sqsubset)}(R)$ has an injective resolution $Y \xrightarrow{\simeq} I$ with $I_\ell = 0$ for $\ell > \sup Y$.*

(P&F) *Every complex $X \in \mathcal{C}_{(\sqsupset)}(R)$ has a projective, and thereby a flat, resolution $X \xleftarrow{\simeq} P$ with $P_\ell = 0$ for $\ell < \inf X$.*

(L) *Every complex $Z \in \mathcal{C}^{(\mathrm{f})}_{(\sqsupset)}(R)$ has a resolution by finite free modules $Z \xleftarrow{\simeq} L$ with $L_\ell = 0$ for $\ell < \inf X$.*

*Proof.* Various versions can be found in [13,41,43,56]; see [7, 1.7] for further guidance.

(A.3.3) **Lemma.** *If $0 \to Y' \to Y \to Y'' \to 0$ is a short exact sequence in $\mathcal{C}_{(\sqsubset)}(R)$, then there exists a short exact sequence $0 \to I' \to I \to I'' \to 0$ in $\mathcal{C}^{\mathrm{I}}_{\sqsubset}(R)$, where $I'$, $I$, and $I''$ are injective resolutions of, respectively, $Y'$, $Y$, and $Y''$.*

*Proof.* [43, Proposition 6.10].

(A.3.4) **Lemma.** *If $0 \to X' \to X \to X'' \to 0$ is a short exact sequence in $\mathcal{C}_{(\sqsupset)}(R)$, then there exists a short exact sequence $0 \to P' \to P \to P'' \to 0$ in $\mathcal{C}^{\mathrm{P}}_{\sqsupset}(R)$, where $P'$, $P$, and $P''$ are projective resolutions of, respectively, $X'$, $X$, and $X''$.*

*Proof.* See Proposition 6.10° ('°' is for 'opposite') on page 67 in [43].

Complexes of injective and projective modules have convenient lifting properties described by the next two lemmas.

(A.3.5) **Lemma.** *If $Y$ and $I$ are equivalent complexes, and $I$ belongs to $\mathcal{C}^{\mathrm{I}}_{\sqsubset}(R)$, then there exists a quasi-isomorphism $Y \xrightarrow{\simeq} I$; that is, $I$ is an injective resolution of $Y$.*

*Proof.* See [7, 1.1.I and 1.4.I].

(A.3.6) **Lemma.** *If $X$ and $P$ are equivalent complexes, and $P$ belongs to $\mathcal{C}^{\mathrm{P}}_{\sqsupset}(R)$, then there exists a quasi-isomorphism $P \xrightarrow{\simeq} X$; that is, $P$ is a projective resolution of $X$.*

*Proof.* See [7, 1.1.P and 1.4.P].

In (A.3.9)–(A.3.10) we define the **standard homological dimensions** for complexes, and it is **not** done **by way of resolutions** but rather by way of equivalence. Now, by the last two lemmas this is of no importance for the projective and injective dimensions, but it still takes a short argument to see that the new definitions agree with the usual ones for modules; this argument is given in (A.3.11).

(A.3.7) **Definitions.** The full subcategories $\mathcal{P}(R)$, $\mathcal{I}(R)$, and $\mathcal{F}(R)$ of $\mathcal{C}(R)$ are defined as follows:

$$\begin{aligned} Y \in \mathcal{I}(R) \quad &\Longleftrightarrow \quad \exists\, I \in \mathcal{C}^{\mathrm{I}}_{\square}(R) : Y \simeq I; \\ X \in \mathcal{P}(R) \quad &\Longleftrightarrow \quad \exists\, P \in \mathcal{C}^{\mathrm{P}}_{\square}(R) : X \simeq P; \quad \text{and} \\ X \in \mathcal{F}(R) \quad &\Longleftrightarrow \quad \exists\, F \in \mathcal{C}^{\mathrm{F}}_{\square}(R) : X \simeq F. \end{aligned}$$

We also use the notation $\mathcal{P}(R)$, $\mathcal{I}(R)$, and $\mathcal{F}(R)$ with superscripts and subscripts following the general rules from (A.1.10). See also (A.3.12).

(A.3.8) **Definition.** For $Y \in \mathcal{C}_{(\sqsubset)}(R)$ the *injective dimension*, $\mathrm{id}_R\, Y$, is defined as

$$\mathrm{id}_R\, Y = \inf\{\sup\{\ell \in \mathbb{Z} \mid I_{-\ell} \neq 0\} \mid Y \simeq I \in \mathcal{C}^{\mathrm{I}}_{\sqsubset}(R)\}.$$

Note that

$$\begin{aligned} &\mathrm{id}_R\, Y \in \{-\infty\} \cup \mathbb{Z} \cup \{\infty\}; \\ &\mathrm{id}_R\, Y \geq -\inf Y; \\ &\mathrm{id}_R\, Y = -\infty \quad \Longleftrightarrow \quad Y \simeq 0; \quad \text{and} \\ &\mathrm{id}_R\, Y < \infty \quad \Longleftrightarrow \quad Y \in \mathcal{I}(R). \end{aligned}$$

(A.3.9) **Definition.** For $X \in \mathcal{C}_{(\sqsupset)}(R)$ the *projective dimension*, $\mathrm{pd}_R\, X$, is defined as

$$\mathrm{pd}_R\, X = \inf\{\sup\{\ell \in \mathbb{Z} \mid P_\ell \neq 0\} \mid X \simeq P \in \mathcal{C}^{\mathrm{P}}_{\sqsupset}(R)\}.$$

Note that

$$\begin{aligned} &\mathrm{pd}_R\, X \in \{-\infty\} \cup \mathbb{Z} \cup \{\infty\}; \\ &\mathrm{pd}_R\, X \geq \sup X; \\ &\mathrm{pd}_R\, X = -\infty \quad \Longleftrightarrow \quad X \simeq 0; \quad \text{and} \\ &\mathrm{pd}_R\, X < \infty \quad \Longleftrightarrow \quad X \in \mathcal{P}(R). \end{aligned}$$

(A.3.10) **Definition.** For $X \in \mathcal{C}_{(\sqsupset)}(R)$ the *flat dimension*, $\mathrm{fd}_R X$, is defined as

$$\mathrm{fd}_R X = \inf\{\sup\{\ell \in \mathbb{Z} \mid F_\ell \neq 0\} \mid X \simeq F \in \mathcal{C}_{\sqsupset}^{\mathrm{F}}(R)\}.$$

Note that

$$\begin{aligned} &\mathrm{fd}_R X \in \{-\infty\} \cup \mathbb{Z} \cup \{\infty\}; \\ \mathrm{pd}_R X \geq\ &\mathrm{fd}_R X \geq \sup X; \\ &\mathrm{fd}_R X = -\infty \iff X \simeq 0; \text{ and} \\ &\mathrm{fd}_R X < \infty \iff X \in \mathcal{F}(R). \end{aligned}$$

The last result of this section shows that the definitions above extend the usual notions of projective, injective, and flat dimension of modules.

(A.3.11) **Proposition.** *The following hold:*

(I) *If the $R$–module $N$ is equivalent to $I \in \mathcal{C}_{\sqsubset}^{\mathrm{I}}(R)$, then the truncated complex*

$$\subset_0 I \;=\; 0 \to \mathrm{C}_0^I \to I_{-1} \to I_{-2} \to \cdots \to I_\ell \to \cdots$$

*is a (usual) injective resolution of $N$.*

(P) *If the $R$–module $M$ is equivalent to $P \in \mathcal{C}_{\sqsupset}^{\mathrm{P}}(R)$, then the truncated complex*

$$P_0 \supset \;=\; \cdots \to P_\ell \to \cdots \to P_2 \to P_1 \to \mathrm{Z}_0^P \to 0$$

*is a (usual) projective resolution of $M$.*

(F) *If the $R$–module $M$ is equivalent to $F \in \mathcal{C}_{\sqsupset}^{\mathrm{F}}(R)$, then the truncated complex*

$$F_0 \supset \;=\; \cdots \to F_\ell \to \cdots \to F_2 \to F_1 \to \mathrm{Z}_0^F \to 0$$

*is a (usual) flat resolution of $M$.*

*Proof.* If $N$ is equivalent to $I \in \mathcal{C}_{\sqsubset}^{\mathrm{I}}(R)$, then $\sup I = \sup N \leq 0$ so, by (A.1.14.2), $N$ is also equivalent to $\subset_0 I$. That is, $N$ is isomorphic to the kernel $\mathrm{Ker}(\mathrm{C}_0^I \to I_{-1})$, and we just have to prove that $\mathrm{C}_0^I$ is injective. But this is immediate: set $u = \sup\{\ell \in \mathbb{Z} \mid I_\ell \neq 0\}$, and consider the exact sequence

$$0 \to I_u \to I_{u-1} \to \cdots \to I_0 \to \mathrm{C}_0^I \to 0.$$

The modules $I_u, I_{u-1}, \ldots, I_0$ are injective, and hence so is the cokernel $\mathrm{C}_0^I$. This proves (I), and the proofs of (P) and (F) are similar. □

(A.3.12) **Remark.** It follows by the Proposition that, e.g., $\mathcal{F}_0(R)$ is (naturally identified with) the full subcategory of modules of finite flat dimension, and $\mathcal{P}_0^{\mathrm{f}}(R)$ is (naturally identified with) the full subcategory of finite modules of finite projective dimension.

## A.4 (Almost) Derived Functors

The notations $\mathbf{R}\mathrm{Hom}_R(-,-)$ and $-\otimes^{\mathbf{L}}_R-$ are usually used for, respectively, the right and left **derived functors** of the homomorphism and tensor product functors for $R$–complexes. For objects (complexes) $X$ and $Y$ in the derived category $\mathbf{R}\mathrm{Hom}_R(X,Y)$ and $X\otimes^{\mathbf{L}}_R Y$ are uniquely determined up to isomorphism in $\mathcal{D}(R)$. In this book we use this very same notation for certain **equivalence classes** of $R$–complexes; they are defined in (A.4.2) and (A.4.11) below. Not only is this permissible, in as much as we never use the true derived functors, but it is also very convenient, because experienced users of the derived category will certainly want to think of these gadgets as "real complexes". Likewise these readers will interpret equalities of equivalence classes of $R$–complexes as isomorphisms in the derived category $\mathcal{D}(R)$.

The first paragraph collects a series of results from [43, Part I] and [7, Section 1].

(A.4.1) ***Preservation of Quasi-Isomorphisms and Equivalences.*** If $P\in\mathcal{C}^{\mathrm{P}}_{\sqsupset}(R)$ and $I\in\mathcal{C}^{\mathrm{I}}_{\sqsubset}(R)$, then the functors $\mathrm{Hom}_R(P,-)$ and $\mathrm{Hom}_R(-,I)$ preserve quasi isomorphisms and, thereby, equivalences:

$$X\xrightarrow[\alpha]{\simeq}Y \implies \mathrm{Hom}_R(P,X)\xrightarrow[\mathrm{Hom}_R(P,\alpha)]{\simeq}\mathrm{Hom}_R(P,Y);$$
$$X\simeq Y \implies \mathrm{Hom}_R(P,X)\simeq\mathrm{Hom}_R(P,Y);$$
$$X\xrightarrow[\alpha]{\simeq}Y \implies \mathrm{Hom}_R(Y,I)\xrightarrow[\mathrm{Hom}_R(\alpha,I)]{\simeq}\mathrm{Hom}_R(X,I);\quad\text{and}$$
$$X\simeq Y \implies \mathrm{Hom}_R(X,I)\simeq\mathrm{Hom}_R(Y,I).$$

For fixed complexes $V$ and $W$ the restrictions of the functors $\mathrm{Hom}_R(V,-)$ and $\mathrm{Hom}_R(-,W)$ to, respectively, $\mathcal{C}^{\mathrm{I}}_{\sqsubset}(R)$ and $\mathcal{C}^{\mathrm{P}}_{\sqsupset}(R)$ also preserve quasi-isomorphisms and equivalences. That is, for $I,I'\in\mathcal{C}^{\mathrm{I}}_{\sqsubset}(R)$ there are implications:

$$I\xrightarrow[\iota]{\simeq}I' \implies \mathrm{Hom}_R(V,I)\xrightarrow[\mathrm{Hom}_R(V,\iota)]{\simeq}\mathrm{Hom}_R(V,I');\quad\text{and}$$
$$I\simeq I' \implies \mathrm{Hom}_R(V,I)\simeq\mathrm{Hom}_R(V,I').$$

Similarly for $P,P'\in\mathcal{C}^{\mathrm{P}}_{\sqsupset}(R)$:

$$P\xrightarrow[\pi]{\simeq}P' \implies \mathrm{Hom}_R(P',W)\xrightarrow[\mathrm{Hom}_R(\pi,W)]{\simeq}\mathrm{Hom}_R(P,W);\quad\text{and}$$
$$P\simeq P' \implies \mathrm{Hom}_R(P,W)\simeq\mathrm{Hom}_R(P',W).$$

If $F\in\mathcal{C}^{\mathrm{F}}_{\sqsupset}(R)$, then the functor $F\otimes_R-$ (and the isomorphic $-\otimes_R F$) preserves quasi-isomorphisms and, therefore, equivalences:

$$X\xrightarrow[\alpha]{\simeq}Y \implies F\otimes_R X\xrightarrow[F\otimes_R\alpha]{\simeq}F\otimes_R Y;\quad\text{and}$$
$$X\simeq Y \implies F\otimes_R X\simeq F\otimes_R Y.$$

For a fixed complex $V$ the restriction of the functor $V \otimes_R -$ (and the isomorphic $- \otimes_R V$) to $\mathcal{C}^{\mathrm{F}}_{\sqsupset}(R)$ also preserves quasi-isomorphisms and equivalences. That is, if $F, F' \in \mathcal{C}^{\mathrm{F}}_{\sqsupset}(R)$, then

$$F \xrightarrow[\varphi]{\simeq} F' \quad \Longrightarrow \quad V \otimes_R F \xrightarrow[V \otimes_R \varphi]{\simeq} V \otimes_R F'; \text{ and}$$
$$F \simeq F' \quad \Longrightarrow \quad V \otimes_R F \simeq V \otimes_R F'.$$

(A.4.2) **Definition.** Let $X$ and $Y$ be $R$–complexes and assume that $X \in \mathcal{C}_{(\sqsupset)}(R)$ and/or $Y \in \mathcal{C}_{(\sqsubset)}(R)$. By $\mathbf{R}\mathrm{Hom}_R(X,Y)$ we denote the equivalence class of $R$–complexes represented by $\mathrm{Hom}_R(P,Y)$ and/or $\mathrm{Hom}_R(X,I)$, where $X \simeq P \in \mathcal{C}^{\mathrm{P}}_{\sqsupset}(R)$ and/or $Y \simeq I \in \mathcal{C}^{\mathrm{I}}_{\sqsubset}(R)$. The equivalence preserving properties described above ensure that $\mathbf{R}\mathrm{Hom}_R(X,Y)$ is well-defined and only depends on the equivalence classes of $X$ and $Y$.

(A.4.3) ***Ext Modules***. For $R$–modules $M, N$ and $m \in \mathbb{Z}$ the isomorphism class $\mathrm{H}_{-m}(\mathbf{R}\mathrm{Hom}_R(M,N))$ is known as the *m-th Ext module*, i.e.,

$$\mathrm{Ext}^m_R(M,N) = \mathrm{H}_{-m}(\mathbf{R}\mathrm{Hom}_R(M,N)).$$

(A.4.4) **Lemma.** *If $X \in \mathcal{C}^{(\mathrm{f})}_{(\sqsupset)}(R)$ and $Y \in \mathcal{C}^{(\mathrm{f})}_{(\sqsubset)}(R)$, then also $\mathbf{R}\mathrm{Hom}_R(X,Y)$ belongs to $\mathcal{C}^{(\mathrm{f})}_{(\sqsubset)}(R)$.*

*Proof.* An easy consequences of the definition and (A.2.2), see also [8, (1.2.2)].

(A.4.5) **Lemma.** *If $X \in \mathcal{C}^{(\mathrm{f})}_{(\sqsupset)}(R)$, $Y \in \mathcal{C}_{(\sqsubset)}(R)$, and $\mathfrak{p}$ is a prime ideal in $R$, then there is an equality of equivalence classes of $R_\mathfrak{p}$–complexes:*

$$\mathbf{R}\mathrm{Hom}_R(X,Y)_\mathfrak{p} = \mathbf{R}\mathrm{Hom}_{R_\mathfrak{p}}(X_\mathfrak{p}, Y_\mathfrak{p}).$$

*Proof.* The equality is an easy consequences of the definition and (A.2.3), see also [7, Lemma 5.2(b)].

It makes sense to talk about the supremum of an equivalence class, cf. (A.1.12), and the next result is very useful.

(A.4.6) **Proposition.** *If $X \in \mathcal{C}_{(\sqsupset)}(R)$ and $Y \in \mathcal{C}_{(\sqsubset)}(R)$, then $\mathbf{R}\mathrm{Hom}_R(X,Y)$ belongs to $\mathcal{C}_{(\sqsubset)}(R)$ and there is an inequality:*

(A.4.6.1) $$\sup(\mathbf{R}\mathrm{Hom}_R(X,Y)) \leq \sup Y - \inf X.$$

*Furthermore, assume that both $X$ and $Y$ are homologically non-trivial, and set $s = \sup Y$ and $i = \inf X$. Then $\mathrm{H}_{s-i}(\mathbf{R}\mathrm{Hom}_R(X,Y))$ is represented by the module $\mathrm{Hom}_R(\mathrm{H}_i(X), \mathrm{H}_s(Y))$, so*

(A.4.6.2) $$\sup(\mathbf{R}\mathrm{Hom}_R(X,Y)) = \sup Y - \inf X \iff \mathrm{Hom}_R(\mathrm{H}_i(X), \mathrm{H}_s(Y)) \neq 0.$$

*Proof.* [35, Lemma 2.1(1)].

(A.4.7) **Lemma.** *Let $0 \to K \to H \to C \to 0$ be a short exact sequence of $R$–modules. If $X \in \mathcal{C}_{(\sqsupset)}(R)$, then there is a long exact sequence of homology modules:*

$$\cdots \to \mathrm{H}_\ell(\mathbf{R}\mathrm{Hom}_R(X,K)) \to \mathrm{H}_\ell(\mathbf{R}\mathrm{Hom}_R(X,H)) \to \mathrm{H}_\ell(\mathbf{R}\mathrm{Hom}_R(X,C)) \to \mathrm{H}_{\ell-1}(\mathbf{R}\mathrm{Hom}_R(X,K)) \to \cdots$$

*Proof.* Choose a projective resolution $X \xleftarrow{\simeq} P \in \mathcal{C}^{\mathrm{P}}_{\sqsupset}(R)$, then

$$0 \to \mathrm{Hom}_R(P,K) \to \mathrm{Hom}_R(P,H) \to \mathrm{Hom}_R(P,C) \to 0$$

is a short exact sequence of complexes, and the associated long exact sequence

$$\cdots \to \mathrm{H}_\ell(\mathrm{Hom}_R(P,K)) \to \mathrm{H}_\ell(\mathrm{Hom}_R(P,H)) \to \mathrm{H}_\ell(\mathrm{Hom}_R(P,C)) \to \mathrm{H}_{\ell-1}(\mathrm{Hom}_R(P,K)) \to \cdots$$

is the desired one. □

(A.4.8) **Lemma.** *Let $0 \to K \to H \to C \to 0$ be a short exact sequence of $R$–modules. If $Y \in \mathcal{C}_{(\sqsubset)}(R)$, then there is a long exact sequence of homology modules:*

$$\cdots \to \mathrm{H}_\ell(\mathbf{R}\mathrm{Hom}_R(C,Y)) \to \mathrm{H}_\ell(\mathbf{R}\mathrm{Hom}_R(H,Y)) \to \mathrm{H}_\ell(\mathbf{R}\mathrm{Hom}_R(K,Y)) \to \mathrm{H}_{\ell-1}(\mathbf{R}\mathrm{Hom}_R(C,Y)) \to \cdots$$

*Proof.* Similar to the proof of (A.4.7), only this time choose an injective resolution of $Y$. □

(A.4.9) **Remarks.** If $X = M$ and $Y = N$ are modules, then the long exact sequences in (A.4.7) and (A.4.8) are just the usual long exact sequences of Ext modules:

$$\cdots \to \mathrm{Ext}^\ell_R(M,K) \to \mathrm{Ext}^\ell_R(M,H) \to \mathrm{Ext}^\ell_R(M,C) \to \mathrm{Ext}^{\ell+1}_R(M,K) \to \cdots$$

and

$$\cdots \to \mathrm{Ext}^\ell_R(C,N) \to \mathrm{Ext}^\ell_R(H,N) \to \mathrm{Ext}^\ell_R(K,N) \to \mathrm{Ext}^{\ell+1}_R(C,N) \to \cdots.$$

(A.4.10) ***Faithfully Injective Modules.*** If $E$ is an injective $R$–module, then, for any $R$–complex $X$, the equivalence class $\mathbf{R}\mathrm{Hom}_R(X,E)$ is represented by $\mathrm{Hom}_R(X,E)$. If $E$ is faithfully injective (i.e., the functor $\mathrm{Hom}_R(-,E)$ is faithful and exact), then we have the following special case of (A.1.20.4):

$$\sup(\mathbf{R}\mathrm{Hom}_R(X,E)) = \sup(\mathrm{Hom}_R(X,E)) = -\inf X$$

and

$$\inf(\mathbf{R}\mathrm{Hom}_R(X,E)) = \inf(\mathrm{Hom}_R(X,E)) = -\sup X.$$

If $(R, \mathfrak{m}, k)$ is local, then $\mathrm{E}_R(k)$, the injective hull of the residue field, is a faithfully injective $R$-module, and the (module) functor $\mathrm{Hom}_R(-, \mathrm{E}_R(k))$ (as well as the induced functor on complexes) is called the *Matlis duality functor*.

Every ring $R$ admits a faithfully injective module $E$, e.g.,

$$E = \prod_{\mathfrak{m} \in \mathrm{Max}\, R} \mathrm{E}_R(R/\mathfrak{m}).$$

(A.4.11) **Definition.** Let $X$ and $Y$ be $R$-complexes and assume that $X \in \mathcal{C}_{(\sqsupset)}(R)$ and/or $Y \in \mathcal{C}_{(\sqsupset)}(R)$. By $X \otimes^{\mathbf{L}}_R Y$ we denote the equivalence class of $R$-complexes represented by $F \otimes_R Y$ and/or $X \otimes_R F'$, where $X \simeq F \in \mathcal{C}^{\mathrm{F}}_{\sqsupset}(R)$ and/or $Y \simeq F' \in \mathcal{C}^{\mathrm{F}}_{\sqsupset}(R)$. The equivalence preserving properties described in (A.4.1) ensure that $X \otimes^{\mathbf{L}}_R Y$ is well-defined and only depends on the equivalence classes of $X$ and $Y$.

(A.4.12) ***Tor Modules***. For $R$-modules $M, N$ and $m \in \mathbb{Z}$ the isomorphism class $\mathrm{H}_m(M \otimes^{\mathbf{L}}_R N)$ is known as the *m-th Tor module*, i.e.,

$$\mathrm{Tor}^R_m(M, N) = \mathrm{H}_m(M \otimes^{\mathbf{L}}_R N).$$

(A.4.13) **Lemma.** *If $X, Y \in \mathcal{C}^{(\mathrm{f})}_{(\sqsupset)}(R)$, then also $X \otimes^{\mathbf{L}}_R Y \in \mathcal{C}^{(\mathrm{f})}_{(\sqsupset)}(R)$.*

*Proof.* An easy consequences of the definition and (A.2.6), see also [8, (1.2.1)].

(A.4.14) **Lemma.** *Let $X$ and $Y$ be $R$-complexes and $\mathfrak{p}$ be a prime ideal in $R$. If one of the complexes is homologically bounded to the right, then there is an equality of equivalence classes of $R_\mathfrak{p}$-complexes:*

$$(X \otimes^{\mathbf{L}}_R Y)_\mathfrak{p} = X_\mathfrak{p} \otimes^{\mathbf{L}}_{R_\mathfrak{p}} Y_\mathfrak{p}.$$

*Proof.* The equality is an easy consequences of the definition and (A.2.5), see also [7, Lemma 5.2(a)].

(A.4.15) **Proposition.** *If $X, Y \in \mathcal{C}_{(\sqsupset)}(R)$, then also $X \otimes^{\mathbf{L}}_R Y \in \mathcal{C}_{(\sqsupset)}(R)$ and there is an inequality:*

(A.4.15.1) $$\inf (X \otimes^{\mathbf{L}}_R Y) \geq \inf X + \inf Y.$$

*Furthermore, assume that $X$ and $Y$ are both homologically non-trivial, and set $i = \inf X$ and $j = \inf Y$. Then $\mathrm{H}_{i+j}(X \otimes^{\mathbf{L}}_R Y)$ is represented by the module $\mathrm{H}_i(X) \otimes_R \mathrm{H}_j(Y)$, so*

(A.4.15.2) $$\inf (X \otimes^{\mathbf{L}}_R Y) = \inf X + \inf Y \iff \mathrm{H}_i(X) \otimes_R \mathrm{H}_j(Y) \neq 0.$$

*Proof.* [35, Lemma 2.1(2)].

The next corollary is an immediate consequence of Nakayama's lemma and (A.4.15.2); it is sometimes called 'Nakayama's lemma for complexes'.

(A.4.16) **Corollary.** *Let $(R, \mathfrak{m}, k)$ be local. If $X$ and $Y$ belong to $\mathcal{C}^{(f)}_{(\sqsupset)}(R)$, then*

$$\inf (X \otimes^{\mathbf{L}}_R Y) = \inf X + \inf Y.$$

(A.4.17) **Lemma.** *Let $0 \to K \to H \to C \to 0$ be a short exact sequence of $R$–modules. If $X \in \mathcal{C}_{(\sqsupset)}(R)$, then there is a long exact sequence of homology modules:*

$$\cdots \to \mathrm{H}_{\ell+1}(C \otimes^{\mathbf{L}}_R X) \to \mathrm{H}_\ell(K \otimes^{\mathbf{L}}_R X) \to \mathrm{H}_\ell(H \otimes^{\mathbf{L}}_R X) \to \mathrm{H}_\ell(C \otimes^{\mathbf{L}}_R X) \to \cdots$$

*Proof.* Similar to the proof of (A.4.7). □

(A.4.18) **Remark.** If $X = M$ is a module, then the long exact sequence in (A.4.17) is just the usual long exact sequence of Tor modules:

$$\cdots \to \mathrm{Tor}^R_{\ell+1}(C, M) \to \mathrm{Tor}^R_\ell(K, M) \to \mathrm{Tor}^R_\ell(H, M) \to \mathrm{Tor}^R_\ell(C, M) \to \cdots$$

The **standard isomorphisms of complexes** (A.2.4.1) and (A.2.7)–(A.2.11) **induce** six **identities of equivalence classes**; these are described below in (A.4.19)–(A.4.24). As usual $S$ is an $R$–algebra; and for an equivalence class $\boldsymbol{X}$ of $R$–complexes we write $\boldsymbol{X} \in \mathcal{C}(S)$ **if $\boldsymbol{X}$ has a representative in $\mathcal{C}(S)$.**

(A.4.19) **Commutativity.** *Assume that $X \in \mathcal{C}_{(\sqsupset)}(R)$ and $Y \in \mathcal{C}(R)$. Then there is an identity of equivalence classes of $R$–complexes:*

$$X \otimes^{\mathbf{L}}_R Y = Y \otimes^{\mathbf{L}}_R X.$$

(A.4.20) **Associativity.** *Assume that $Z \in \mathcal{C}_{(\sqsupset)}(S)$, $Y \in \mathcal{C}(S)$, and $X$ belongs to $\mathcal{C}_{(\sqsupset)}(R)$. Then $Z \otimes^{\mathbf{L}}_S Y \in \mathcal{C}(R)$ and $Y \otimes^{\mathbf{L}}_R X \in \mathcal{C}(S)$, and there is an identity of equivalence classes of $S$–complexes:*

$$(Z \otimes^{\mathbf{L}}_S Y) \otimes^{\mathbf{L}}_R X = Z \otimes^{\mathbf{L}}_S (Y \otimes^{\mathbf{L}}_R X).$$

(A.4.21) **Adjointness.** *Assume that $Z \in \mathcal{C}_{(\sqsupset)}(S)$, $Y \in \mathcal{C}(S)$, and $X \in \mathcal{C}_{(\sqsubset)}(R)$. Then $Z \otimes^{\mathbf{L}}_S Y \in \mathcal{C}(R)$ and $\mathbf{R}\mathrm{Hom}_R(Y, X) \in \mathcal{C}(S)$, and there is an identity of equivalence classes of $S$–complexes:*

$$\mathbf{R}\mathrm{Hom}_R(Z \otimes^{\mathbf{L}}_S Y, X) = \mathbf{R}\mathrm{Hom}_S(Z, \mathbf{R}\mathrm{Hom}_R(Y, X)).$$

(A.4.22) **Swap.** *Assume that $Z \in \mathcal{C}_{(\sqsupset)}(S)$, $Y \in \mathcal{C}(S)$, and $X$ belongs to $\mathcal{C}_{(\sqsupset)}(R)$. Then $\mathbf{R}\mathrm{Hom}_R(X, Y) \in \mathcal{C}(S)$ and $\mathbf{R}\mathrm{Hom}_S(Z, Y) \in \mathcal{C}(R)$, and there is an identity of equivalence classes of $S$–complexes:*

$$\mathbf{R}\mathrm{Hom}_S(Z, \mathbf{R}\mathrm{Hom}_R(X, Y)) = \mathbf{R}\mathrm{Hom}_R(X, \mathbf{R}\mathrm{Hom}_S(Z, Y)).$$

(A.4.23) **Tensor Evaluation.** *Assume that* $Z \in \mathcal{C}^{(\mathrm{f})}_{(\sqsupset)}(S)$, $Y \in \mathcal{C}_{(\square)}(S)$, *and* $X \in \mathcal{C}_{(\sqsupset)}(R)$. *Then* $\mathbf{R}\mathrm{Hom}_S(Z,Y) \in \mathcal{C}(R)$ *and* $Y \otimes^{\mathbf{L}}_R X \in \mathcal{C}(S)$, *and there is an identity of equivalence classes of S–complexes:*

$$\mathbf{R}\mathrm{Hom}_S(Z,Y) \otimes^{\mathbf{L}}_R X = \mathbf{R}\mathrm{Hom}_S(Z, Y \otimes^{\mathbf{L}}_R X),$$

*provided that* $Z \in \mathcal{P}^{(\mathrm{f})}(S)$ *or* $X \in \mathcal{F}(R)$.

(A.4.24) **Hom Evaluation.** *Assume that* $Z \in \mathcal{C}^{(\mathrm{f})}_{(\sqsupset)}(S)$, $Y \in \mathcal{C}_{(\square)}(S)$, *and X belongs to* $\mathcal{C}_{(\sqsubset)}(R)$. *Then* $\mathbf{R}\mathrm{Hom}_S(Z,Y) \in \mathcal{C}(R)$ *and* $\mathbf{R}\mathrm{Hom}_R(Y,X) \in \mathcal{C}(S)$, *and there is an identity of equivalence classes of S–complexes:*

$$Z \otimes^{\mathbf{L}}_S \mathbf{R}\mathrm{Hom}_R(Y,X) = \mathbf{R}\mathrm{Hom}_R(\mathbf{R}\mathrm{Hom}_S(Z,Y),X),$$

*provided that* $Z \in \mathcal{P}^{(\mathrm{f})}(S)$ *or* $X \in \mathcal{I}(R)$.

*Proof of (A.4.19)–(A.4.24).* The equalities are straightforward consequences of the standard morphisms, but the reader may want to check with [7, Lemma 4.4].

## A.5 Homological Dimensions

The standard homological dimensions were defined in section A.3. In this section we collect a number of results that allow us to compute the dimensions in terms of (almost) derived functors.

(A.5.1) **ID Theorem.** *Let* $Y \in \mathcal{C}_{(\sqsubset)}(R)$ *and* $n \in \mathbb{Z}$. *The following are equivalent:*

- *(i)* *Y is equivalent to a complex* $I \in \mathcal{C}^{\mathrm{I}}_{\square}(R)$ *concentrated in degrees at least* $-n$; *and I can be chosen with* $I_\ell = 0$ *for* $\ell > \sup Y$.
- *(ii)* $\mathrm{id}_R Y \leq n$.
- *(iii)* $n \geq -\sup U - \inf(\mathbf{R}\mathrm{Hom}_R(U,Y))$ *for all* $U \not\simeq 0$ *in* $\mathcal{C}_{(\square)}(R)$.
- *(iv)* $n \geq -\inf Y$ *and* $\mathrm{H}_{-(n+1)}(\mathbf{R}\mathrm{Hom}_R(T,Y)) = 0$ *for all cyclic modules T.*
- *(v)* $n \geq -\inf Y$ *and the module* $\mathrm{Z}^I_{-n}$ *is injective whenever* $Y \simeq I \in \mathcal{C}^{\mathrm{I}}_{\sqsubset}(R)$.

*Proof.* See Theorem 2.4.I and Corollary 2.7.I in [7].

(A.5.2) **ID Corollary.** *For $Y \in \mathcal{C}_{(\sqsubset)}(R)$ there are equalities:*

$$\text{(A.5.2.1)}\quad \begin{aligned} \operatorname{id}_R Y &= \sup\{-\sup U - \inf(\mathbf{R}\mathrm{Hom}_R(U,Y)) \mid U \in \mathcal{C}_{(\square)}(R) \wedge U \not\simeq 0\} \\ &= \sup\{-\inf(\mathbf{R}\mathrm{Hom}_R(T,Y)) \mid T \in \mathcal{C}_0(R) \text{ cyclic}\}; \end{aligned}$$

*and the following are equivalent:*

(*i*) $Y \in \mathcal{I}(R)$.

(*ii*) $\mathbf{R}\mathrm{Hom}_R(U,Y) \in \mathcal{C}_{(\square)}(R)$ *for all* $U \in \mathcal{C}_{(\square)}(R)$.

(*iii*) $\mathbf{R}\mathrm{Hom}_R(T,Y) \in \mathcal{C}_{(\square)}(R)$ *for all* $T \in \mathcal{C}_0(R)$.

*Proof.* See section 2.I in [7].

(A.5.3) **PD Theorem.** *Let $X \in \mathcal{C}_{(\sqsupset)}(R)$ and $n \in \mathbb{Z}$. The following are equivalent:*

(*i*) *$X$ is equivalent to a complex $P \in \mathcal{C}^{\mathrm{P}}_{\square}(R)$ concentrated in degrees at most $n$; and $P$ can be chosen with $P_\ell = 0$ for $\ell < \inf X$.*

(*ii*) $\operatorname{pd}_R X \leq n$.

(*iii*) $n \geq \inf U - \inf(\mathbf{R}\mathrm{Hom}_R(X,U))$ *for all* $U \not\simeq 0$ *in* $\mathcal{C}_{(\square)}(R)$.

(*iv*) $n \geq \sup X$ *and* $\mathrm{H}_{-(n+1)}(\mathbf{R}\mathrm{Hom}_R(X,T)) = 0$ *for all $R$-modules $T$.*

(*v*) $n \geq \sup X$ *and the module $\mathrm{C}^P_n$ is projective whenever $X \simeq P \in \mathcal{C}^{\mathrm{P}}_{\sqsupset}(R)$.*

*Proof.* See Theorem 2.4.P and Corollary 2.7.P in [7].

(A.5.4) **PD Corollary.** *For $X \in \mathcal{C}_{(\sqsupset)}(R)$ there are equalities:*

$$\text{(A.5.4.1)}\quad \begin{aligned} \operatorname{pd}_R X &= \sup\{\inf U - \inf(\mathbf{R}\mathrm{Hom}_R(X,U)) \mid U \in \mathcal{C}_{(\square)}(R) \wedge U \not\simeq 0\} \\ &= \sup\{-\inf(\mathbf{R}\mathrm{Hom}_R(X,T)) \mid T \in \mathcal{C}_0(R)\}; \end{aligned}$$

*and the following are equivalent:*

(*i*) $X \in \mathcal{P}(R)$.

(*ii*) $\mathbf{R}\mathrm{Hom}_R(X,U) \in \mathcal{C}_{(\square)}(R)$ *for all* $U \in \mathcal{C}_{(\square)}(R)$.

(*iii*) $\mathbf{R}\mathrm{Hom}_R(X,T) \in \mathcal{C}_{(\square)}(R)$ *for all* $T \in \mathcal{C}_0(R)$.

*Furthermore, the following hold if $X \in \mathcal{C}^{(\mathrm{f})}_{(\sqsupset)}(R)$:*

$$\text{(A.5.4.2)}\qquad X \in \mathcal{P}^{(\mathrm{f})}(R) \iff \exists\, P \in \mathcal{C}^{\mathrm{fP}}_{\square}(R) : X \simeq P;$$

*and*

$$\text{(A.5.4.3)}\qquad \operatorname{pd}_R X = \sup\{-\inf(\mathbf{R}\mathrm{Hom}_R(X,T)) \mid T \in \mathcal{C}^{\mathrm{f}}_0(R)\}.$$

*Proof.* See section 2.P and Proposition 5.3.P in [7].

(A.5.5) **FD Theorem.** *Let $X \in \mathcal{C}_{(\sqsupset)}(R)$ and $n \in \mathbb{Z}$. The following are equivalent*

(*i*) *$X$ is equivalent to a complex $F \in \mathcal{C}^{\mathrm{F}}_{\square}(R)$ concentrated in degrees at most $n$; and $F$ can be chosen with $F_\ell = 0$ for $\ell < \inf X$.*
(*ii*) $\operatorname{fd}_R X \leq n$.
(*iii*) $\sup(U \otimes^{\mathbf{L}}_R X) - \sup U \leq n$ *for all* $U \not\simeq 0$ *in* $\mathcal{C}_{(\square)}(R)$.
(*iv*) $n \geq \sup X$ *and* $\mathrm{H}_{n+1}(T \otimes^{\mathbf{L}}_R X) = 0$ *for all cyclic modules* $T$.
(*v*) $n \geq \sup X$ *and the module* $\mathrm{C}^F_n$ *is flat whenever* $X \simeq F \in \mathcal{C}^{\mathrm{F}}_{\sqsupset}(R)$.

*Proof.* See Theorem 2.4.F and Corollary 2.7.F in [7].

(A.5.6) **FD Corollary.** *For $X \in \mathcal{C}_{(\sqsupset)}(R)$ there are equalities:*

$$\text{(A.5.6.1)}\qquad \begin{aligned}\operatorname{fd}_R X &= \sup\{\sup(U \otimes^{\mathbf{L}}_R X) - \sup U \mid U \in \mathcal{C}_{(\square)}(R) \wedge U \not\simeq 0\}\\ &= \sup\{\sup(T \otimes^{\mathbf{L}}_R X) \mid T \in \mathcal{C}_0(R) \textit{ cyclic}\};\end{aligned}$$

*and the following are equivalent:*

(*i*) $X \in \mathcal{F}(R)$.
(*ii*) $U \otimes^{\mathbf{L}}_R X \in \mathcal{C}_{(\square)}(R)$ *for all* $U \in \mathcal{C}_{(\square)}(R)$.
(*iii*) $T \otimes^{\mathbf{L}}_R X \in \mathcal{C}_{(\square)}(R)$ *for all* $T \in \mathcal{C}_0(R)$.

*Proof.* See section 2.F in [7].

(A.5.7) **Theorem (Homological Dimensions over Local Rings).** *Let $(R, \mathfrak{m}, k)$ be local. There is an equality of full subcategories:*

$$\text{(A.5.7.1)}\qquad \mathcal{P}^{(\mathrm{f})}(R) = \mathcal{F}^{(\mathrm{f})}(R);$$

*and the following hold for $X \in \mathcal{C}^{(\mathrm{f})}_{(\sqsupset)}(R)$ and $Y \in \mathcal{C}^{(\mathrm{f})}_{(\sqsubset)}(R)$:*

$$\text{(A.5.7.2)}\qquad \operatorname{pd}_R X = \operatorname{fd}_R X = \sup(X \otimes^{\mathbf{L}}_R k);$$

$$\text{(A.5.7.3)}\qquad \operatorname{pd}_R X = -\inf(\mathbf{R}\mathrm{Hom}_R(X, k)); \quad \textit{and}$$

$$\text{(A.5.7.4)}\qquad \operatorname{id}_R Y = -\inf(\mathbf{R}\mathrm{Hom}_R(k, Y)).$$

*Proof.* See Corollary 2.10.F and Proposition 5.5 in [7].

(A.5.8) **Theorem (Stability).** *If $X, X' \in \mathcal{F}(R)$ and $Y, Y' \in \mathcal{I}(R)$, then*

$$\text{(A.5.8.1)}\qquad \operatorname{fd}_R(X \otimes^{\mathbf{L}}_R X') \leq \operatorname{fd}_R X + \operatorname{fd}_R X';$$

$$\text{(A.5.8.2)}\qquad \operatorname{id}_R(\mathbf{R}\mathrm{Hom}_R(X, Y)) \leq \operatorname{fd}_R X + \operatorname{id}_R Y;$$

$$\text{(A.5.8.3)}\qquad \operatorname{id}_R(Y \otimes^{\mathbf{L}}_R X) \leq \operatorname{id}_R Y - \inf X; \quad \textit{and}$$

$$\text{(A.5.8.4)}\qquad \operatorname{fd}_R(\mathbf{R}\mathrm{Hom}_R(Y, Y')) \leq \operatorname{id}_R Y + \sup Y'.$$

*Proof.* See Theorems 4.1 and 4.5 in [7].

**(A.5.9) Lemma.** *If $X \in \mathcal{C}_{(\square)}(R)$ is equivalent to $P \in \mathcal{C}^{\mathrm{P}}_{\sqsupset}(R)$, $n \geq \sup X$, and $N \in \mathcal{C}_0(R)$, then*

$$\mathrm{Ext}^1_R(\mathrm{C}^P_n, N) = \mathrm{H}_{-(n+1)}(\mathbf{R}\mathrm{Hom}_R(X, N)).$$

*Proof.* Since $n \geq \sup X = \sup P$ we have $P_n \sqsupset \simeq \Sigma^n \mathrm{C}^P_n$, cf. (A.1.14.3), and $\mathbf{R}\mathrm{Hom}_R(\mathrm{C}^P_n, N)$ is, therefore, represented by $\mathrm{Hom}_R(\Sigma^{-n}(P_n\sqsupset), N)$. The isomorphism class $\mathrm{Ext}^1_R(\mathrm{C}^P_n, N)$ is then represented by

$$\begin{aligned}
\mathrm{H}_{-1}(\mathrm{Hom}_R(\Sigma^{-n}(P_n\sqsupset), N)) &= \mathrm{H}_{-1}(\Sigma^n \mathrm{Hom}_R(P_n\sqsupset, N)) \\
&= \mathrm{H}_{-(n+1)}(\mathrm{Hom}_R(P_n\sqsupset, N)) \\
&= \mathrm{H}_{-(n+1)}(\sqsubset_{-n} \mathrm{Hom}_R(P, N)) \\
&= \mathrm{H}_{-(n+1)}(\mathrm{Hom}_R(P, N));
\end{aligned}$$

cf. (A.2.1.3), (A.1.3.1), and (A.1.20.2). Since the complex $\mathrm{Hom}_R(P, N)$ represents $\mathbf{R}\mathrm{Hom}_R(X, N)$, we have $\mathrm{Ext}^1_R(\mathrm{C}^P_n, N) = \mathrm{H}_{-(n+1)}(\mathbf{R}\mathrm{Hom}_R(X, N))$ as wanted. □

## A.6 Depth and Width

The invariants depth and width for modules (the latter is sometimes called codepth or Tor–depth) have been extended to complexes by Foxby [36] and Yassemi [63]. Depth for complexes has also been studied by Iyengar in [44].

(A.6.1) ***Depth.*** If $(R, \mathfrak{m}, k)$ is local and $Y \in \mathcal{C}_{(\sqsubset)}(R)$, then the ***depth*** of $Y$ is defined as:

$$\operatorname{depth}_R Y = -\sup(\mathbf{R}\mathrm{Hom}_R(k, Y)).$$

For finite modules this definition agrees with the classical one (the maximal length of a regular sequence).

For every prime ideal $\mathfrak{p}$ in $R$ (not necessarily a local ring) and every complex $Y \in \mathcal{C}_{(\sqsubset)}(R)$ there are inequalities:

(A.6.1.1) $$\operatorname{depth}_{R_\mathfrak{p}} Y_\mathfrak{p} \geq -\sup Y_\mathfrak{p} \geq -\sup Y.$$

Furthermore, if $Y$ is homologically non-trivial and $s = \sup Y$, then

(A.6.1.2) $$\mathfrak{p} \in \mathrm{Ass}_R(\mathrm{H}_s(Y)) \iff \operatorname{depth}_{R_\mathfrak{p}} Y_\mathfrak{p} = -\sup Y.$$

**(A.6.2) Lemma.** *If $R$ is local and $Y \in \mathcal{C}^{(\mathrm{f})}_{(\sqsubset)}(R)$, then*

$$\operatorname{depth}_R Y \leq \operatorname{depth}_{R_\mathfrak{p}} Y_\mathfrak{p} + \dim R/\mathfrak{p}$$

*for every $\mathfrak{p} \in \operatorname{Spec} R$.*

*Proof.* For finite modules the inequality is a consequence of [11, Lemma (3.1)]; a proof for complexes is given in [33, Chapter 13].

(A.6.3) ***Width.*** If $(R, \mathfrak{m}, k)$ is local and $X \in \mathcal{C}_{(\sqsupset)}(R)$, then the *width* of $X$ is defined as:

$$\operatorname{width}_R X = \inf (X \otimes^{\mathbf{L}}_R k).$$

For modules this is in agreement with the usual definition.

For every prime ideal $\mathfrak{p}$ in $R$ (not necessarily a local ring) and every complex $X \in \mathcal{C}_{(\sqsupset)}(R)$ there are inequalities:

$$\operatorname{width}_{R_\mathfrak{p}} X_\mathfrak{p} \geq \inf X_\mathfrak{p} \geq \inf X. \tag{A.6.3.1}$$

If $R$ is local and $X \in \mathcal{C}^{(\mathrm{f})}_{(\sqsupset)}(R)$, then

$$\operatorname{width}_R X = \inf X. \tag{A.6.3.2}$$

(A.6.4) **Lemma.** *Let $R$ be local. If $X \in \mathcal{C}_{(\sqsupset)}(R)$ and $Y \in \mathcal{C}_{(\sqsubset)}(R)$, then*

$$\operatorname{depth}_R(\mathbf{R}\mathrm{Hom}_R(X, Y)) = \operatorname{width}_R X + \operatorname{depth}_R Y.$$

*Proof.* [63, Theorem 2.4(a)].

(A.6.5) **Lemma.** *Let $R$ be local. If $X, Y \in \mathcal{C}_{(\sqsupset)}(R)$, then*

$$\operatorname{width}_R(X \otimes^{\mathbf{L}}_R Y) = \operatorname{width}_R X + \operatorname{width}_R Y.$$

*Proof.* [63, Theorem 2.4(b)].

(A.6.6) **Theorem.** *If $R$ is local and $X \in \mathcal{C}_{(\square)}(R)$, then*

$$\operatorname{width}_R X < \infty \iff \operatorname{depth}_R X < \infty.$$

*Proof.* Follows by [36, Proposition 2.8].

(A.6.7) **Theorem.** *If $R$ is local, $U \in \mathcal{F}(R)$, and $X \in \mathcal{C}_{(\square)}(R)$, then*

$$\operatorname{depth}_R(U \otimes^{\mathbf{L}}_R X) = \operatorname{depth}_R U + \operatorname{depth}_R X - \operatorname{depth} R.$$

*Proof.* See [37, Lemma 2.1] or [44, Theorem 4.1].

## A.7 Numerical and Formal Invariants

In this section we review a practical technique for manipulating certain invariants for complexes over local rings.

(A.7.1) ***Betti Numbers***. Let $X$ be any $R$–complex. If $(R, \mathfrak{m}, k)$ is local, then

$$\beta_m^R(X) = \operatorname{rank}_k(\mathrm{H}_m(X \otimes_R^{\mathbf{L}} k))$$

is the *$m$-th Betti number of* $X$. Note that if $X \in \mathcal{C}_{(\sqsupset)}^{(\mathrm{f})}(R)$, then $\beta_m^R(X) \in \mathbb{N}_0$ for all $m \in \mathbb{Z}$, cf. (A.4.13).

For $M \in \mathcal{C}_0(R)$ the definition reads: $\beta_m^R(M) = \operatorname{rank}_k(\operatorname{Tor}_m^R(M, k))$; and if $M \in \mathcal{C}_0^{\mathrm{f}}(R)$, then $\beta_0^R(M) = \operatorname{rank}_k M/\mathfrak{m}M$ is the minimal number of generators for $M$.

In general, for $\mathfrak{p} \in \operatorname{Spec} R$ the *$m$-th Betti number of $X$ at* $\mathfrak{p}$ is $\beta_m^R(\mathfrak{p}, X) = \beta_m^{R_\mathfrak{p}}(X_\mathfrak{p})$. Note that if $(R, \mathfrak{m}, k)$ is local, then $\beta_m^R(\mathfrak{m}, X) = \beta_m^R(X)$.

(A.7.2) **Proposition.** *For $X \in \mathcal{C}_{(\sqsupset)}(R)$ there is an equality:*

$$\operatorname{fd}_R X = \sup\{m \in \mathbb{Z} \mid \exists\, \mathfrak{p} \in \operatorname{Spec} R : \beta_m^R(\mathfrak{p}, X) \neq 0\}.$$

*Proof.* [7, Proposition 5.3.F].

(A.7.3) ***Bass Numbers***. Let $Y$ be any $R$–complex. If $(R, \mathfrak{m}, k)$ is local, then

$$\mu_R^m(Y) = \operatorname{rank}_k(\mathrm{H}_{-m}(\mathbf{R}\mathrm{Hom}_R(k, Y)))$$

is the *$m$-th Bass number of* $Y$. Note that if $Y \in \mathcal{C}_{(\sqsubset)}^{(\mathrm{f})}(R)$, then $\mu_R^m(Y) \in \mathbb{N}_0$ for all $m \in \mathbb{Z}$, cf. (A.4.4).

For $N \in \mathcal{C}_0(R)$ the definition reads $\mu_R^m(N) = \operatorname{rank}_k(\operatorname{Ext}_R^m(k, N))$. For brevity we set $\mu_R^m = \mu_R^m(R)$.

In general, for $\mathfrak{p} \in \operatorname{Spec} R$ the *$m$-th Bass number of $Y$ at* $\mathfrak{p}$ is $\mu_R^m(\mathfrak{p}, Y) = \mu_{R_\mathfrak{p}}^m(Y_\mathfrak{p})$. Note that if $(R, \mathfrak{m}, k)$ is local, then $\mu_R^m(\mathfrak{m}, Y) = \mu_R^m(Y)$.

(A.7.4) ***Poincaré series***. Let $(R, \mathfrak{m}, k)$ be local. The *Poincaré series*, $\mathrm{P}_X^R(t)$, of a complex $X \in \mathcal{C}_{(\sqsupset)}^{(\mathrm{f})}(R)$ is defined as:

$$\mathrm{P}_X^R(t) = \sum_{m \in \mathbb{Z}} \beta_m^R(X) t^m;$$

it is a formal Laurant series with non-negative integer coefficients.

It follows by (A.5.7.2) and (A.4.16) that

(A.7.4.1) $$\deg \mathrm{P}_X^R(t) = \operatorname{pd}_R X \quad \text{and} \quad \operatorname{ord} \mathrm{P}_X^R(t) = \inf X.$$

(A.7.5) ***Bass series.*** Let $(R, \mathfrak{m}, k)$ be local. The *Bass series*, $\mathrm{I}_R^Y(t)$, of a complex $Y \in \mathcal{C}_{(\sqsubset)}^{(\mathrm{f})}(R)$ is defined as:

$$\mathrm{I}_R^X(t) = \sum_{m \in \mathbb{Z}} \mu_R^m(Y) t^m;$$

it is a formal Laurent series with non-negative integer coefficients.

It follows by (A.5.7.4) and the definition of depth that

(A.7.5.1) $$\deg \mathrm{I}_R^Y(t) = \mathrm{id}_R Y \quad \text{and} \quad \mathrm{ord}\, \mathrm{I}_R^Y(t) = \mathrm{depth}_R Y.$$

(A.7.6) **Theorem.** *If $R$ is local and $X, Y \in \mathcal{C}_{(\sqsupset)}^{(\mathrm{f})}(R)$, then there is an equality of formal Laurent series:*

$$\mathrm{P}_{X \otimes_R^{\mathbf{L}} Y}^R(t) = \mathrm{P}_X^R(t)\, \mathrm{P}_Y^R(t).$$

*Proof.* See [35, Theorem 4.2(a)] or [8, Lemma (1.5.3)(a)].

(A.7.7) **Theorem.** *If $R$ is local, $X \in \mathcal{C}_{(\sqsupset)}^{(\mathrm{f})}(R)$, and $Y \in \mathcal{C}_{(\sqsubset)}^{(\mathrm{f})}(R)$, then there is an equality of formal Laurent series:*

$$\mathrm{I}_R^{\mathbf{R}\mathrm{Hom}_R(X,Y)}(t) = \mathrm{P}_X^R(t)\, \mathrm{I}_R^Y(t).$$

*Proof.* See [35, Theorem 4.1(a)] or [8, Lemma (1.5.3)(b)].

(A.7.8) **Theorem.** *If $R$ is local, $X \in \mathcal{P}^{(\mathrm{f})}(R)$, and $Y \in \mathcal{C}_{(\square)}^{(\mathrm{f})}(R)$, then there is an equality of formal Laurent series:*

$$\mathrm{P}_{\mathbf{R}\mathrm{Hom}_R(X,Y)}^R(t) = \mathrm{P}_X^R(t^{-1})\, \mathrm{P}_Y^R(t).$$

*In particular,*

$$\inf (\mathbf{R}\mathrm{Hom}_R(X, Y)) = \inf Y - \mathrm{pd}_R X.$$

*Proof.* [15, Corollary (2.14)]. (The result is due to Foxby.)

The proofs of (A.7.6)–(A.7.8) all use what Foxby calls 'accounting principles'.

(A.7.9) **Lemma (Accounting Principles).** *Let $R$ be a local ring with residue field $k$. If $X \in \mathcal{C}_{(\sqsupset)}(R)$ and $V \in \mathcal{C}(k)$, then*

(A.7.9.1) $$\sup (V \otimes_R^{\mathbf{L}} X) = \sup V + \sup (X \otimes_R^{\mathbf{L}} k); \quad \textit{and}$$

(A.7.9.2) $$\inf (V \otimes_R^{\mathbf{L}} X) = \inf V + \inf (X \otimes_R^{\mathbf{L}} k).$$

*If $Y \in \mathcal{C}_{(\sqsubset)}(R)$ and $V \in \mathcal{C}(k)$, then*

(A.7.9.3) $$\sup (\mathbf{R}\mathrm{Hom}_R(V, Y)) = \sup (\mathbf{R}\mathrm{Hom}_R(k, Y)) - \inf V; \quad \textit{and}$$

(A.7.9.4) $$\inf (\mathbf{R}\mathrm{Hom}_R(V, Y)) = \inf (\mathbf{R}\mathrm{Hom}_R(k, Y)) - \sup V.$$

*Proof.* See, e.g., the proofs of Theorems 4.1 and 4.2 in [35].

## A.8 Dualizing Complexes

The definition of dualizing complexes goes back to [41]. In the literature these complexes are usually taken to be bounded complexes of injective modules, cf. part (1) in the definition below; for the proofs in this book, however, we usually need the (infinite) projective resolution instead, and this accounts for the formulation of part (2).

(A.8.1) **Definition.** Let $R$ be a local ring. A complex $D \in \mathcal{C}_{(\square)}^{(\mathrm{f})}(R)$ is ***dualizing*** for $R$ if and only if

(1) it has finite injective dimension, i.e., $D \in \mathcal{I}^{(\mathrm{f})}(R)$; and
(2) if $P \in \mathcal{C}_{\sqsupset}^{\mathrm{P}}(R)$ is a projective resolution of $D$, then the homothety morphism $\chi_P^R \colon R \to \mathrm{Hom}_R(P,P)$ is a quasi-isomorphism.

(A.8.2) **Remark.** To see that this definition of dualizing complexes makes sense, take two projective resolutions $P, P' \in \mathcal{C}_{\sqsupset}^{\mathrm{P}}(R)$ of $R$. By (A.3.6) there is then a quasi-isomorphism $\pi\colon P \xrightarrow{\simeq} P'$, and using the quasi-isomorphism preserving properties described in (A.4.1) we establish a commutative diagram

$$\begin{array}{ccc} R & \xrightarrow{\chi_{P'}^R} & \mathrm{Hom}_R(P',P') \\ \Big\downarrow{\scriptstyle \chi_P^R} & & \simeq\Big\downarrow{\scriptstyle \mathrm{Hom}_R(\pi,P')} \\ \mathrm{Hom}_R(P,P) & \xrightarrow[\simeq]{\mathrm{Hom}_R(P,\pi)} & \mathrm{Hom}_R(P,P') \end{array}$$

which shows that $\chi_P^R$ is a quasi-isomorphism if and only if $\chi_{P'}^R$ is so.

(A.8.3) **Theorem (Existence and Uniqueness).** *Let $R$ be a local ring. The following hold:*

(A.8.3.1) *$R$ is Gorenstein if and only if the $R$–module $R$ is dualizing for $R$.*
(A.8.3.2) *If $R$ is a homomorphic image of a Gorenstein ring, then $R$ has a dualizing complex.*
(A.8.3.3) *If both $D$ and $D'$ are dualizing complexes for $R$, then $D \sim D'$.*
(A.8.3.4) *If $D$ is a dualizing complex for $R$ and $\mathfrak{p} \in \operatorname{Spec} R$, then $D_\mathfrak{p}$ is dualizing for $R_\mathfrak{p}$.*

*Proof.* See, respectively, Proposition 3.4, §10, Theorem 3.1, and §8 in [41, Chapter V] for the original results; or refer to sections 2 and 6 in [15].

(A.8.4) ***Support and Dimension for Complexes.*** Let $X$ be any $R$–complex; the *support* of $X$ is the set

$$\mathrm{Supp}_R X = \{\mathfrak{p} \in \operatorname{Spec} R \mid X_\mathfrak{p} \not\simeq 0\} = \bigcup_{\ell \in \mathbb{Z}} \mathrm{Supp}_R(\mathrm{H}_\ell(X)).$$

For modules this agrees with the usual definition.

Note that

(A.8.4.1) $$X \not\simeq 0 \iff \operatorname{Supp}_R X \neq \emptyset.$$

The (*Krull*) *dimension* of an $R$–complex $X$ is defined as

$$\begin{aligned}\dim_R X &= \sup\{\dim R/\mathfrak{p} - \inf X_\mathfrak{p} \mid \mathfrak{p} \in \operatorname{Spec} R\}\\ &= \sup\{\dim R/\mathfrak{p} - \inf X_\mathfrak{p} \mid \mathfrak{p} \in \operatorname{Supp}_R X\}.\end{aligned}$$

Also this definition [36, Section 3] extends the usual concept for modules.

(A.8.5) **Theorem (Biduality).** *Let $R$ be a local ring and assume that $D$ is a dualizing complex for $R$. For every complex $Z \in \mathcal{C}_{(\square)}^{(f)}(R)$ the biduality morphism*

$$\delta_Z^D : Z \longrightarrow \mathbf{R}\mathrm{Hom}_R(\mathbf{R}\mathrm{Hom}_R(Z, D), D)$$

*is then a quasi-isomorphism, and the following equalities hold:*

(A.8.5.1) $$\inf(\mathbf{R}\mathrm{Hom}_R(Z, D)) = \operatorname{depth}_R Z - \operatorname{depth}_R D; \quad \textit{and}$$

(A.8.5.2) $$\sup(\mathbf{R}\mathrm{Hom}_R(Z, D)) = \dim_R Z - \operatorname{depth}_R D.$$

*In particular, we have*

(A.8.5.3) $$\operatorname{amp} D = \dim R - \operatorname{depth} R = \operatorname{cmd} R.$$

*Proof.* See [41, Proposition V.2.1] and [36, Proposition 3.14], or refer to section 3 in [15].

(A.8.6) **Remark.** If $R$ is a local ring, and $D$ is a dualizing complex for $R$, then it follows by (A.8.3.4) that

(A.8.6.1) $$\operatorname{Supp}_R D = \operatorname{Spec} R.$$

In our applications of the next two lemmas it is always a dualizing complex that plays the role of $Z$.

(A.8.7) **Lemma.** *Assume that $Z \in \mathcal{C}_{(\sqsupset)}^{(f)}(R)$ and $Y \in \mathcal{C}_{(\sqsubset)}(R)$ are both homologically non-trivial. If* $\operatorname{Ass}_R(\mathrm{H}_{\sup Y}(Y)) \subseteq \operatorname{Supp}_R Z$ *(e.g.,* $\operatorname{supp}_R Z = \operatorname{Spec} R$*), then*

$$\sup Y - \sup Z \leq \sup(\mathbf{R}\mathrm{Hom}_R(Z, Y)).$$

*Proof.* See [35, Proposition 2.2].

(A.8.8) **Lemma.** *Assume that $Z \in \mathcal{C}_{(\sqsupset)}^{(f)}(R)$ and $X \in \mathcal{C}_{(\sqsupset)}(R)$ are both homologically non-trivial. If* $\operatorname{Supp}_R Z = \operatorname{Spec} R$*, then*

$$\sup Z + \inf X \geq \inf(Z \otimes_R^{\mathbf{L}} X).$$

*Proof.* Follows from (A.8.7); see [15, Lemma (4.11)].

In the rest of this section we work to establish three lemmas — (A.8.11), (A.8.12), and (A.8.13) — which allow us to conclude that a morphism $\alpha$, between appropriately bounded complexes, is invertible in the derived category if $\mathbf{R}\mathrm{Hom}_R(\alpha, R)$, $D \otimes^{\mathbf{L}}_R \alpha$, or $\mathbf{R}\mathrm{Hom}_R(D, \alpha)$ is so. Those who are familiar with the derived category can skip the rest of this section and refer to [8, Lemma (1.2.3)(b)] instead.

For the proofs of (A.8.11)–(A.8.13) we need some extra properties of depth and width.

(A.8.9) **Lemma.** *If $Y \in \mathcal{C}^{(\mathrm{f})}_{(\sqsubset)}(R)$ and $\mathfrak{p} \in \operatorname{Spec} R$, then*

$$\mathfrak{p} \in \operatorname{Supp}_R Y \iff \operatorname{depth}_{R_\mathfrak{p}} Y_\mathfrak{p} < \infty.$$

*Proof.* Follows, e.g., by [16, Corollary (5.2)].

(A.8.10) **Lemma.** *If $X \in \mathcal{C}_{(\sqsupset)}(R)$, then*

$$X \not\simeq 0 \iff \{\mathfrak{p} \in \operatorname{Spec} R \mid \operatorname{width}_{R_\mathfrak{p}} X_\mathfrak{p} < \infty\} \neq \emptyset.$$

*Proof.* Follows, e.g., by [36, Lemma 2.6].

(A.8.11) **Lemma.** *Assume that $Y \in \mathcal{C}^{(\mathrm{f})}_{(\sqsubset)}(R)$ with $\operatorname{Supp}_R Y = \operatorname{Spec} R$, and let $I \in \mathcal{C}^{\mathrm{I}}_{\sqsubset}(R)$ be an injective resolution of $Y$. If $\alpha \colon V \to V'$ is a morphism in $\mathcal{C}^{(\mathrm{f})}_{(\sqsupset)}(R)$, and $\mathrm{Hom}_R(\alpha, I)$ is a quasi-isomorphism, then $\alpha$ is a quasi-isomorphism.*

*Proof.* If $\mathrm{Hom}_R(\alpha, I)$ is a quasi-isomorphism, then the mapping cone $\mathcal{M}(\mathrm{Hom}_R(\alpha, I))$ is homologically trivial, cf. (A.1.19), and by (A.2.1.4) also the complex $\mathrm{Hom}_R(\mathcal{M}(\alpha), I)$ is homologically trivial. We want to prove that $\mathcal{M}(\alpha) \simeq 0$, so we assume that $\operatorname{Supp}_R \mathcal{M}(\alpha) \neq \emptyset$ and work to establish a contradiction, cf. (A.8.4.1). The complex $\mathrm{Hom}_R(\mathcal{M}(\alpha), I)$ represents $\mathbf{R}\mathrm{Hom}_R(\mathcal{M}(\alpha), Y)$, and $\mathcal{M}(\alpha)$ belongs to $\mathcal{C}^{(\mathrm{f})}_{(\sqsupset)}(R)$ because both $V$ and $V'$ do so. For $\mathfrak{p} \in \operatorname{Supp}_R \mathcal{M}(\alpha)$ it now follows by (A.4.5), (A.6.4), (A.6.3.2), and (A.8.9) that

$$\begin{aligned}
\operatorname{depth}_{R_\mathfrak{p}}(\mathbf{R}\mathrm{Hom}_R(\mathcal{M}(\alpha), Y))_\mathfrak{p} &= \operatorname{depth}_{R_\mathfrak{p}}(\mathbf{R}\mathrm{Hom}_{R_\mathfrak{p}}(\mathcal{M}(\alpha)_\mathfrak{p}, Y_\mathfrak{p}))\\
&= \operatorname{width}_{R_\mathfrak{p}} \mathcal{M}(\alpha)_\mathfrak{p} + \operatorname{depth}_{R_\mathfrak{p}} Y_\mathfrak{p}\\
&= \inf \mathcal{M}(\alpha)_\mathfrak{p} + \operatorname{depth}_{R_\mathfrak{p}} Y_\mathfrak{p}\\
&< \infty.
\end{aligned}$$

But $\mathbf{R}\mathrm{Hom}_R(\mathcal{M}(\alpha), Y)_\mathfrak{p}$ is homologically trivial, and this means, in particular, that $\operatorname{depth}_{R_\mathfrak{p}}(\mathbf{R}\mathrm{Hom}_R(\mathcal{M}(\alpha), Y))_\mathfrak{p} = \infty$, whence the desired contradiction has been obtained. □

(A.8.12) **Lemma.** *Assume that $Z \in \mathcal{C}_{(\sqsupset)}^{(\mathrm{f})}(R)$ with $\operatorname{Supp}_R Z = \operatorname{Spec} R$, and let $P \in \mathcal{C}_{\sqsupset}^{\mathrm{P}}(R)$ be a projective resolution of $Z$. If $\alpha\colon V \to V'$ is a morphism in $\mathcal{C}_{(\sqsupset)}(R)$, and $P \otimes_R \alpha$ is a quasi-isomorphism, then $\alpha$ is a quasi-isomorphism.*

*Proof.* If $P \otimes_R \alpha$ is a quasi-isomorphism, then the mapping cone $\mathcal{M}(P \otimes_R \alpha)$ is homologically trivial, cf. (A.1.19). By (A.2.4.4) we have $\mathcal{M}(P \otimes_R \alpha) \cong P \otimes_R \mathcal{M}(\alpha)$, and the latter complex represents $Z \otimes_R^{\mathbf{L}} \mathcal{M}(\alpha)$. The mapping cone of $\alpha$ belongs to $\mathcal{C}_{(\sqsupset)}(R)$ because both $V$ and $V'$ do so. Assume that $\mathcal{M}(\alpha) \not\simeq 0$ and choose by (A.8.10) a prime ideal $\mathfrak{p}$ such that $\operatorname{width}_{R_\mathfrak{p}} \mathcal{M}(\alpha)_\mathfrak{p} < \infty$. It now follows by (A.4.14), (A.6.5), and (A.6.3.2) that

$$\begin{aligned}\operatorname{width}_{R_\mathfrak{p}}(Z \otimes_R^{\mathbf{L}} \mathcal{M}(\alpha))_\mathfrak{p} &= \operatorname{width}_{R_\mathfrak{p}}(Z_\mathfrak{p} \otimes_{R_\mathfrak{p}}^{\mathbf{L}} \mathcal{M}(\alpha)_\mathfrak{p})\\ &= \operatorname{width}_{R_\mathfrak{p}} Z_\mathfrak{p} + \operatorname{width}_{R_\mathfrak{p}} \mathcal{M}(\alpha)_\mathfrak{p}\\ &= \inf Z_\mathfrak{p} + \operatorname{width}_{R_\mathfrak{p}} \mathcal{M}(\alpha)_\mathfrak{p}\\ &< \infty.\end{aligned}$$

But $Z \otimes_R^{\mathbf{L}} \mathcal{M}(\alpha)$ is homologically trivial, so $\operatorname{width}_{R_\mathfrak{p}}(Z \otimes_R^{\mathbf{L}} \mathcal{M}(\alpha))_\mathfrak{p} = \infty$. Thus, we have reached a contradiction, and we conclude that $\mathcal{M}(\alpha)$ is homologically trivial as wanted. □

(A.8.13) **Lemma.** *Assume that $Z \in \mathcal{C}_{(\sqsupset)}^{(\mathrm{f})}(R)$ with $\operatorname{Supp}_R Z = \operatorname{Spec} R$, and let $P \in \mathcal{C}_{\sqsupset}^{\mathrm{P}}(R)$ be a projective resolution of $Z$. If $\alpha\colon W \to W'$ is a morphism in $\mathcal{C}_{(\sqsubset)}(R)$, and $\operatorname{Hom}_R(P, \alpha)$ is a quasi-isomorphism, then $\alpha$ is a quasi-isomorphism.*

*Proof.* If $\operatorname{Hom}_R(P,\alpha)$ is a quasi-isomorphism, then the mapping cone $\mathcal{M}(\operatorname{Hom}_R(P,\alpha))$ is homologically trivial, cf. (A.1.19), and by (A.2.1.2) we have $\mathcal{M}(\operatorname{Hom}_R(P,\alpha)) = \operatorname{Hom}_R(P, \mathcal{M}(\alpha))$. The latter complex represents $\mathbf{R}\operatorname{Hom}_R(Z, \mathcal{M}(\alpha))$, and $\mathcal{M}(\alpha)$ belongs to $\mathcal{C}_{(\sqsubset)}(R)$ as both $W$ and $W'$ do so. If $\mathcal{M}(\alpha)$ is homologically non-trivial, we can choose a prime ideal $\mathfrak{p}$ associated to the top homology module; by (A.4.5), (A.6.4), (A.6.3.2), and (A.6.1.2) we then have

$$\begin{aligned}\operatorname{depth}_{R_\mathfrak{p}}(\mathbf{R}\operatorname{Hom}_R(Z, \mathcal{M}(\alpha)))_\mathfrak{p} &= \operatorname{depth}_{R_\mathfrak{p}}(\mathbf{R}\operatorname{Hom}_{R_\mathfrak{p}}(Z_\mathfrak{p}, \mathcal{M}(\alpha)_\mathfrak{p}))\\ &= \operatorname{width}_{R_\mathfrak{p}} Z_\mathfrak{p} + \operatorname{depth}_{R_\mathfrak{p}} \mathcal{M}(\alpha)_\mathfrak{p}\\ &= \inf Z_\mathfrak{p} - \sup \mathcal{M}(\alpha)\\ &< \infty.\end{aligned}$$

But $\mathbf{R}\operatorname{Hom}_R(Z, \mathcal{M}(\alpha))_\mathfrak{p}$ is homologically trivial and has, therefore, infinite depth. Thus, we have reached a contradiction, and we conclude that $\mathcal{M}(\alpha)$ is homologically trivial, i.e., $\alpha$ is a quasi-isomorphism. □

# Bibliography


[1] Maurice Auslander, *Anneaux de Gorenstein, et torsion en algèbre commutative*, Secrétariat mathématique, Paris, 1967, Séminaire d'Algèbre Commutative dirigé par Pierre Samuel, 1966/67. Texte rédigé, d'après des exposés de Maurice Auslander, par Marquerite Mangeney, Christian Peskine et Lucien Szpiro. École Normale Supérieure de Jeunes Filles.

[2] Maurice Auslander and Mark Bridger, *Stable module theory*, American Mathematical Society, Providence, R.I., 1969, Memoirs of the American Mathematical Society, no. 94.

[3] Maurice Auslander and David A. Buchsbaum, *Homological dimension in Noetherian rings*, Proc. Nat. Acad. Sci. U.S.A. **42** (1956), 36–38.

[4] ———, *Homological dimension in local rings*, Trans. Amer. Math. Soc. **85** (1957), 390–405.

[5] ———, *Homological dimension in Noetherian rings. II*, Trans. Amer. Math. Soc. **88** (1958), 194–206.

[6] Luchezar L. Avramov, Ragnar-Olaf Buchweitz, Alex Martsinkovsky, and Idun Reiten, *Stable cohomological algebra*, in preparation.

[7] Luchezar L. Avramov and Hans-Bjørn Foxby, *Homological dimensions of unbounded complexes*, J. Pure Appl. Algebra **71** (1991), no. 2-3, 129–155.

[8] ———, *Ring homomorphisms and finite Gorenstein dimension*, Proc. London Math. Soc. (3) **75** (1997), no. 2, 241–270.

[9] ———, *Cohen-Macaulay properties of ring homomorphisms*, Adv. Math. **133** (1998), no. 1, 54–95.

[10] Hyman Bass, *Injective dimension in Noetherian rings*, Trans. Amer. Math. Soc. **102** (1962), 18–29.

[11] ———, *On the ubiquity of Gorenstein rings*, Math. Z. **82** (1963), 8–28.

[12] Winfried Bruns and Jürgen Herzog, *Cohen-Macaulay rings*, revised ed., Cambridge University Press, Cambridge, 1998, Cambridge studies in advanced mathematics, vol. 39.

[13] Henri Cartan and Samuel Eilenberg, *Homological algebra*, Princeton University Press, Princeton, N. J., 1956.

[14] Leo G. Chouinard, II, *On finite weak and injective dimension*, Proc. Amer. Math. Soc. **60** (1976), 57–60 (1977).

[15] Lars W. Christensen, *Semi-dualizing complexes and their Auslander categories*, to appear in Trans. Amer. Math. Soc.

[16] ———, *Sequences for complexes*, to appear in Math. Scand.

[17] ———, *Gorenstein dimensions*, Master's thesis, Matematisk Institut, Københavns Universitet, 1996.

[18] ———, *Functorial dimensions*, Ph.D. thesis, University of Copenhagen, 1999.

[19] Lars W. Christensen and Anders Frankild, *Functorial dimensions and Cohen-Macaulayness of rings*, in preparation.

[20] David Eisenbud, *Commutative algebra with a view toward algebraic geometry*, Springer-Verlag, New York, 1995, Graduate Texts in Mathematics, vol. 150.

[21] Edgar E. Enochs, *Injective and flat covers, envelopes and resolvents*, Israel J. Math. **39** (1981), no. 3, 189–209.

[22] Edgar E. Enochs and Overtoun M. G. Jenda, *On Gorenstein injective modules*, Comm. Algebra **21** (1993), no. 10, 3489–3501.

[23] ———, *Mock finitely generated Gorenstein injective modules and isolated singularities*, J. Pure Appl. Algebra **96** (1994), no. 3, 259–269.

[24] ———, *Gorenstein flat preenvelopes and resolvents*, Nanjing Daxue Xuebao Shuxue Bannian Kan **12** (1995), no. 1, 1–9.

[25] ———, *Gorenstein injective and projective modules*, Math. Z. **220** (1995), no. 4, 611–633.

[26] ———, *Resolutions by Gorenstein injective and projective modules and modules of finite injective dimension over Gorenstein rings*, Comm. Algebra **23** (1995), no. 3, 869–877.

[27] ———, *Gorenstein injective and flat dimensions*, Math. Japon. **44** (1996), no. 2, 261–268.

[28] ———, *Coliftings and Gorenstein injective modules*, J. Math. Kyoto Univ. **38** (1998), no. 2, 241–254.

[29] ———, *Gorenstein injective modules over Gorenstein rings*, Comm. Algebra **26** (1998), no. 11, 3489–3496.

[30] ———, *Gorenstein injective dimension and Tor-depth of modules*, Arch. Math. (Basel) **72** (1999), no. 2, 107–117.

[31] Edgar E. Enochs, Overtoun M. G. Jenda, and Blas Torrecillas, *Gorenstein flat modules*, Nanjing Daxue Xuebao Shuxue Bannian Kan **10** (1993), no. 1, 1–9.

[32] Edgar E. Enochs, Overtoun M. G. Jenda, and Jinzhong Xu, *Foxby duality and Gorenstein injective and projective modules*, Trans. Amer. Math. Soc. **348** (1996), no. 8, 3223–3234.

[33] Hans-Bjørn Foxby, *Hyperhomological algebra & commutative rings*, notes in preparation.

[34] ———, *Gorenstein modules and related modules*, Math. Scand. **31** (1972), 267–284 (1973).

[35] ———, *Isomorphisms between complexes with applications to the homological theory of modules*, Math. Scand. **40** (1977), no. 1, 5–19.

[36] ———, *Bounded complexes of flat modules*, J. Pure Appl. Algebra **15** (1979), no. 2, 149–172.

[37] ———, *Homological dimensions of complexes of modules*, Séminaire d'Algèbre Paul Dubreil et Marie-Paule Malliavin, 32ème année (Paris, 1979), Springer, Berlin, 1980, Lecture Notes in Mathematics, no. 795, pp. 360–368.

[38] ———, *A homological theory for complexes of modules*, Preprint Series 1981 nos. 19a&b, Matematisk Institut, Københavns Universitet, 1981.

[39] ———, *Gorenstein dimension over Cohen-Macaulay rings*, Proceedings of international conference on commutative algebra (W. Bruns, ed.), Universität Onsabrück, 1994.

[40] Evgeniy S. Golod, *G-dimension and generalized perfect ideals*, Trudy Mat. Inst. Steklov. **165** (1984), 62–66, Algebraic geometry and its applications.

[41] Robin Hartshorne, *Residues and duality*, Springer-Verlag, Berlin, 1966, Lecture notes of a seminar on the work of A. Grothendieck, given at Harvard 1963/64. With an appendix by P. Deligne. Lecture Notes in Mathematics, no. 20.

[42] Takeshi Ishikawa, *On injective modules and flat modules*, J. Math. Soc. Japan **17** (1965), 291–296.

[43] Birger Iversen, *Cohomology of sheaves*, Springer-Verlag, Berlin, 1986, Universitext.

[44] Srikanth Iyengar, *Depth for complexes, and intersection theorems*, Math. Z. **230** (1999), no. 3, 545–567.

[45] Christian U. Jensen, *On the vanishing of* $\varprojlim^{(i)}$, J. Algebra **15** (1970), 151–166.

[46] Daniel Lazard, *Autour de la platitude*, Bull. Soc. Math. France **97** (1969), 81–128.

[47] Saunders MacLane, *Categories for the working mathematician*, Springer-Verlag, New York, 1971, Graduate Texts in Mathematics, vol. 5.

[48] Eben Matlis, *Applications of duality*, Proc. Amer. Math. Soc. **10** (1959), 659–662.

[49] Hideyuki Matsumura, *Commutative ring theory*, second ed., Cambridge University Press, Cambridge, 1989, Cambridge studies in advanced mathematics, vol. 8.

[50] Hõ Dình Duâ'n, *A note on Gorenstein dimension and the Auslander-Buchsbaum formula*, Kodai Math. J. **17** (1994), no. 3, 390–394, Workshop on Geometry and Topology (Hanoi, 1993).

[51] Christian Peskine and Lucien Szpiro, *Dimension projective finie et cohomologie locale. Applications à la démonstration de conjectures de M. Auslander, H. Bass et A. Grothendieck*, Inst. Hautes Études Sci. Publ. Math. (1973), no. 42, 47–119.

[52] Michel Raynaud and Laurent Gruson, *Critères de platitude et de projectivité. Techniques de "platification" d'un module*, Invent. Math. **13** (1971), 1–89.

[53] Paul Roberts, *Le théorème d'intersection*, C. R. Acad. Sci. Paris Sér. I Math. **304** (1987), no. 7, 177–180.

[54] Jean-Pierre Serre, *Sur la dimension homologique des anneaux et des modules noethériens*, Proceedings of the international symposium on algebraic number theory, Tokyo & Nikko, 1955 (Tokyo), Science Council of Japan, 1956, pp. 175–189.

[55] Rodney Y. Sharp, *Finitely generated modules of finite injective dimension over certain Cohen-Macaulay rings*, Proc. London Math. Soc. (3) **25** (1972), 303–328.

[56] Nicolas Spaltenstein, *Resolutions of unbounded complexes*, Compositio Math. **65** (1988), no. 2, 121–154.

[57] Yasuji Takeuchi, *On finite Gorenstein dimension*, Kobe J. Math. **4** (1987), no. 1, 103–106.

[58] ———, *On weakly Gorenstein dimension*, Kobe J. Math. **5** (1988), no. 2, 271–277.

[59] Jean-Louis Verdier, *Catégories dérivées. quelques résultats (état 0)*, SGA $4\frac{1}{2}$, Springer, Berlin Heidelberg New York, 1977, pp. 262–311.

[60] Charles A. Weibel, *An introduction to homological algebra*, Cambridge University Press, Cambridge, 1994, Cambridge studies in advanced mathematics, vol. 38.

[61] Jinzhong Xu, *Flat covers of modules*, Springer-Verlag, Berlin, 1996, Lecture Notes in Mathematics, no. 1634.

[62] Siamak Yassemi, *G-dimension*, Math. Scand. **77** (1995), no. 2, 161–174.

[63] ———, *Width of complexes of modules*, Acta Math. Vietnam. **23** (1998), no. 1, 161–169.

# List of Symbols


$\mathbb{Z}$, 10
$\mathbb{N}$, 10
$\mathbb{N}_0$, 10
$(x_1, \dots, x_t)$, 10
$(R, \mathfrak{m}, k)$, 10
$k$, 10
$k(\mathfrak{p})$, 10
$\mathrm{E}_R(M)$, 10
$\mathrm{z}_R\, M$, 10
$\mathrm{z}\, R$, 10
$\mathrm{Ann}_R\, M$, 10
$\sigma_{PNM}$, 11
$\rho_{PNM}$, 12
$\varsigma_{PMN}$, 12
$\omega_{PNM}$, 12
$\theta_{PNM}$, 12
$\delta_M$, 17
$\mathrm{G}(R)$, 18
$M^*$, 18
$M^{**}$, 18
$-^*$, 18
$M_{\mathrm{T}}$, 18
$\text{G-dim}_R\, M$, 22
$\mathrm{depth}_R\, M$, 32
$\delta_X^Y$, 42
$\chi_Y^R$, 42
$\mathcal{R}(R)$, 44
$X^*$, 44
$\mathcal{C}^{\mathrm{G}}(R)$, 52
$\text{G-dim}_R\, X$, 52
$\gamma_X^Z$, 66
$\mathcal{A}(R)$, 67
$\xi_Y^Z$, 72
$\mathcal{B}(R)$, 73
$\mathcal{C}^{\mathrm{GP}}(R)$, 104
$\mathrm{Gpd}_R\, X$, 106
$\mathcal{C}^{\mathrm{GF}}(R)$, 120
$\mathrm{Gfd}_R\, X$, 120
$\mathrm{Td}_R\, X$, 127
$\mathcal{C}^{\mathrm{GI}}(R)$, 139
$\mathrm{Gid}_R\, Y$, 141
$\partial_\ell^X$, 160
$|x|$, 160
$\mathrm{Z}_\ell^X$, 160
$\mathrm{B}_\ell^X$, 160
$\mathrm{C}_\ell^X$, 160
$\mathrm{H}_\ell(X)$, 160
$\mathrm{H}(X)$, 160
$\Sigma^m X$, 161
$r_X$, 161
$1_X$, 161
$\cong$, 161
$\simeq$, 162
$\sup X$, 162
$\inf X$, 162
$\mathrm{amp}\, X$, 162
$\mathcal{C}(R)$, 162
$\simeq$, 164
$\sim$, 164
$\mathcal{D}(R)$, 165
$\sqsubset_u X$, 165
$X_v\sqsupset$, 165
$\subset_u X$, 165
$X_v\supset$, 165
$\mathrm{Hom}_R(X, Y)$, 168
$X \otimes_R Y$, 169
$\sigma_{ZYX}$, 170
$\rho_{ZYX}$, 170
$\varsigma_{ZXY}$, 170
$\omega_{ZYX}$, 170

$\theta_{ZYX}$, 171
$\mathcal{P}(R)$, 173
$\mathcal{I}(R)$, 173
$\mathcal{F}(R)$, 173
$\mathrm{id}_R\, Y$, 173
$\mathrm{pd}_R\, X$, 173
$\mathrm{fd}_R\, X$, 174
$\mathbf{R}\mathrm{Hom}_R(X, Y)$, 176
$\mathrm{Ext}_R^m(M, N)$, 176
$X \otimes_R^{\mathbf{L}} Y$, 178
$\mathrm{Tor}_m^R(M, N)$, 178
$\mathrm{depth}_R\, Y$, 183
$\mathrm{width}_R\, X$, 184
$\beta_m^R(X)$, 185
$\mu_R^m(Y)$, 185
$\mu_R^m$, 185
$\mathrm{Supp}_R\, X$, 187

# Index


AB formula, **5**
  for flat dimension, 132
  for G–dimension, 56
    of finite modules, 35
  for Gorenstein flat dimension, 134
  for projective dimension of finite modules, 13
  for restricted Tor–dimension, 129
    of complexes with finite homology, 130
accounting principles, 186
adjointness
  isomorphism of complexes, **170**
    identity induced by, **179**
  isomorphism of modules, **12**
amplitude inequalities, 81
amplitude of complex, **162**, 163
annihilator, **10**
associativity
  isomorphism of complexes, **170**
    identity induced by, **179**
  isomorphism of modules, **11**
Auslander, Maurice, 1–3, 17, 63, 97
Auslander class, **67**, 158
  and finite flat dimension, 68
  and finite G–dimension, 70
  and finite Gorenstein flat dimension, 122
    of modules, 125
  and finite Gorenstein projective dimension, 107
    of modules, 110
  and localization, 68
  and short exact sequences, 70, 85
  for Gorenstein rings, 70, 80, 88
  modules in, 85
    and short exact sequences, 85
    bounded complex of, 71
Auslander's zero-divisor conjecture, 39
Auslander–Bridger formula, 35, *see also* AB formula
Auslander–Buchsbaum formula, 13, *see also* AB formula
Avramov, Luchezar L., 52, 58, 99, 159

Bass, Hyman, 5, 7
Bass class, **73**, 158
  and finite Gorenstein injective dimension, 143
    of modules, 145
  and finite injective dimension, 73
  and localization, 73
  and short exact sequences, 76, 87
  for Gorenstein rings, 75, 80, 88
  modules in, 86
    and short exact sequences, 87
    bounded complex of, 76
Bass conjecture, 7, 147
Bass formula, 13

for Gorenstein injective dimension, 147
Bass numbers, **185**
Bass series, **186**
Betti numbers, **185**
bidual module, **18**
biduality, 18, 19, 22, 42, 43, 50, 188
homomorphism of modules, **17**
morphism of complexes, **42**
Bridger, Mark, 2, 17, 63
Buchsbaum, David A., 3
Buchweitz, Ragner–Olaf, 2

canonical module, *see* dualizing module
canonical representation
of $\mathbf{R}\mathrm{Hom}_R(\mathbf{R}\mathrm{Hom}_R(X,R),R)$, 43, **43**, 68
of $\mathbf{R}\mathrm{Hom}_R(D, D\otimes^{\mathbf{L}}_R X)$, **66**, 67, 68
of $D\otimes^{\mathbf{L}}_R \mathbf{R}\mathrm{Hom}_R(D,Y)$, 72, **72**
category of $R$–complexes, **162**
full subcategories of, 162–164
Change of Rings Theorem for G–dimension, 56
of finite modules, 39, 51
Chouinard, Leo G., II, 5, 134, 147
CM rings, *see* Cohen–Macaulay rings
codepth, *see* width
Cohen–Macaulay defect, **14**
at most 1, 131
Cohen–Macaulay rings, **14**, 40, 63, 81, 112, 130, 131
commutativity
isomorphism of complexes, **169**
identity induced by, **179**
complete flat resolution, **113**, 114, 115, 118, 153, 156
complete injective resolution, **135**, 136, 153, 156
complete projective resolution, **97**, 98, 114, 118, 156
complete resolution by finite free modules, 93, **93**, 94, 118
complex(es), **160**
amplitude of, **162**
bounded to the left/right, **162**
concentrated in certain degrees, **160**
equivalent, **164**
up to a shift, **164**
homologically bounded (to the left/right), **163**
homologically trivial, **161**, 164
infimum of, **162**
isomorphic, **161**
modules considered as, 160
shifted, **161**
short exact sequence of, **166**
supremum of, **162**
truncated, **165**
with finite homology, **163**

degree of element in complex, **160**
depth, 10, **183**
and Bass series, 186
of modules, **32**
derived category, 82, **165**, 175
differential, **160**
dimension, *see* Krull dimension
dual module, **18**, 19
duality functor (algebraic), **18**, 44
dualizing, **18**
dualizing complex, 65, 81, **187**
dualizing module, 76, **83**, *see also* dualizing complex

Enochs, Edgar E., 1, 2, 91, 99, 100, 105, 113, 115, 120, 126, 135, 136, 147, 148, 152
equivalence of complexes, **164**
preservation of, 168, 175
up to a shift, **164**
exact sequence, 161
Ext modules, **176**
long exact sequence of, 177

faithful functor, 168

faithfully injective module, **177**
finer invariant, **3**
finite free modules
belong to the G–class, 18
finite module, **10**
flat dimension, 68, 89, 123, 131, **174**, 182
and Betti numbers, 185
of complexes with finite homology, 182
of modules, 13
flat modules, 14
are Gorenstein flat, 113
flat preenvelope, **100**
flat resolution, **172**
existence of, 172
of module, 174
Foxby, Hans-Bjørn, 1, 2, 5, 52, 58, 76, 82, 00, 112, 113, 126, 131, 134, 158, 159, 183, 186
Foxby equivalence, 1, **76**
over Cohen–Macaulay rings, 87
over Gorenstein rings, 79

G–class, **18**
and localization, 30
dual modules in, 18
finite free modules in, 18
finite projective modules in, 21
modules in, 94
and short exact sequences, 20
are Gorenstein flat, 119
are Gorenstein projective, 98
are reflexive as complexes, 48
are torsion-free, 18
modulo regular sequence, 33
non-projective modules in, 21, 95
of residue class ring, 31
G–dimension, **52**, 52–63, 70, 108, 132
and change of rings, 56
and localization, 55
and reflexive complexes, 54
of finite modules, **22**, 48, 57, 59–61, 63
and change of rings, 38, 51
and localization, 30
in short exact sequence, 27, 50
modulo regular sequence, 34
over Cohen–Macaulay rings, 63
over residue class ring, 32
refinement of projective dimension, 28
over Gorenstein rings, 36, 56
refinement projective dimension, 55
G–resolution, **22**, 57
existence of, 22
length of, **22**
GD Theorem, 54
for finite modules, 24
$\mathcal{R}$ version, 57
GD–GPD equality, 108
GD–PD inequality, 55
for finite modules, 29
GFD Corollary, 122
for modules, 125
GFD Theorem, 121
for modules, 124
GFD–FD inequality, 123
GFD–GPD inequality, 123
GID Corollary, 143
for modules, 145
GID Theorem, 142
for modules, 145
GID–ID inequality, 143
Golod, Evgeniy, S., 2, 40, 52
Gorenstein flat dimension, **120**, 120–125, 133–134, 148–151, 155–157
and localization, 122
and the Auslander class, 122, 125
finer invariant than Gorenstein projective dimension, 122

of complexes with finite homology, 123
of modules, 124, 125, 134
over Gorenstein rings, 124
refinement of flat dimension, 122
Gorenstein flat modules, **113**, 124, 149, 150, 153, 156
and localization, 114
and short exact sequences, 117
and the Auslander class, 115, 120
finite, 119
Gorenstein flat resolution, 124, **124**
Gorenstein injective dimension, **141**, 141–152, 155–157
and localization, 146
and the Bass class, 143, 145
of complexes with finite homology, 147
of finite modules, 147
of modules, 145, 147, 152
over Gorenstein rings, 144
refinement of injective dimension, 143
Gorenstein injective modules, **135**, 145, 149, 150, 153, 156
and localization, 135, 146
and short exact sequences, 139
and the Bass class, 137, 139
width of, 150
Gorenstein injective resolution, 144, **144**, 145
Gorenstein projective dimension, **106**, 106–112, 156
and localization, 110
and the Auslander class, 107, 110
of complexes with finite homology, 108, 123
of modules, 109, 110, 112
over Gorenstein rings, 108
refinement of projective dimension, 108
Gorenstein projective modules, **97**, 109, 156
and localization, 97, 110
and short exact sequences, 103
and the Auslander class, 100, 105
are Gorenstein flat, 114
finite, 98
Gorenstein projective resolution, 109, **109**
Gorenstein rings, 6, **14**, 70, 75, 80, 88, 187
and finite flat dimension, 6, 79, 148
and finite G–dimension, 36, 56
and finite Gorenstein flat dimension, 124, 148
and finite Gorenstein injective dimension, 144, 148
and finite Gorenstein projective dimension, 108
and finite injective dimension, 6, 14, 79, 148
and Foxby equivalence, 79
Gorenstein Theorem
$\mathcal{A}$ version, 70
$\mathcal{B}$ version, 75
Foxby equivalence version, 79
GD version, 36
GFD version, 124
GFD/GID version, 148
GID version, 144
GPD version, 108
PD/ID version, 6
$\mathcal{R}$ version, 56
special complexes version, 80
special modules version, 88
GPD Corollary, 107
for modules, 110
GPD Theorem, 106
for modules, 109
GPD–PD inequality, 108

Hom complex, *see* homomorphism complex
Hom evaluation, 17, 21, 42, 43, 72

homomorphism of modules, **12**
morphism of complexes, **171**
identity induced by, **180**
Hom vanishing corollary, 11
Hom vanishing lemma, 11
homology complex, **160**, 164
homology isomorphism, *see* quasi-isomorphism
homology modules, **160**
long exact sequence of, 166, 177, 179
homomorphism complex, **168**
and mapping cones, 168, 169
and shift, 168, 169
homothety, **161**
homothety morphism, **42**, 187

identity morphism, 161
induced functor on complexes, 167
infimum
of complex, **162**, 163
of empty set, **10**
injective dimension, 73, 143, **173**, 180–182
of complexes with finite homology, 182
and Bass series, 186
of modules, 13
injective hull, **10**
injective modules, 14
are Gorenstein injective, 135
faithfully so, **177**
injective precover, **136**
injective resolution, **171**, 172
existence of, 172
of module, 174
intersection theorems, 39
Ishikawa, Takeshi, 7, 149
isomorphism
of complexes, **161**
of modules, 161
Iyengar, Srikanth, 29, 183

Jenda, Overtoun M. G., 2, 91, 99, 113, 135, 147, 148, 152
Jensen, Christian U., 90

Krull dimension, 10
of complexes, **188**
of rings, 13, 63

ladder, **10**
local ring, **10**
localization at prime ideal, 168

$M$-regular element, **30**, 33
$M$-sequence, **30**, 33
mapping cone, 26, 29, 167, **167**
Matlis duality, **178**
modules considered as complexes, 160, 163
morphism of complexes, **161**
identity, 161
induced in homology, 161

Nakayama's lemma, **11**
for complexes, 178
natural map, **10**

Peskine, Christian, 7, 40, 52
Poincaré series, **185**
preservation of quasi-isomorphisms and equivalences, 168, 175
projective dimension, 44, 45, 55, 89, 108, **173**, 181
of complexes with finite homology, 182
and Poincaré series, 185
of finite modules, 13, 29
and change of rings, 38
of modules, 13
over regular rings, 3
projective modules, 14
are Gorenstein projective, 97
finite, 14
belong to the G–class, 21
projective resolution, **172**, 173
existence of, 172
of module, 174

quasi-isomorphism(s), **162**
and mapping cones, 167
preservation of, 168, 175

*R*–complex, *see* complex
*R*–sequence, 33
refinement, **4**
reflexive complex(es), **44**
and finite G–dimension, 48, 54, 61, 63
and finite projective dimension, 44
and localization, 45
and short exact sequences, 46
belong to the Auslander class, 69
over Gorenstein rings, 56
reflexive module, **44**, 50
regular element, **30**
regular rings, **14**, 37
and finite projective dimension, 3, 14
Regularity Theorem, 3
Reiten, Idun, 2
residue field, **10**
resolution by finite free modules, **172**
existence of, 172
restricted Tor–dimension, **127**, 127–131
and localization, 128
of complexes with finite homology, 130, 132
over Cohen–Macaulay rings, 130
over Gorenstein rings, 128
refinement of flat dimension, 131
refinement of G–dimension, 132
refinement of Gorenstein flat dimension, 134
Roberts, Paul C., 7

Serre, Jean–Pierre, 3
Sharp, Rodney Y., 76, 90
snake lemma, 11
stability, **7**, 14, 45, 58, 182
standard homomorphisms, 11–12
standard identities, 179–180
standard morphisms, 170–171
support, **187**
supremum
of complex, **162**, 163
of empty set, **10**
swap
isomorphism of complexes, **170**
identity induced by, **179**
isomorphism of modules, **12**
Szpiro, Lucien, 7, 40, 52

Takeuchi, Yasuji, 63
TD–FD inequality, 131
TD–GD inequality, 132
TD–GFD inequality, 134
tensor evaluation, 67
homomorphism of modules, **12**
morphism of complexes, **170**
identity induced by, **180**
tensor product complex, **169**
and mapping cones, 169
and shift, 169
Tor modules, **178**
long exact sequence of, 179
Tor–depth, *see* width
Torrecillas, Blas, 2, 113
torsion, **18**
-free, **18**, 19, 33
-less, **19**
truncations, **165**

width, **184**

Xu, Jinzhong, 2, 99, 105, 120, 126, 135

Yassemi, Siamak, 2, 41, 58, 147, 183

zero-complex, 160, **164**
zero-divisors, **10**, 39

# Lecture Notes in Mathematics

**For information about Vols. 1–1560**
**please contact your bookseller or Springer-Verlag**

Vol. 1561: I. S. Molchanov, Limit Theorems for Unions of Random Closed Sets. X, 157 pages. 1993.

Vol. 1562: G. Harder, Eisensteinkohomologie und die Konstruktion gemischter Motive. XX, 184 pages. 1993.

Vol. 1563: E. Fabes, M. Fukushima, L. Gross, C. Kenig, M. Röckner, D. W. Stroock, Dirichlet Forms. Varenna, 1992. Editors: G. Dell'Antonio, U. Mosco. VII, 245 pages. 1993.

Vol. 1564: J. Jorgenson, S. Lang, Basic Analysis of Regularized Series and Products. IX, 122 pages. 1993.

Vol. 1565: L. Boutet de Monvel, C. De Concini, C. Procesi, P. Schapira, M. Vergne. D-modules, Representation Theory, and Quantum Groups. Venezia, 1992. Editors: G. Zampieri, A. D'Agnolo. VII, 217 pages. 1993.

Vol. 1566: B. Edixhoven, J.-H. Evertse (Eds.), Diophantine Approximation and Abelian Varieties. XIII, 127 pages. 1993.

Vol. 1567: R. L. Dobrushin, S. Kusuoka, Statistical Mechanics and Fractals. VII, 98 pages. 1993.

Vol. 1568: F. Weisz, Martingale Hardy Spaces and their Application in Fourier Analysis. VIII, 217 pages. 1994.

Vol. 1569: V. Totik, Weighted Approximation with Varying Weight. VI, 117 pages. 1994.

Vol. 1570: R. deLaubenfels, Existence Families, Functional Calculi and Evolution Equations. XV, 234 pages. 1994.

Vol. 1571: S. Yu. Pilyugin, The Space of Dynamical Systems with the $C^0$-Topology. X, 188 pages. 1994.

Vol. 1572: L. Göttsche, Hilbert Schemes of Zero-Dimensional Subschemes of Smooth Varieties. IX, 196 pages. 1994.

Vol. 1573: V. P. Havin, N. K. Nikolski (Eds.), Linear and Complex Analysis – Problem Book 3 – Part I. XXII, 489 pages. 1994.

Vol. 1574: V. P. Havin, N. K. Nikolski (Eds.), Linear and Complex Analysis – Problem Book 3 – Part II. XXII, 507 pages. 1994.

Vol. 1575: M. Mitrea, Clifford Wavelets, Singular Integrals, and Hardy Spaces. XI, 116 pages. 1994.

Vol. 1576: K. Kitahara, Spaces of Approximating Functions with Haar-Like Conditions. X, 110 pages. 1994.

Vol. 1577: N. Obata, White Noise Calculus and Fock Space. X, 183 pages. 1994.

Vol. 1578: J. Bernstein, V. Lunts, Equivariant Sheaves and Functors. V, 139 pages. 1994.

Vol. 1579: N. Kazamaki, Continuous Exponential Martingales and *BMO*. VII, 91 pages. 1994.

Vol. 1580: M. Milman, Extrapolation and Optimal Decompositions with Applications to Analysis. XI, 161 pages. 1994.

Vol. 1581: D. Bakry, R. D. Gill, S. A. Molchanov, Lectures on Probability Theory. Editor: P. Bernard. VIII, 420 pages. 1994.

Vol. 1582: W. Balser, From Divergent Power Series to Analytic Functions. X, 108 pages. 1994.

Vol. 1583: J. Azéma, P. A. Meyer, M. Yor (Eds.), Séminaire de Probabilités XXVIII. VI, 334 pages. 1994.

Vol. 1584: M. Brokate, N. Kenmochi, I. Müller, J. F. Rodriguez, C. Verdi, Phase Transitions and Hysteresis. Montecatini Terme, 1993. Editor: A. Visintin. VII, 291 pages. 1994.

Vol. 1585: G. Frey (Ed.), On Artin's Conjecture for Odd 2-dimensional Representations. VIII, 148 pages. 1994.

Vol. 1586: R. Nillsen, Difference Spaces and Invariant Linear Forms. XII, 186 pages. 1994.

Vol. 1587: N. Xi, Representations of Affine Hecke Algebras. VIII, 137 pages. 1994.

Vol. 1588: C. Scheiderer, Real and Étale Cohomology. XXIV, 273 pages. 1994.

Vol. 1589: J. Bellissard, M. Degli Esposti, G. Forni, S. Graffi, S. Isola, J. N. Mather, Transition to Chaos in Classical and Quantum Mechanics. Montecatini Terme, 1991. Editor: 2S. Graffi. VII, 192 pages. 1994.

Vol. 1590: P. M. Soardi, Potential Theory on Infinite Networks. VIII, 187 pages. 1994.

Vol. 1591: M. Abate, G. Patrizio, Finsler Metrics – A Global Approach. IX, 180 pages. 1994.

Vol. 1592: K. W. Breitung, Asymptotic Approximations for Probability Integrals. IX, 146 pages. 1994.

Vol. 1593: J. Jorgenson & S. Lang, D. Goldfeld, Explicit Formulas for Regularized Products and Series. VIII, 154 pages. 1994.

Vol. 1594: M. Green, J. Murre, C. Voisin, Algebraic Cycles and Hodge Theory. Torino, 1993. Editors: A. Albano, F. Bardelli. VII, 275 pages. 1994.

Vol. 1595: R.D.M. Accola, Topics in the Theory of Riemann Surfaces. IX, 105 pages. 1994.

Vol. 1596: L. Heindorf, L. B. Shapiro, Nearly Projective Boolean Algebras. X, 202 pages. 1994.

Vol. 1597: B. Herzog, Kodaira-Spencer Maps in Local Algebra. XVII, 176 pages. 1994.

Vol. 1598: J. Berndt, F. Tricerri, L. Vanhecke, Generalized Heisenberg Groups and Damek-Ricci Harmonic Spaces. VIII, 125 pages. 1995.

Vol. 1599: K. Johannson, Topology and Combinatorics of 3-Manifolds. XVIII, 446 pages. 1995.

Vol. 1600: W. Narkiewicz, Polynomial Mappings. VII, 130 pages. 1995.

Vol. 1601: A. Pott, Finite Geometry and Character Theory. VII, 181 pages. 1995.

Vol. 1602: J. Winkelmann, The Classification of Three-dimensional Homogeneous Complex Manifolds. XI, 230 pages. 1995.

Vol. 1603: V. Ene, Real Functions – Current Topics. XIII, 310 pages. 1995.

Vol. 1604: A. Huber, Mixed Motives and their Realization in Derived Categories. XV, 207 pages. 1995.

Vol. 1605: L. B. Wahlbin, Superconvergence in Galerkin Finite Element Methods. XI, 166 pages. 1995.

Vol. 1606: P.-D. Liu, M. Qian, Smooth Ergodic Theory of Random Dynamical Systems. XI, 221 pages. 1995.

Vol. 1607: G. Schwarz, Hodge Decomposition – A Method for Solving Boundary Value Problems. VII, 155 pages. 1995.

Vol. 1608: P. Biane, R. Durrett, Lectures on Probability Theory. Editor: P. Bernard. VII, 210 pages. 1995.

Vol. 1609: L. Arnold, C. Jones, K. Mischaikow, G. Raugel, Dynamical Systems. Montecatini Terme, 1994. Editor: R. Johnson. VIII, 329 pages. 1995.

Vol. 1610: A. S. Üstünel, An Introduction to Analysis on Wiener Space. X, 95 pages. 1995.

Vol. 1611: N. Knarr, Translation Planes. VI, 112 pages. 1995.

Vol. 1612: W. Kühnel, Tight Polyhedral Submanifolds and Tight Triangulations. VII, 122 pages. 1995.

Vol. 1613: J. Azéma, M. Emery, P. A. Meyer, M. Yor (Eds.), Séminaire de Probabilités XXIX. VI, 326 pages. 1995.

Vol. 1614: A. Koshelev, Regularity Problem for Quasilinear Elliptic and Parabolic Systems. XXI, 255 pages. 1995.

Vol. 1615: D. B. Massey, Le Cycles and Hypersurface Singularities. XI, 131 pages. 1995.

Vol. 1616: I. Moerdijk, Classifying Spaces and Classifying Topoi. VII, 94 pages. 1995.

Vol. 1617: V. Yurinsky, Sums and Gaussian Vectors. XI, 305 pages. 1995.

Vol. 1618: G. Pisier, Similarity Problems and Completely Bounded Maps. VII, 156 pages. 1996.

Vol. 1619: E. Landvogt, A Compactification of the Bruhat-Tits Building. VII, 152 pages. 1996.

Vol. 1620: R. Donagi, B. Dubrovin, E. Frenkel, E. Previato, Integrable Systems and Quantum Groups. Montecatini Terme, 1993. Editors:M. Francaviglia, S. Greco. VIII, 488 pages. 1996.

Vol. 1621: H. Bass, M. V. Otero-Espinar, D. N. Rockmore, C. P. L. Tresser, Cyclic Renormalization and Auto-morphism Groups of Rooted Trees. XXI, 136 pages. 1996.

Vol. 1622: E. D. Farjoun, Cellular Spaces, Null Spaces and Homotopy Localization. XIV, 199 pages. 1996.

Vol. 1623: H.P. Yap, Total Colourings of Graphs. VIII, 131 pages. 1996.

Vol. 1624: V. Brınzanescu, Holomorphic Vector Bundles over Compact Complex Surfaces. X, 170 pages. 1996.

Vol.1625: S. Lang, Topics in Cohomology of Groups. VII, 226 pages. 1996.

Vol. 1626: J. Azéma, M. Emery, M. Yor (Eds.), Séminaire de Probabilités XXX. VIII, 382 pages. 1996.

Vol. 1627: C. Graham, Th. G. Kurtz, S. Méléard, Ph. E. Protter, M. Pulvirenti, D. Talay, Probabilistic Models for Nonlinear Partial Differential Equations. Montecatini Terme, 1995. Editors: D. Talay, L. Tubaro. X, 301 pages. 1996.

Vol. 1628: P.-H. Zieschang, An Algebraic Approach to Association Schemes. XII, 189 pages. 1996.

Vol. 1629: J. D. Moore. Lectures on Seiberg-Witten Invariants. VII. 105 pages. 1996.

Vol. 1630: D. Neuenschwander. Probabilities on the Heisenberg Group: Limit Theorems and Brownian Motion. VIII, 139 pages. 1996.

Vol. 1631: K. Nishioka. Mahler Functions and Transcendence.VIII, 185 pages.1996.

Vol. 1632: A. Kushkuley. Z. Balanov. Geometric Methods in Degree Theory for Equivariant Maps. VII. 136 pages. 1996.

Vol.1633: H. Aikawa, M. Essén, Potential Theory – Selected Topics. IX, 200 pages.1996.

Vol. 1634: J. Xu, Flat Covers of Modules. IX. 161 pages. 1996.

Vol. 1635: E. Hebey. Sobolev Spaces on Riemannian Manifolds. X, 116 pages. 1996.

Vol. 1636: M. A. Marshall, Spaces of Orderings and Abstract Real Spectra. VI, 190 pages. 1996.

Vol. 1637: B. Hunt. The Geometry of some special Arithmetic Quotients. XIII, 332 pages. 1996.

Vol. 1638: P. Vanhaecke. Integrable Systems in the realm of Algebraic Geometry. VIII. 218 pages. 1996.

Vol. 1639: K. Dekimpe. Almost-Bieberbach Groups: Affine and Polynomial Structures. X, 259 pages. 1996.

Vol. 1640: G. Boillat, C. M. Dafermos. P. D. Lax, T. P. Liu, Recent Mathematical Methods in Nonlinear Wave Propagation. Montecatini Terme, 1994. Editor: T. Ruggeri. VII, 142 pages. 1996.

Vol. 1641: P. Abramenko, Twin Buildings and Applications to S-Arithmetic Groups. IX, 123 pages. 1996.

Vol. 1642: M. Puschnigg, Asymptotic Cyclic Cohomology. XXII, 138 pages. 1996.

Vol. 1643: J. Richter-Gebert, Realization Spaces of Polytopes. XI, 187 pages. 1996.

Vol. 1644: A. Adler, S. Ramanan, Moduli of Abelian Varieties. VI, 196 pages. 1996.

Vol. 1645: H. W. Broer, G. B. Huitema, M. B. Sevryuk, Quasi-Periodic Motions in Families of Dynamical Systems. XI, 195 pages. 1996.

Vol. 1646: J.-P. Demailly, T. Peternell, G. Tian, A. N. Tyurin, Transcendental Methods in Algebraic Geometry. Cetraro, 1994. Editors: F. Catanese, C. Ciliberto. VII. 257 pages. 1996.

Vol. 1647: D. Dias, P. Le Barz, Configuration Spaces over Hilbert Schemes and Applications. VII. 143 pages. 1996.

Vol. 1648: R. Dobrushin, P. Groeneboom, M. Ledoux, Lectures on Probability Theory and Statistics. Editor: P. Bernard. VIII, 300 pages. 1996.

Vol. 1649: S. Kumar, G. Laumon. U. Stuhler, Vector Bundles on Curves – New Directions. Cetraro, 1995. Editor: M. S. Narasimhan. VII, 193 pages. 1997.

Vol. 1650: J. Wildeshaus, Realizations of Polylogarithms. XI, 343 pages. 1997.

Vol. 1651: M. Drmota, R. F. Tichy. Sequences, Discrepancies and Applications. XIII. 503 pages. 1997.

Vol. 1652: S. Todorcevic, Topics in Topology. VIII. 153 pages. 1997.

Vol. 1653: R. Benedetti, C. Petronio. Branched Standard Spines of 3-manifolds. VIII, 132 pages. 1997.

Vol. 1654: R. W. Ghrist, P. J. Holmes, M. C. Sullivan, Knots and Links in Three-Dimensional Flows. X, 208 pages. 1997.

Vol. 1655: J. Azéma, M. Emery, M. Yor (Eds.), Séminaire de Probabilités XXXI. VIII, 329 pages. 1997.

Vol. 1656: B. Biais, T. Björk, J. Cvitanic, N. El Karoui, E. Jouini, J. C. Rochet, Financial Mathematics. Bressanone, 1996. Editor: W. J. Runggaldier. VII, 316 pages. 1997.

Vol. 1657: H. Reimann, The semi-simple zeta function of quaternionic Shimura varieties. IX, 143 pages. 1997.

Vol. 1658: A. Pumarino, J. A. Rodrıguez, Coexistence and Persistence of Strange Attractors. VIII, 195 pages. 1997.

Vol. 1659: V. Kozlov, V. Maz'ya, Theory of a Higher-Order Sturm-Liouville Equation. XI, 140 pages. 1997.

Vol. 1660: M. Bardi, M. G. Crandall, L. C. Evans, H. M. Soner, P. E. Souganidis, Viscosity Solutions and Applications. Montecatini Terme, 1995. Editors: I. Capuzzo Dolcetta, P. L. Lions. IX, 259 pages. 1997.

Vol. 1661: A. Tralle, J. Oprea, Symplectic Manifolds with no Kähler Structure. VIII, 207 pages. 1997.

Vol. 1662: J. W. Rutter, Spaces of Homotopy Self-Equivalences – A Survey. IX, 170 pages. 1997.

Vol. 1663: Y. E. Karpeshina; Perturbation Theory for the Schrödinger Operator with a Periodic Potential. VII, 352 pages. 1997.

Vol. 1664. M. Väth, Ideal Spaces. V, 146 pages. 1997.

Vol. 1665: E. Giné, G. R. Grimmett, L. Saloff-Coste, Lectures on Probability Theory and Statistics 1996. Editor: P. Bernard. X, 424 pages, 1997.

Vol. 1666: M. van der Put, M. F. Singer, Galois Theory of Difference Equations. VII, 179 pages. 1997.

Vol. 1667: J. M. F. Castillo, M. González, Three-space Problems in Banach Space Theory. XII, 267 pages. 1997.

Vol. 1668: D. B. Dix, Large-Time Behavior of Solutions of Linear Dispersive Equations. XIV, 203 pages. 1997.

Vol. 1669: U. Kaiser, Link Theory in Manifolds. XIV, 167 pages. 1997.

Vol. 1670: J. W. Neuberger, Sobolev Gradients and Differential Equations. VIII, 150 pages. 1997.

Vol. 1671: S. Bouc, Green Functors and *G*-sets. VII, 342 pages. 1997.

Vol. 1672: S. Mandal, Projective Modules and Complete Intersections. VIII, 114 pages. 1997.

Vol. 1673: F. D. Grosshans, Algebraic Homogeneous Spaces and Invariant Theory. VI, 148 pages. 1997.

Vol. 1674: G. Klaas, C. R. Leedham-Green, W. Plesken, Linear Pro-*p*-Groups of Finite Width. VIII, 115 pages. 1997.

Vol. 1675: J. E. Yukich, Probability Theory of Classical Euclidean Optimization Problems. X, 152 pages. 1998.

Vol. 1676: P. Cembranos, J. Mendoza, Banach Spaces of Vector-Valued Functions. VIII, 118 pages. 1997.

Vol. 1677: N. Proskurin, Cubic Metaplectic Forms and Theta Functions. VIII, 196 pages. 1998.

Vol. 1678: O. Krupková, The Geometry of Ordinary Variational Equations. X, 251 pages. 1997.

Vol. 1679: K.-G. Grosse-Erdmann, The Blocking Technique. Weighted Mean Operators and Hardy's Inequality. IX, 114 pages. 1998.

Vol. 1680: K.-Z. Li, F. Oort, Moduli of Supersingular Abelian Varieties. V, 116 pages. 1998.

Vol. 1681: G. J. Wirsching, The Dynamical System Generated by the 3n+1 Function. VII,158 pages. 1998.

Vol. 1682: H.-D. Alber, Materials with Memory. X, 166 pages. 1998.

Vol. 1683: A. Pomp, The Boundary-Domain Integral Method for Elliptic Systems. XVI, 163 pages. 1998.

Vol. 1684: C. A. Berenstein, P. F. Ebenfelt, S. G. Gindikin, S. Helgason, A. E. Tumanov, Integral Geometry, Radon Transforms and Complex Analysis. Firenze, 1996. Editors: E. Casadio Tarabusi, M. A. Picardello, G. Zampieri. VII, 160 pages. 1998

Vol. 1685: S. König, A. Zimmermann, Derived Equivalences for Group Rings. X, 146 pages. 1998.

Vol. 1686: J. Azéma, M. Émery, M. Ledoux, M. Yor (Eds.), Séminaire de Probabilités XXXII. VI, 440 pages. 1998.

Vol. 1687: F. Bornemann, Homogenization in Time of Singularly Perturbed Mechanical Systems. XII, 156 pages. 1998.

Vol. 1688: S. Assing, W. Schmidt, Continuous Strong Markov Processes in Dimension One. XII, 137 page. 1998.

Vol. 1689: W. Fulton, P. Pragacz, Schubert Varieties and Degeneracy Loci. XI, 148 pages. 1998.

Vol. 1690: M. T. Barlow, D. Nualart, Lectures on Probability Theory and Statistics. Editor: P. Bernard. VIII, 237 pages. 1998.

Vol. 1691: R. Bezrukavnikov, M. Finkelberg, V. Schechtman, Factorizable Sheaves and Quantum Groups. X, 282 pages. 1998.

Vol. 1692: T. M. W. Eyre, Quantum Stochastic Calculus and Representations of Lie Superalgebras. IX, 138 pages. 1998.

Vol. 1694: A. Braides, Approximation of Free-Discontinuity Problems. XI, 149 pages. 1998.

Vol. 1695: D. J. Hartfiel, Markov Set-Chains. VIII, 131 pages. 1998.

Vol. 1696: E. Bouscaren (Ed.): Model Theory and Algebraic Geometry. XV, 211 pages. 1998.

Vol. 1697: B. Cockburn, C. Johnson, C.-W. Shu, E. Tadmor, Advanced Numerical Approximation of Nonlinear Hyperbolic Equations. Cetraro, Italy, 1997. Editor: A. Quarteroni. VII, 390 pages. 1998.

Vol. 1698: M. Bhattacharjee, D. Macpherson, R. G. Möller, P. Neumann, Notes on Infinite Permutation Groups. XI, 202 pages. 1998.

Vol. 1699: A. Inoue,Tomita-Takesaki Theory in Algebras of Unbounded Operators. VIII, 241 pages. 1998.

Vol. 1700: W. A. Woyczyński, Burgers-KPZ Turbulence,XI, 318 pages. 1998.

Vol. 1701: Ti-Jun Xiao, J. Liang, The Cauchy Problem of Higher Order Abstract Differential Equations. XII, 302 pages. 1998.

Vol. 1702: J. Ma, J. Yong, Forward-Backward Stochastic Differential Equations and Their Applications. XIII, 270 pages. 1999.

Vol. 1703: R. M. Dudley, R. Norvaiša, Differentiability of Six Operators on Nonsmooth Functions and p-Variation. VIII, 272 pages. 1999.

Vol. 1704: H. Tamanoi, Elliptic Genera and Vertex Operator Super-Algebras. VI, 390 pages. 1999.

Vol. 1705: I. Nikolaev, E. Zhuzhoma, Flows in 2-dimensional Manifolds. XIX, 294 pages. 1999.

Vol. 1706: S. Yu. Pilyugin, Shadowing in Dynamical Systems. XVII, 271 pages. 1999.

Vol. 1707: R. Pytlak, Numerical Methods for Optimal Control Problems with State Constraints. XV, 215 pages. 1999.

Vol. 1708: K. Zuo, Representations of Fundamental Groups of Algebraic Varieties. VII, 139 pages. 1999.

Vol. 1709: J. Azéma, M. Émery, M. Ledoux, M. Yor (Eds), Séminaire de Probabilités XXXIII. VIII, 418 pages. 1999.

Vol. 1710: M. Koecher, The Minnesota Notes on Jordan Algebras and Their Applications. IX, 173 pages. 1999.

Vol. 1711: W. Ricker, Operator Algebras Generated by Commuting Projections: A Vector Measure Approach. XVII, 159 pages. 1999.

Vol. 1712: N. Schwartz, J. J. Madden, Semi-algebraic Function Rings and Reflectors of Partially Ordered Rings. XI, 279 pages. 1999.

Vol. 1713: F. Bethuel, G. Huisken, S. Müller, K. Steffen, Calculus of Variations and Geometric Evolution Problems. Cetraro, 1996. Editors: S. Hildebrandt, M. Struwe. VII, 293 pages. 1999.

Vol. 1714: O. Diekmann, R. Durrett, K. P. Hadeler, P. K. Maini, H. L. Smith, Mathematics Inspired by Biology. Martina Franca, 1997. Editors: V. Capasso, O. Diekmann. VII, 268 pages. 1999.

Vol. 1715: N. V. Krylov, M. Röckner, J. Zabczyk, Stochastic PDE's and Kolmogorov Equations in Infinite Dimensions. Cetraro, 1998. Editor: G. Da Prato. VIII, 239 pages. 1999.

Vol. 1716: J. Coates, R. Greenberg, K. A. Ribet, K. Rubin, Arithmetic Theory of Elliptic Curves. Cetraro, 1997. Editor: C. Viola. VIII, 260 pages. 1999.

Vol. 1717: J. Bertoin, F. Martinelli, Y. Peres, Lectures on Probability Theory and Statistics. Saint-Flour, 1997. Editor: P. Bernard. IX, 291 pages. 1999.

Vol. 1718: A. Eberle, Uniqueness and Non-Uniqueness of Semigroups Generated by Singular Diffusion Operators. VIII, 262 pages. 1999.

Vol. 1719: K. R. Meyer, Periodic Solutions of the N-Body Problem. IX, 144 pages. 1999.

Vol. 1720: D. Elworthy, Y. Le Jan, X-M. Li, On the Geometry of Diffusion Operators and Stochastic Flows. IV, 118 pages. 1999.

Vol. 1721: A. Iarrobino, V. Kanev, Power Sums, Gorenstein Algebras, and Determinantal Loci. XXVII, 345 pages. 1999.

Vol. 1722: R. McCutcheon, Elemental Methods in Ergodic Ramsey Theory. VI, 160 pages. 1999.

Vol. 1723: J. P. Croisille, C. Lebeau, Diffraction by an Immersed Elastic Wedge. VI, 134 pages. 1999.

Vol. 1724: V. N. Kolokoltsov, Semiclassical Analysis for Diffusions and Stochastic Processes. VIII, 347 pages. 2000.

Vol. 1725: D. A. Wolf-Gladrow, Lattice-Gas Cellular Automata and Lattice Boltzmann Models. IX, 308 pages. 2000.

Vol. 1726: V. Marić, Regular Variation and Differential Equations. X, 127 pages. 2000.

Vol. 1727: P. Kravanja, M. Van Barel, Computing the Zeros of Analytic Functions. VII, 111 pages. 2000.

Vol. 1728: K. Gatermann, Computer Algebra Methods for Equivariant Dynamical Systems. XV, 153 pages. 2000.

Vol. 1729: J. Azéma, M. Émery, M. Ledoux, M. Yor, Séminaire de Probabilités XXXIV. VI, 431 pages. 2000.

Vol. 1730: S. Graf, H. Luschgy. Foundations of Quantization for Probability Distributions. X, 230 pages. 2000.

Vol. 1731: T. Hsu, Quilts: Central Extensions, Braid Actions, and Finite Groups,. XII, 185 pages. 2000.

Vol. 1732: K. Keller, Invariant Factors, Julia Equivalences and the (Abstract) Mandelbrot Set. X, 206 pages. 2000.

Vol. 1733: K. Ritter, Average-Case Analysis of NumericalProblems. IX, 254 pages. 2000.

Vol. 1734: M. Espedal, A. Fasano, A. Mikelić, Filtration in Porous Media and Industrial Applications. Cetraro 1998. Editor: A. Fasano. 2000.

Vol. 1735: D. Yafaev, Scattering Theory: Some Old and New Problems. XVI, 169 pages. 2000.

Vol. 1736: B. O. Turesson, Nonlinear Potential Theory and Weighted Sobolev Spaces. XIV, 173 pages. 2000.

Vol. 1737: S. Wakabayashi, Classical Microlocal Analysis in the Space of Hyperfunctions. VIII, 367 pages. 2000.

Vol. 1738: M. Emery, A. Nemirovski, D. Voiculescu, Lectures on Probability Theory and Statistics. XI, 356 pages. 2000.

Vol. 1739: R. Burkard, P. Deuflhard, A. Jameson, J.-L. Lions, G. Strang, Computational Mathematics Driven by Industrial Problems. Martina Franca, 1999. Editors: V. Capasso, H. Engl, J. Periaux. VII, 418 pages. 2000.

Vol. 1740: B. Kawohl, O. Pironneau, L. Tartar, J.-P. Zolesio, Optimal Shape Design. Tróia, Portugal 1999. Editors: A. Cellina, A. Ornelas. IX, 388 pages. 2000.

Vol. 1741: E. Lombardi, Oscillatory Integrals and Phenomena Beyond all Algebraic Orders. XV, 413 pages. 2000.

Vol. 1742: A. Unterberger, Quantization and Non-holomorphic Modular Forms. VIII, 253 pages. 2000.

Vol. 1743: L. Habermann, Riemannian Metrics of Constant Mass and Moduli Spaces of Conformal Structures. XII, 116 pages. 2000.

Vol. 1744: M. Kunze, Non-Smooth Dynamical Systems. X, 228 pages. 2000.

Vol. 1745: V. D. Milman, G. Schechtman, Geometric Aspects of Functional Analysis. VIII, 289 pages. 2000.

Vol. 1746: A. Degtyarev, I. Itenberg, V. Kharlamov, Real Enriques Surfaces. XVI, 259 pages. 2000.

Vol. 1747: L. W. Christensen, Gorenstein Dimensions. VIII, 204 pages. 2000.

4. Lecture Notes are printed by photo-offset from the master-copy delivered in camera-ready form by the authors. Springer-Verlag provides technical instructions for the preparation of manuscripts. Macro packages in TEX, LATEX2e, LATEX2.09 are available from Springer's web-pages at

http://www.springer.de/math/authors/b-tex.html.

Careful preparation of the manuscripts will help keep production time short and ensure satisfactory appearance of the finished book.

The actual production of a Lecture Notes volume takes approximately 12 weeks.

5. Authors receive a total of 50 free copies of their volume, but no royalties. They are entitled to a discount of 33.3 % on the price of Springer books purchase for their personal use, if ordering directly from Springer-Verlag.

Commitment to publish is made by letter of intent rather than by signing a formal contract. Springer-Verlag secures the copyright for each volume. Authors are free to reuse material contained in their LNM volumes in later publications: A brief written (or e-mail) request for formal permission is sufficient.

**Addresses:**

Professor F. Takens, Mathematisch Instituut,
Rijksuniversiteit Groningen, Postbus 800,
9700 AV Groningen, The Netherlands
E-mail: F.Takens@math.rug.nl

Professor B. Teissier
Université Paris 7
UFR de Mathématiques
Equipe Géométrie et Dynamique
Case 7012
2 place Jussieu
75251 Paris Cedex 05
E-mail: Teissier@math.jussieu.fr

Springer-Verlag, Mathematics Editorial, Tiergartenstr. 17,
D-69121 Heidelberg, Germany,
Tel.: *49 (6221) 487-701
Fax: *49 (6221) 487-355
E-mail: lnm@Springer.de